Finance Construction-11, *Tim Asikin, Steve Asikin, Indra Senihardja*

FINANCE

Construction-11

Corporate IFRS-GAAP (B/S-I/S) Engineering

Technologies No. 7,501-8,000 of 111,111 Laws

by

Tim Asikin, Steve Asikin,

Indra Senihardja

0

Corporate IFRS-GAAP (B/S-I/S), ISBN-13: **978-1720792789**, ISBN-10: **172079278X**

FINANCE CONSTRUCTION-11

Corporate IFRS-GAAP (B/S-I/S) Engineering
Technology No. 7,501-8,000 of 111,111 Laws
By Tim Asikin, Steve Asikin, Indra Senihardja

Creative Team:Steve Asikin
            & Asikin Business Quant, New York
An Amazon Paperback
First Published in United States 2018
By Amazon Createspace
This 1st edition published in 2018
AMZN, PO. Box 81226, Seattle
WA 98108-1226, United States

ISBN-13: **978-1720792789**

ISBN-10: **172079278X**

PREFACE

This one is the only Real Finance Engineering Book, written by Engineering & Business Experts with strong Accounting Maths. It comes from a very long world wide experience, analyzing more than 100,000 Corporate Balancesheet and Income Statement reports, in conservative IFRS (International Finance Reporting Standard) and GAAP (Generally Accepted Accounting Principles) of 9,000 days works as Financing Banker, plus Business CEO expert & fresh Engineering views.

This is a manual for healthy Corporate Governance assurance dan practical business management. It is best for Government Authorities, Accounting Regulators, Business Owners & Small Public Investors (all majority & minority share holders as well).

This is not just a tool for making debt instrument unrelated to daily Corporate Finance. This only good for sincere Directors & Executive, and truly bad for the lazy and /or incapable one.

The nicest thing about it, is that only requires simple Junior Highschool Algebra. We just need to be consistent with its less number treatment of constellations. Its saddest part is that many professors and doctors are no longer able to und erstand Junior Highschool Algebra, and hiding in Accounting jargons. Please go back to its Introduction for glossary, at anywhere inside the book, so then we can still understand what are those mean with simple Algebraic Rule there. We try to avoid some sentences multiplied, added, eliminated divided, rooted or powered by another sentences, that make many traditional finance book successful (delivering few simple things, seen as difficult and scary), but less useful.

Corporate IFRS-GAAP (B/S-I/S), ISBN-13: **978-1720792789**, ISBN-10: **172079278X**

Finance Construction-11, *Tim Asikin, Steve Asikin, Indra Senihardja*

TABLE OF CONTENTS

Corporate IFRS-GAAP (B/S-I/S), ISBN-13: **978-1720792789**, ISBN-10: **172079278X**

INTRODUCTION:
The Math Fin Law Terminologies

Income Statement

S	V				
	M	F			
		O	I		
			B	T	
				A	D	Y
						H
				R		

Balance Sheet

K	C	W	L	X	G
J					Z
P				Q	
	N		E	U	
				R	

Please carefully look that all of 26 characters (**A** to **Z**) used once, except the **R**, that in Accountancy must be appeared in both diagrams at same identical value, with glossary (of well known abbreviations) in the next page. Only 2 (two) of them are historical as **U=E'** and **S'** means carried forward legal numbers from previous year.

The 13 (thirteen) common measures (**c, d, f, g, i, j, l, p, q, s, t, v**, and **y**), are put up front as desired criteria along the way, so then later analyst could be surprised.

In Balance Sheet, we found that **W= C+N= K+J+P+N= L+E= X+Q+U+R= G+Z+Q+U+R** as well, so then we put **W** in the middle of its T-Account balance.

In Income Statement, same applied as **S= V+F+I +T+Y+H+R= V+M= V+F+ O= V+F+I+B= V+F+I+T+A= V+F+I+B= V+F+I+T+D+R**.

Corporate IFRS-GAAP (B/S-I/S), ISBN-13: **978-1720792789**, ISBN-10: **172079278X**

A= After Tax Income;

B= Before Tax Income,

C= Current Assets, c= C/X= Current Ratio,

D= Dividend Paid, d= D/A=Dividend Portion or Payout,

E= Equity or Capital, E'= Equity of Past Year,

F= Fixed Cost, f= F/S=Fixed Portion,

G= Goods or Trade Payables,

　　　g= 360G/V= Goods Payable Days,

H= Home Taken Dividend,

I= Interest Expense, i= I/S= Interest Portion,

J= Jobs or Account Receivable,

　　　j= 360J/S= Jobs or Trade Account ReceivableDays,

K= Kind of Cash,

L= Liabilities or Debt, l= L/E= Leverage or Gearing Ratio,

M= Margin of Contribution,

N= Non Current Assets,

O= Operational Surplus,

P= Procured Inventories,

　　　p= 360P/V= Procured Inventory Days,

Q= Quoted Longterm Liabilities,

　　　q= [C-P]/X= Quick or Acid Test Ratio,

R= Retained Earnings,

S= Sales or Revenues, S'= Sales of Past Year,

　　　s= [S/S']-1= Sales Growth,

T= Tax Paid, t= T/B= Tax Rate,

U= Utilized or Starting Capital,

V= Variable Cost, v= V/S= Variable Portion,

W= Wealth or Total Assets,

X= Xpress or Current Debt,

Y= Yielding Tax of Dividend,

　　　y= Y/D= Yielding Tax Rate of Dividend,

Z= Zero Trade Current Debt.

Corporate IFRS-GAAP (B/S-I/S), ISBN-13: **978-1720792789**, ISBN-10: **172079278X**

CHAPTER-20:
Q= QUOTED Longterm Debt Optimization

Based on 12/31/15 Pfizer Annual Report
(http://www.nasdaq.com/symbol/pfe/financials?query=income-statement)

Continuing the Laws No.1-7,500
at Finance Construction 1-10

Law-7501:

If both (**U**= Utilized or Starting Capital= 64,720 millions USD), (**I**= Interest Expense= 513 millions USD), (**F**= Fixed Cost= 29,740 millions USD), (**M**= Margin of Contribution= 41,548 millions USD), (**d**= D/A= Dividend Portion or Payout= 30.00%), (**c**= C/X= Current Ratio= 1.20 times), (**q**= [C-P]/X= Quick or Acid Test Ratio= 1.01 times), (**p**= 360 P/V= Procured Inventory Days= 240 Days), (**V**= Variable Cost= 9,746 millions USD), (**T**= Tax Paid= 3,385 millions USD), and **Q**= Quoted Longterm Liabilities= 35,351 millions USD), then it's (**I** = Leverage or Gearing Ratio planned), is:

$$I = (\mathbf{Q}+\mathbf{Vp}/\{360[\mathbf{c\text{-}q}]\})$$
$$/\{\mathbf{U}+[\mathbf{M\text{-}F\text{-}I\text{-}T}][1\text{-}\mathbf{d}]\}$$
$$= (35,351+9,746*240/\{360[1.2000$$
$$-1.0100]\})/\{64,720+[41,548$$
$$-29,750-513-3,385]$$
$$[1-0.3000]\}$$
$$= 99.00\%$$

Corporate IFRS-GAAP (B/S-I/S), ISBN-13: **978-1720792789**, ISBN-10: **172079278X**

Law-7502:

If both (**I**= L/E= Leverage or Gearing Ratio= 99.00%), (**I**= Interest Expense= 513 millions USD), (**F**= Fixed Cost= 29,740 millions USD), (**M**= Margin of Contribution= 41,548 millions USD), (**d**= D/A= Dividend Portion or Payout= 30.00%), (**c**= C/X= Current Ratio= 1.20 times), (**q**= [C-P]/X= Quick or Acid Test Ratio= 1.01 times), (**p**= 360 P/V= Procured Inventory Days= 240 Days), (**V**= Variable Cost= 9,746 millions USD), (**T**= Tax Paid= 3,385 millions USD), and (**Q**= Quoted Longterm Liabilities= 35,351 millions USD), then it's (**U**= Utilized or Starting Capital planned), is:

$$U= (Q+Vp/\{360[c-q]\})/I -[M-F-I-T][1-d]$$
$$= (35,351+9,746*240/\{360[1.2000$$
$$-1.0100]\})/0.9900-[41,548$$
$$-29,750-513-3,385]$$
$$[1-0.3000]$$
$$= \underline{64,720} \text{ millions USD}$$

Corporate IFRS-GAAP (B/S-I/S), ISBN-13: **978-1720792789**, ISBN-10: **172079278X**

Law-7503:

If both (I= L/E= Leverage or Gearing Ratio= 99.00%), (I= Interest Expense= 513 millions USD), (F= Fixed Cost= 29,740 millions USD), (U= Utilized or Starting Capital= 64,720 millions USD), (d= D/A= Dividend Portion or Payout= 30.00%), (c= C/X= Current Ratio= 1.20 times), (q= [C-P]/X= Quick or Acid Test Ratio= 1.01 times), (p= 360 P/V= Procured Inventory Days= 240 Days), (V= Variable Cost= 9,746 millions USD), (T= Tax Paid= 3,385 millions USD), and (Q= Quoted Longterm Liabilities= 35,351 millions USD), then it's (M= Margin of Contribution planned), is:

$$M= F+I+T+[(Q+Vp/\{360[c-q]\})/I-U]/[1-d]$$
$$= 29,750+513+3,385+[(35,351$$
$$+9,746*240/\{360[1.2000$$
$$-1.0100]\})/0.9900-64,720]$$
$$/[1-0.3000]$$
$$= 41,548 \text{ millions USD}$$

Corporate IFRS-GAAP (B/S-I/S), ISBN-13: **978-1720792789**, ISBN-10: **172079278X**

<u>Law-7504</u>:

If both (**I**= L/E= Leverage or Gearing Ratio=
<u>99.00%</u>), (**I**= Interest Expense= <u>513</u> millions USD),
(**M**= Margin of Contribution= <u>41,548</u> millions USD),
(**U**= Utilized or Starting Capital= <u>64,720</u> millions
USD), (**d**= D/A= Dividend Portion or Payout=
<u>30.00%</u>), (**c**= C/X= Current Ratio= <u>1.20</u> times), (**q**=
[C-P]/X= Quick or Acid Test Ratio= <u>1.01</u> times), (**p**=
360 P/V= Procured Inventory Days= <u>240</u> Days), (**V**=
Variable Cost= <u>9,746</u> millions USD), (**T**= Tax Paid=
<u>3,385</u> millions USD), and (**Q**= Quoted Longterm
Liabilities= <u>35,351</u> millions USD), then it's (**F**=
Fixed Cost planned), is:

$$F= M\text{-}I\text{-}T\text{-}[(Q+Vp/\{360[c\text{-}q]\})/I\text{-}U]/[1\text{-}d]$$
$$= 41,548\text{-}513\text{-}3,385\text{-}[(35,351+9,746$$
$$*240/\{360[1.2000\text{-}1.0100]\})$$
$$/0.9900\text{-}64,720]/[1\text{-}0.3000]$$
$$= \underline{29,750} \text{ millions USD}$$

Corporate IFRS-GAAP (B/S-I/S), ISBN-13: **978-1720792789**, ISBN-10: **172079278X**

Law-7505:

If both (**I**= L/E= Leverage or Gearing Ratio= 99.00%), (**F**= Fixed Cost= 29,740 millions USD), (**M**= Margin of Contribution= 41,548 millions USD), (**U**= Utilized or Starting Capital= 64,720 millions USD), (**d**= D/A= Dividend Portion or Payout= 30.00%), (**c**= C/X= Current Ratio= 1.20 times), (**q**= [C-P]/X= Quick or Acid Test Ratio= 1.01 times), (**p**= 360 P/V= Procured Inventory Days= 240 Days), (**V**= Variable Cost= 9,746 millions USD), (**T**= Tax Paid= 3,385 millions USD), and (**Q**= Quoted Longterm Liabilities= 35,351 millions USD), then it's (**I**= Interest Expense planned), is:

$$\mathbf{I= M\text{-}F\text{-}T\text{-}[(Q+Vp}/\{360[\mathbf{c\text{-}q}]\})/\mathbf{I}\text{ -}\mathbf{U}]/[1\text{-}\mathbf{d}]$$

$$= 41,548\text{-}29,750\text{-}3,385\text{-}[(35,351$$
$$+9,746*240/\{360[1.2000$$
$$-1.0100]\})/0.9900\text{-}64,720]$$
$$/[1\text{-}0.3000]$$

$$= \underline{513} \text{ millions USD}$$

Corporate IFRS-GAAP (B/S-I/S), ISBN-13: **978-1720792789**, ISBN-10: **172079278X**

Law-7506:

If both (**I**= L/E= Leverage or Gearing Ratio= 99.00%), (**F**= Fixed Cost= 29,740 millions USD), (**M**= Margin of Contribution= 41,548 millions USD), (**U**= Utilized or Starting Capital= 64,720 millions USD), (**d**= D/A= Dividend Portion or Payout= 30.00%), (**c**= C/X= Current Ratio= 1.20 times), (**q**= [C-P]/X= Quick or Acid Test Ratio= 1.01 times), (**p**= 360 P/V= Procured Inventory Days= 240 Days), (**V**= Variable Cost= 9,746 millions USD), (**I**= Interest Expense= 513 millions USD), and (**Q**= Quoted Longterm Liabilities= 35,351 millions USD), then it's (**T**= Tax Paid planned), is:

$$T = M\text{-}F\text{-}I\text{-}[(Q+Vp/\{360[c\text{-}q]\})/I\text{-}U]/[1\text{-}d]$$

$$= 41{,}548\text{-}29{,}750\text{-}513\text{-}[(35{,}351 +9{,}746*240/\{360[1.2000 -1.0100]\})/0.9900\text{-}64{,}720]/[1\text{-}0.3000]$$

$$= \underline{3{,}385} \text{ millions USD}$$

Corporate IFRS-GAAP (B/S-I/S), ISBN-13: **978-1720792789**, ISBN-10: **172079278X**

Law-7507:

If both (**I**= L/E= Leverage or Gearing Ratio= 99.00%), (**F**= Fixed Cost= 29,740 millions USD), (**M**= Margin of Contribution= 41,548 millions USD), (**U**= Utilized or Starting Capital= 64,720 millions USD), (**T**= Tax Paid= 3,385 millions USD), (**c**= C/X= Current Ratio= 1.20 times), (**q**= [C-P]/X= Quick or Acid Test Ratio= 1.01 times), (**p**= 360 P/V= Procured Inventory Days= 240 Days), (**V**= Variable Cost= 9,746 millions USD), (**I**= Interest Expense= 513 millions USD), and (**Q**= Quoted Longterm Liabilities= 35,351 millions USD), then it's (**d**= Dividend Portion or Payout planned), is:

$$d= 1-[(\mathbf{Q}+\mathbf{V}\mathbf{p}/\{360[\mathbf{c}\text{-}\mathbf{q}]\})/\mathbf{I}\text{-}\mathbf{U}]$$
$$/[\mathbf{M}\text{-}\mathbf{F}\text{-}\mathbf{I}\text{-}\mathbf{T}]$$
$$= 1-[(35,351+9,746*240/\{360[1.2000$$
$$-1.0100]\})/0.9900\text{-}64,720]$$
$$/[41,548\text{-}29,750\text{-}513\text{-}3,385]$$
$$= 30.00\%$$

Corporate IFRS-GAAP (B/S-I/S), ISBN-13: **978-1720792789**, ISBN-10: **172079278X**

Law-7508:

If both (I= L/E= Leverage or Gearing Ratio= 99.00%), (F= Fixed Cost= 29,740 millions USD), (M= Margin of Contribution= 41,548 millions USD), (U= Utilized or Starting Capital= 64,720 millions USD), (T= Tax Paid= 3,385 millions USD), (c= C/X= Current Ratio= 1.20 times), (q= [C-P]/X= Quick or Acid Test Ratio= 1.01 times), (p= 360 P/V= Procured Inventory Days= 240 Days), (d= D/A= Dividend Portion or Payout= 30.00%), (I= Interest Expense= 513 millions USD), and (Q= Quoted Longterm Liabilities= 35,351 millions USD), then it's (V= Variable Cost planned), is:

$$V= 360[c\text{-}q](I\{U+[M\text{-}F\text{-}I\text{-}T][1\text{-}d]\}\text{-}Q)/p$$
$$= 360[1.2000\text{-}1.0100](0.9900\{64,720$$
$$+[41,548\text{-}29,750\text{-}513\text{-}3,385]$$
$$[1\text{-}0.3000]\}\text{-}35,351)/240$$
$$= 9,746 \text{ millions USD}$$

Corporate IFRS-GAAP (B/S-I/S), ISBN-13: **978-1720792789**, ISBN-10: **172079278X**

Law-7509:

If both (**I**= L/E= Leverage or Gearing Ratio=
99.00%), (**F**= Fixed Cost= 29,740 millions USD),
(**M**= Margin of Contribution= 41,548 millions USD),
(**U**= Utilized or Starting Capital= 64,720 millions
USD), (**T**= Tax Paid= 3,385 millions USD), (**c**= C/X=
Current Ratio= 1.20 times), (**q**= [C-P]/X= Quick or
Acid Test Ratio= 1.01 times), (**V**= Variable Cost=
9,746 millions USD), (**d**= D/A= Dividend Portion or
Payout= 30.00%), (**I**= Interest Expense= 513 millions
USD), and (**Q**= Quoted Longterm Liabilities= 35,351
millions USD), then it's (**p**= Procured Inventory
Days planned), is:

$$p = 360[c\text{-}q](I\,\{U+[M\text{-}F\text{-}I\text{-}T][1\text{-}d]\}\text{-}Q)/V$$
$$= 360[1.2000\text{-}1.0100](0.9900\{64,720$$
$$+[41,548\text{-}29,750\text{-}513\text{-}3,385]$$
$$[1\text{-}0.3000]\}\text{-}35,351)/9,746$$
$$= \underline{240}\text{ days}$$

Corporate IFRS-GAAP (B/S-I/S), ISBN-13: **978-1720792789**, ISBN-10: **172079278X**

Law-7510:

If both (**I**= L/E= Leverage or Gearing Ratio= 99.00%), (**F**= Fixed Cost= 29,740 millions USD), (**M**= Margin of Contribution= 41,548 millions USD), (**U**= Utilized or Starting Capital= 64,720 millions USD), (**T**= Tax Paid= 3,385 millions USD), (**p**= 360 P/V= Procured Inventory Days= 240 Days), (**q**= Quick or Acid Test Ratio= [C-P]/X= 1.01 times), (**V**= Variable Cost= 9,746 millions USD), (**d**= D/A= Dividend Portion or Payout= 30.00%), (**I**= Interest Expense= 513 millions USD), and (**Q**= Quoted Longterm Liabilities= 35,351 millions USD), then it's (**c**= Current Ratio planned), is:

$$c = q + Vp / [360(I\{U + [M-F-I-T][1-d]\} - Q)]$$
$$= 1.0100 + 9,746 * 240 / [360(0.9900$$
$$\{64,720 + [41,548 - 29,750$$
$$-513 - 3,385][1 - 0.3000]\}$$
$$-35,351)]$$
$$= 1.20 \text{ times}$$

Corporate IFRS-GAAP (B/S-I/S), ISBN-13: **978-1720792789**, ISBN-10: **172079278X**

Law-7511:

If both (**I**= L/E= Leverage or Gearing Ratio= 99.00%), (**F**= Fixed Cost= 29,740 millions USD), (**M**= Margin of Contribution= 41,548 millions USD), (**U**= Utilized or Starting Capital= 64,720 millions USD), (**T**= Tax Paid= 3,385 millions USD), (**p**= 360 P/V= Procured Inventory Days= 240 Days), (**c**= C/X= Current Ratio= 1.20 times), (**V**= Variable Cost= 9,746 millions USD), (**d**= D/A= Dividend Portion or Payout= 30.00%), (**I**= Interest Expense= 513 millions USD), and (**Q**= Quoted Longterm Liabilities= 35,351 millions USD), then it's (**q**= Quick or Acid Test Ratio planned), is:

$$q = c - Vp/[360(I\{U+[M-F-I-T][1-d]\}-Q)]$$
$$= 1.2000 - 9,746*240/[360(0.9900$$
$$\{64,720+[41,548-29,750$$
$$-513-3,385][1-0.3000]\}$$
$$-35,351)]$$

$$= 1.01 \text{ times}$$

Corporate IFRS-GAAP (B/S-I/S), ISBN-13: **978-1720792789**, ISBN-10: **172079278X**

<u>Law-7512</u>:

If both (**I**= L/E= Leverage or Gearing Ratio= 99.00%), (**F**= Fixed Cost= <u>29,740</u> millions USD), (**M**= Margin of Contribution= <u>41,548</u> millions USD), (**U**= Utilized or Starting Capital= <u>64,720</u> millions USD), (**T**= Tax Paid= <u>3,385</u> millions USD), (**p**= 360 P/V= Procured Inventory Days= <u>240</u> Days), (**c**= C/X= Current Ratio= <u>1.20</u> times), (**v**= V/S= Variable Portion= <u>19.00%</u>), (**S**= Sales or Revenues= <u>51,294</u> millions USD), (**d**= D/A= Dividend Portion or Payout= <u>30.00%</u>), (**I**= Interest Expense= <u>513</u> millions USD), and (**q**= [C-P]/X= Quick or Acid Test Ratio= <u>1.01</u> times), then it's (**Q**= Quoted Longterm Liabilities planned), is:

$$Q= I\{U+[M\text{-}F\text{-}I\text{-}T][1\text{-}d]\}\text{-}Svp/\{360[c\text{-}q]\}$$
$$= 0.9900\{64,720+[41,548\text{-}29,750$$
$$-513\text{-}3,385][1\text{-}0.3000]\}$$
$$-51,294*0.1900*240$$
$$/\{360[1.2000\text{-}1.0100]\}$$
$$= \underline{35,351} \text{ millions USD}$$

Corporate IFRS-GAAP (B/S-I/S), ISBN-13: **978-1720792789**, ISBN-10: **172079278X**

<u>Law-7513</u>:

If both (**Q**= Quoted Longterm Liabilities= <u>35,351</u> millions USD), (**F**= Fixed Cost= <u>29,740</u> millions USD), (**M**= Margin of Contribution= <u>41,548</u> millions USD), (**U**= Utilized or Starting Capital= <u>64,720</u> millions USD), (**T**= Tax Paid= <u>3,385</u> millions USD), (**p**= 360 P/V= Procured Inventory Days= <u>240</u> Days), (**c**= C/X= Current Ratio= <u>1.20</u> times), (**v**= V/S= Variable Portion= <u>19.00%</u>), (**S**= Sales or Revenues= <u>51,294</u> millions USD), (**d**= D/A= Dividend Portion or Payout= <u>30.00%</u>), (**I**= Interest Expense= <u>513</u> millions USD), and (**q**= [C-P]/X= Quick or Acid Test Ratio= <u>1.01</u> times), then it's (**I** = Leverage or Gearing Ratio planned), is:

$$I = (Q+Svp/\{360[c-q]\})$$
$$/\{U+[M-F-I-T][1-d]\}$$
$$= (35,351+51,294*0.1900*240$$
$$/\{360[1.2000-1.0100]\})$$
$$/\{64,720+[41,548-29,750$$
$$-513-3,385][1-0.3000]\}$$
$$= \underline{99.00\%}$$

Corporate IFRS-GAAP (B/S-I/S), ISBN-13: **978-1720792789**, ISBN-10: **172079278X**

Law-7514:

If both (**Q**= Quoted Longterm Liabilities= 35,351 millions USD), (**F**= Fixed Cost= 29,740 millions USD), (**M**= Margin of Contribution= 41,548 millions USD), (**I**= L/E= Leverage or Gearing Ratio= 99.00%), (**T**= Tax Paid= 3,385 millions USD), (**p**= 360 P/V= Procured Inventory Days= 240 Days), (**c**= C/X= Current Ratio= 1.20 times), (**v**= V/S= Variable Portion= 19.00%), (**S**= Sales or Revenues= 51,294 millions USD), (**d**= D/A= Dividend Portion or Payout= 30.00%), (**I**= Interest Expense= 513 millions USD), and (**q**= [C-P]/X= Quick or Acid Test Ratio= 1.01 times), then it's (**U**= Utilized or Starting Capital planned), is:

$$U = (Q + Svp/\{360[c-q]\})/I - [M-F-I-T][1-d]$$
$$= (35,351 + 51,294*0.1900*240$$
$$/\{360[1.2000-1.0100]\})$$
$$/0.9900 - [41,548-29,750$$
$$-513-3,385][1-0.3000]$$
$$= \underline{64,720} \text{ millions USD}$$

Corporate IFRS-GAAP (B/S-I/S), ISBN-13: **978-1720792789**, ISBN-10: **172079278X**

Law-7515:

If both (**Q**= Quoted Longterm Liabilities= 35,351 millions USD), (**F**= Fixed Cost= 29,740 millions USD), (**U**= Utilized or Starting Capital= 64,720 millions USD), (**l**= L/E= Leverage or Gearing Ratio= 99.00%), (**T**= Tax Paid= 3,385 millions USD), (**p**= 360 P/V= Procured Inventory Days= 240 Days), (**c**= C/X= Current Ratio= 1.20 times), (**v**= V/S= Variable Portion= 19.00%), (**s**= Sales or Revenues= 51,294 millions USD), (**d**= D/A= Dividend Portion or Payout= 30.00%), (**l**= Interest Expense= 513 millions USD), and (**q**= [C-P]/X= Quick or Acid Test Ratio= 1.01 times), then it's (**M**= Margin of Contribution planned), is:

$$M= F+l+T+[(Q+Svp/\{360[c\text{-}q]\})/l \text{-}U]$$
$$/[1\text{-}d]$$
$$= 29,750+513+3,385+[(35,351$$
$$+51,294*0.1900*240$$
$$/\{360[1.2000\text{-}1.0100]\})$$
$$/0.9900\text{-}64,720]$$
$$/[1\text{-}0.3000]$$
$$= 41,548 \text{ millions USD}$$

Corporate IFRS-GAAP (B/S-I/S), ISBN-13: **978-1720792789**, ISBN-10: **172079278X**

Law-7516:

If both (**Q**= Quoted Longterm Liabilities= 35,351 millions USD), (**M**= Margin of Contribution= 41,548 millions USD), (**U**= Utilized or Starting Capital= 64,720 millions USD), (**l**= L/E= Leverage or Gearing Ratio= 99.00%), (**T**= Tax Paid= 3,385 millions USD), (**p**= 360 P/V= Procured Inventory Days= 240 Days), (**c**= C/X= Current Ratio= 1.20 times), (**v**= V/S= Variable Portion= 19.00%), (**$**= Sales or Revenues= 51,294 millions USD), (**d**= D/A= Dividend Portion or Payout= 30.00%), (**I**= Interest Expense= 513 millions USD), and (**q**= [C-P]/X= Quick or Acid Test Ratio= 1.01 times), then it's (**F**= Fixed Cost planned), is:

$$F= M-I-T-[(Q+Svp/\{360[c-q]\})/I-U]/[1-d]$$
$$= 41{,}548{-}513{-}3{,}385{-}[(35{,}351{+}51{,}294$$
$$*0.1900*240/\{360[1.2000$$
$$-1.0100]\})/0.9900{-}64{,}720]$$
$$/[1{-}0.3000]$$
$$= 29{,}750 \text{ millions USD}$$

Corporate IFRS-GAAP (B/S-I/S), ISBN-13: **978-1720792789**, ISBN-10: **172079278X**

Law-7517:

If both (**Q**= Quoted Longterm Liabilities= 35,351 millions USD), (**M**= Margin of Contribution= 41,548 millions USD), (**U**= Utilized or Starting Capital= 64,720 millions USD), (**⌐**= L/E= Leverage or Gearing Ratio= 99.00%), (**T**= Tax Paid= 3,385 millions USD), (**p**= 360 P/V= Procured Inventory Days= 240 Days), (**c**= C/X= Current Ratio= 1.20 times), (**v**= V/S= Variable Portion= 19.00%), (**$**= Sales or Revenues= 51,294 millions USD), (**d**= D/A= Dividend Portion or Payout= 30.00%), (**F**= Fixed Cost= 29,740 millions USD), and (**q**= [C-P]/X= Quick or Acid Test Ratio= 1.01 times), then it's (**I**= Interest Expense planned), is:

$$I= M-F-T-[(Q+\$vp/\{360[c-q]\})/\mathit{l}-U]/[1-d]$$

$$= 41,548-29,750-3,385-[(35,351$$
$$+51,294*0.1900*240/\{360$$
$$[1.2000-1.0100]\})/0.9900$$
$$-64,720]/[1-0.3000]$$

$$= \underline{513} \text{ millions USD}$$

Corporate IFRS-GAAP (B/S-I/S), ISBN-13: **978-1720792789**, ISBN-10: **172079278X**

Law-7518:

If both (**Q**= Quoted Longterm Liabilities= 35,351 millions USD), (**M**= Margin of Contribution= 41,548 millions USD), (**U**= Utilized or Starting Capital= 64,720 millions USD), (**I**= L/E= Leverage or Gearing Ratio= 99.00%), (**I**= Interest Expense= 513 millions USD), (**p**= 360 P/V= Procured Inventory Days= 240 Days), (**c**= C/X= Current Ratio= 1.20 times), (**v**= V/S= Variable Portion= 19.00%), (**S**= Sales or Revenues= 51,294 millions USD), (**d**= D/A= Dividend Portion or Payout= 30.00%), (**F**= Fixed Cost= 29,740 millions USD), and (**q**= [C-P]/X= Quick or Acid Test Ratio= 1.01 times), then it's (**T**= Tax Paid planned), is:

$$T= M-F-I-[(Q+Svp/\{360[c-q]\})/I-U]/[1-d]$$
$$= 41,548-29,750-513-[(35,351$$
$$+51,294*0.1900*240/\{360$$
$$[1.2000-1.0100]\})/0.9900$$
$$-64,720]/[1-0.3000]$$
$$= 3,385 \text{ millions USD}$$

Corporate IFRS-GAAP (B/S-I/S), ISBN-13: **978-1720792789**, ISBN-10: **172079278X**

Law-7519:

If both (**Q**= Quoted Longterm Liabilities= 35,351 millions USD), (**M**= Margin of Contribution= 41,548 millions USD), (**U**= Utilized or Starting Capital= 64,720 millions USD), (**l**= L/E= Leverage or Gearing Ratio= 99.00%), (**I**= Interest Expense= 513 millions USD), (**p**= 360 P/V= Procured Inventory Days= 240 Days), (**c**= C/X= Current Ratio= 1.20 times), (**v**= V/S= Variable Portion= 19.00%), (**S**= Sales or Revenues= 51,294 millions USD), (**T**= Tax Paid= 3,385 millions USD), (**F**= Fixed Cost= 29,740 millions USD), and (**q**= [C-P]/X= Quick or Acid Test Ratio= 1.01 times), then it's (**d**= Dividend Portion or Payout planned), is:

$$d= 1\text{-}[(Q+Svp/\{360[c\text{-}q]\})/l\text{-}U]/[M\text{-}F\text{-}I\text{-}T]$$
$$= 1\text{-}[(35,351+51,294*0.1900*240$$
$$/\{360[1.2000\text{-}1.0100]\})$$
$$/0.9900\text{-}64,720]/[41,548$$
$$-29,750\text{-}513\text{-}3,385]$$
$$= 30.00\%$$

Corporate IFRS-GAAP (B/S-I/S), ISBN-13: **978-1720792789**, ISBN-10: **172079278X**

Law-7520:

If both (**Q**= Quoted Longterm Liabilities= 35,351 millions USD), (**M**= Margin of Contribution= 41,548 millions USD), (**U**= Utilized or Starting Capital= 64,720 millions USD), (**I**= L/E= Leverage or Gearing Ratio= 99.00%), (**I**= Interest Expense= 513 millions USD), (**p**= 360 P/V= Procured Inventory Days= 240 Days), (**c**= C/X= Current Ratio= 1.20 times), (**v**= V/S= Variable Portion= 19.00%), (**d**= D/A= Dividend Portion or Payout= 30.00%), (**T**= Tax Paid= 3,385 millions USD), (**F**= Fixed Cost= 29,740 millions USD), and (**q**= [C-P]/X= Quick or Acid Test Ratio= 1.01 times), then it's (**$**= Sales or Revenues planned), is:

$$\$= 360[\mathbf{c}\text{-}\mathbf{q}](\mathbf{I} - \{\mathbf{U}+[\mathbf{M}\text{-}\mathbf{F}\text{-}\mathbf{I}\text{-}\mathbf{T}][1\text{-}\mathbf{d}]\}\text{-}\mathbf{Q})/[\mathbf{vp}]$$
$$= 360[1.2000\text{-}1.0100](0.9900$$
$$\{64,720+[41,548\text{-}29,750$$
$$-513\text{-}3,385][1\text{-}0.3000]\}$$
$$-35,351)/[0.1900*240]$$
$$= \underline{51,294} \text{ millions USD}$$

Corporate IFRS-GAAP (B/S-I/S), ISBN-13: **978-1720792789**, ISBN-10: **172079278X**

Law-7521:

If both (**Q**= Quoted Longterm Liabilities= 35,351 millions USD), (**M**= Margin of Contribution= 41,548 millions USD), (**U**= Utilized or Starting Capital= 64,720 millions USD), (**I**= L/E= Leverage or Gearing Ratio= 99.00%), (**I**= Interest Expense= 513 millions USD), (**p**= 360 P/V= Procured Inventory Days= 240 Days), (**c**= C/X= Current Ratio= 1.20 times), (**\$**= Sales or Revenues= 51,294 millions USD), (**d**= D/A= Dividend Portion or Payout= 30.00%), (**T**= Tax Paid= 3,385 millions USD), (**F**= Fixed Cost= 29,740 millions USD), and (**q**= [C-P]/X= Quick or Acid Test Ratio= 1.01 times), then it's (**v**= Variable Portion planned), is:

$$v= 360[c\text{-}q](I \text{-} \{U+[M\text{-}F\text{-}I\text{-}T][1\text{-}d]\}\text{-}Q)/[\$p]$$
$$= 360[1.2000\text{-}1.0100](0.9900\{64,720$$
$$+[41,548\text{-}29,750\text{-}513\text{-}3,385]$$
$$[1\text{-}0.3000]\}\text{-}35,351)$$
$$/[51,294*240]$$

$$= 19.00\%$$

Corporate IFRS-GAAP (B/S-I/S), ISBN-13: **978-1720792789**, ISBN-10: **172079278X**

Law-7522:

If both (**Q**= Quoted Longterm Liabilities= 35,351 millions USD), (**M**= Margin of Contribution= 41,548 millions USD), (**U**= Utilized or Starting Capital= 64,720 millions USD), (**⌐**= L/E= Leverage or Gearing Ratio= 99.00%), (**I**= Interest Expense= 513 millions USD), (**v**= V/S= Variable Portion= 19.00%), (**c**= C/X= Current Ratio= 1.20 times), (**$**= Sales or Revenues= 51,294 millions USD), (**d**= D/A= Dividend Portion or Payout= 30.00%), (**T**= Tax Paid= 3,385 millions USD), (**F**= Fixed Cost= 29,740 millions USD), and (**q**= [C-P]/X= Quick or Acid Test Ratio= 1.01 times), then it's (**p**= Procured Inventory Days planned), is:

$$p= 360[\mathbf{c}-\mathbf{q}](\mathbf{⌐}-\{\mathbf{U}+[\mathbf{M}-\mathbf{F}-\mathbf{I}-\mathbf{T}][1-\mathbf{d}]\}-\mathbf{Q})/[\mathbf{\$v}]$$
$$= 360[1.2000-1.0100](0.9900\{64,720$$
$$+[41,548-29,750-513-3,385]$$
$$[1-0.3000]\}-35,351)$$
$$/[51,294*0.1900]$$
$$= \underline{240} \text{ days}$$

Corporate IFRS-GAAP (B/S-I/S), ISBN-13: **978-1720792789**, ISBN-10: **172079278X**

Law-7523:

If both (**Q**= Quoted Longterm Liabilities= <u>35,351</u> millions USD), (**M**= Margin of Contribution= <u>41,548</u> millions USD), (**U**= Utilized or Starting Capital= <u>64,720</u> millions USD), (**I**= L/E= Leverage or Gearing Ratio= <u>99.00%</u>), (**I**= Interest Expense= <u>513</u> millions USD), (**v**= V/S= Variable Portion= <u>19.00%</u>), (**p**= 360 P/V= Procured Inventory Days= <u>240</u> Days), (**S**= Sales or Revenues= <u>51,294</u> millions USD), (**d**= D/A= Dividend Portion or Payout= <u>30.00%</u>), (**T**= Tax Paid= <u>3,385</u> millions USD), (**F**= Fixed Cost= <u>29,740</u> millions USD), and (**q**= [C-P]/X= Quick or Acid Test Ratio= <u>1.01</u> times), then it's (**c**= Current Ratio planned), is:

$$c = q + Svp / [360(I - \{U + [M-F-I-T][1-d]\} - Q)]$$
$$= 1.0100 + 51,294*0.1900*240$$
$$/[360(0.9900\{64,720$$
$$+[41,548-29,750-513-3,385]$$
$$[1-0.3000]\}-35,351)]$$

$$= \underline{1.20} \text{ times}$$

Corporate IFRS-GAAP (B/S-I/S), ISBN-13: **978-1720792789**, ISBN-10: **172079278X**

Law-7524:

If both (**Q**= Quoted Longterm Liabilities= 35,351 millions USD), (**M**= Margin of Contribution= 41,548 millions USD), (**U**= Utilized or Starting Capital= 64,720 millions USD), (**I**= L/E= Leverage or Gearing Ratio= 99.00%), (**I**= Interest Expense= 513 millions USD), (**v**= V/S= Variable Portion= 19.00%), (**p**= 360 P/V= Procured Inventory Days= 240 Days), (**S**= Sales or Revenues= 51,294 millions USD), (**d**= D/A= Dividend Portion or Payout= 30.00%), (**T**= Tax Paid= 3,385 millions USD), (**F**= Fixed Cost= 29,740 millions USD), and (**c**= C/X= Current Ratio= 1.20 times), then it's (**q**= Quick or Acid Test Ratio planned), is:

$$q= c\text{-}Svp/[360(I-\{U+[M\text{-}F\text{-}I\text{-}T][1\text{-}d]\}\text{-}Q)]$$
$$= 1.2000\text{-}51,294*0.1900*240/[360$$
$$(0.9900\{64,720+[41,548$$
$$-29,750\text{-}513\text{-}3,385]$$
$$[1\text{-}0.3000]\}\text{-}35,351)]$$
$$= \underline{1.01} \text{ times}$$

Corporate IFRS-GAAP (B/S-I/S), ISBN-13: **978-1720792789**, ISBN-10: **172079278X**

Law-7525:

If both (q= [C-P]/X= Quick or Acid Test Ratio= 1.01 times), (M= Margin of Contribution= 41,548 millions USD), (U= Utilized or Starting Capital= 64,720 millions USD), (I= L/E= Leverage or Gearing Ratio= 99.00%), (I= Interest Expense= 513 millions USD), (v= V/S= Variable Portion= 19.00%), (p= 360 P/V= Procured Inventory Days= 240 Days), (S'= Sales of Past Year= 48,851 millions USD), (s= [S/S']-1= Sales Growth= 5.00%), (d= D/A= Dividend Portion or Payout= 30.00%), (T= Tax Paid= 3,385 millions USD), (F= Fixed Cost= 29,740 millions USD), and (c= C/X= Current Ratio= 1.20 times), then it's (Q= Quoted Longterm Liabilities planned), is:

$$Q= I\{U+[M-F-I-T][1-d]\}-S'vp[1+s]$$
$$/\{360[c-q]\}$$
$$= 0.9900\{64,720+[41,548-29,750$$
$$-513-3,385][1-0.3000]\}$$
$$-48,851*0.1900*240$$
$$[1+0.0500]/\{360[1.2000$$
$$-1.0100]\}$$
$$= 35,351 \text{ millions USD}$$

Corporate IFRS-GAAP (B/S-I/S), ISBN-13: **978-1720792789**, ISBN-10: **172079278X**

Law-7526:

If both (q= [C-P]/X= Quick or Acid Test Ratio= 1.01 times), (M= Margin of Contribution= 41,548 millions USD), (U= Utilized or Starting Capital= 64,720 millions USD), (Q= Quoted Longterm Liabilities= 35,351 millions USD), (I= Interest Expense= 513 millions USD), (v= V/S= Variable Portion= 19.00%), (p= 360 P/V= Procured Inventory Days= 240 Days), (S'= Sales of Past Year= 48,851 millions USD), (s= [S/S']-1= Sales Growth= 5.00%), (d= D/A= Dividend Portion or Payout= 30.00%), (T= Tax Paid= 3,385 millions USD), (F= Fixed Cost= 29,740 millions USD), and (c= C/X= Current Ratio= 1.20 times), then it's (I = Leverage or Gearing Ratio planned), is:

$$I = (Q+S'vp[1+s]/\{360[c-q]\})$$
$$/\{U+[M-F-I-T][1-d]\}$$
$$= (35,351+48,851*0.1900*240$$
$$[1+0.0500]/\{360[1.2000$$
$$-1.0100]\})/\{64,720+[41,548$$
$$-29,750-513-3,385]$$
$$[1-0.3000]\}$$
$$= 99.00\%$$

Corporate IFRS-GAAP (B/S-I/S), ISBN-13: **978-1720792789**, ISBN-10: **172079278X**

<u>Law-7527</u>:

If both (**q**= [C-P]/X= Quick or Acid Test Ratio= <u>1.01</u> times), (**M**= Margin of Contribution= <u>41,548</u> millions USD), (**I**= L/E= Leverage or Gearing Ratio= <u>99.00%</u>), (**Q**= Quoted Longterm Liabilities= <u>35,351</u> millions USD), (**I**= Interest Expense= <u>513</u> millions USD), (**v**= V/S= Variable Portion= <u>19.00%</u>), (**p**= 360 P/V= Procured Inventory Days= <u>240</u> Days), (**S'**= Sales of Past Year= <u>48,851</u> millions USD), (**s**= [S/S']-1= Sales Growth= <u>5.00%</u>), (**d**= D/A= Dividend Portion or Payout= <u>30.00%</u>), (**T**= Tax Paid= <u>3,385</u> millions USD), (**F**= Fixed Cost= <u>29,740</u> millions USD), and (**c**= C/X= Current Ratio= <u>1.20</u> times), then it's (**U**= Utilized or Starting Capital planned), is:

$$U= (Q+S'vp[1+s]/\{360[c-q]\})/I$$
$$-[M-F-I-T][1-d]$$
$$= (35,351+48.851*0.1900*240$$
$$[1+0.0500]/\{360[1.2000$$
$$-1.0100]\})/0.9900-[41,548$$
$$-29,750-513-3,385]$$
$$[1-0.3000]$$
$$= \underline{64,720} \text{ millions USD}$$

Corporate IFRS-GAAP (B/S-I/S), ISBN-13: **978-1720792789**, ISBN-10: **172079278X**

Law-7528:

If both (**q**= [C-P]/X= Quick or Acid Test Ratio= 1.01
times), (**U**= Utilized or Starting Capital= 64,720
millions USD), (**I**= L/E= Leverage or Gearing Ratio=
99.00%), (**Q**= Quoted Longterm Liabilities= 35,351
millions USD), (**I**= Interest Expense= 513 millions
USD), (**v**= V/S= Variable Portion= 19.00%), (**p**= 360
P/V= Procured Inventory Days= 240 Days), (**S'**=
Sales of Past Year= 48,851 millions USD), (**s**=
[S/S']-1= Sales Growth= 5.00%), (**d**= D/A=
Dividend Portion or Payout= 30.00%), (**T**= Tax Paid=
3,385 millions USD), (**F**= Fixed Cost= 29,740
millions USD), and (**c**= C/X= Current Ratio= 1.20
times), then it's (**M**= Margin of Contribution
planned), is:

$$M= F+I+T+[(Q+S'vp[1+s]/\{360[c-q]\})/I-U]$$
$$/[1-d]$$

$$\begin{aligned}
= {}& 29,750+513+3,385+[(35,351 \\
& +48,851*0.1900*240 \\
& [1+0.0500]/\{360[1.2000 \\
& -1.0100]\})/0.9900-64,720] \\
& /[1-0.3000]
\end{aligned}$$

$$= \underline{41,548} \text{ millions USD}$$

Corporate IFRS-GAAP (B/S-I/S), ISBN-13: **978-1720792789**, ISBN-10: **172079278X**

Law-7529:

If both ($q=$ [C-P]/X= Quick or Acid Test Ratio= 1.01 times), ($U=$ Utilized or Starting Capital= 64,720 millions USD), ($l=$ L/E= Leverage or Gearing Ratio= 99.00%), ($Q=$ Quoted Longterm Liabilities= 35,351 millions USD), ($I=$ Interest Expense= 513 millions USD), ($v=$ V/S= Variable Portion= 19.00%), ($p=$ 360 P/V= Procured Inventory Days= 240 Days), ($S'=$ Sales of Past Year= 48,851 millions USD), ($s=$ [S/S']-1= Sales Growth= 5.00%), ($d=$ D/A= Dividend Portion or Payout= 30.00%), ($T=$ Tax Paid= 3,385 millions USD), ($M=$ Margin of Contribution= 41,548 millions USD), and ($c=$ C/X= Current Ratio= 1.20 times), then it's ($F=$ Fixed Cost planned), is:

$$F= M\text{-}I\text{-}T\text{-}[(Q+S'vp[1+s]/\{360[c\text{-}q]\})/l\text{-}U]/[1\text{-}d]$$

$$= 41,548\text{-}513\text{-}3,385\text{-}[(35,351+48,851*0.1900*240[1+0.0500]/\{360[1.2000\text{-}1.0100]\})/0.9900\text{-}64,720]/[1\text{-}0.3000]$$

$$= 29,750 \text{ millions USD}$$

Corporate IFRS-GAAP (B/S-I/S), ISBN-13: **978-1720792789**, ISBN-10: **172079278X**

Law-7530:

If both (**q**= [C-P]/X= Quick or Acid Test Ratio= 1.01 times), (**U**= Utilized or Starting Capital= 64,720 millions USD), (**⌐**= L/E= Leverage or Gearing Ratio= 99.00%), (**Q**= Quoted Longterm Liabilities= 35,351 millions USD), (**F**= Fixed Cost= 29,740 millions USD), (**v**= V/S= Variable Portion= 19.00%), (**p**= 360 P/V= Procured Inventory Days= 240 Days), (**S'**= Sales of Past Year= 48,851 millions USD), (**s**= [S/S']-1= Sales Growth= 5.00%), (**d**= D/A= Dividend Portion or Payout= 30.00%), (**T**= Tax Paid= 3,385 millions USD), (**M**= Margin of Contribution= 41,548 millions USD), and (**c**= C/X= Current Ratio= 1.20 times), then it's (**I**= Interest Extreme planned), is:

$$I= M\text{-}F\text{-}T\text{-}[(Q+S'vp[1+s]/\{360[c\text{-}q]\})/⌐\text{-}U]$$
$$/[1\text{-}d]$$

$$= 41{,}548\text{-}29{,}750\text{-}3{,}385\text{-}[(35{,}351$$
$$+48{,}851*0.1900*240$$
$$[1+0.0500]/\{360[1.2000$$
$$-1.0100]\})/0.9900\text{-}64{,}720]$$
$$/[1\text{-}0.3000]$$

$$= \underline{513} \text{ millions USD}$$

Corporate IFRS-GAAP (B/S-I/S), ISBN-13: **978-1720792789**, ISBN-10: **172079278X**

Law-7531:

If both (**q**= [C-P]/X= Quick or Acid Test Ratio= 1.01
times), (**U**= Utilized or Starting Capital= 64,720
millions USD), (**I**= L/E= Leverage or Gearing Ratio=
99.00%), (**Q**= Quoted Longterm Liabilities= 35,351
millions USD), (**F**= Fixed Cost= 29,740 millions
USD), (**v**= V/S= Variable Portion= 19.00%), (**p**= 360
P/V= Procured Inventory Days= 240 Days), (**S'**=
Sales of Past Year= 48,851 millions USD), (**s**=
[S/S']-1= Sales Growth= 5.00%), (**d**= D/A=
Dividend Portion or Payout= 30.00%), (**I**= Interest
Expense= 513 millions USD), (**M**= Margin of
Contribution= 41,548 millions USD), and (**c**= C/X=
Current Ratio= 1.20 times), then it's (**T**= Tax Paid
planned), is:

$$T= M-F-I-[(Q+S'vp[1+s]/\{360[c-q]\})/I -U]$$
$$/[1-d]$$

$$= 41,548-29,750-513-[(35,351$$
$$+48,851*0.1900*240$$
$$[1+0.0500]/\{360[1.2000$$
$$-1.0100]\})/0.9900-64,720]$$
$$/[1-0.3000]$$

$$= 3,385 \text{ millions USD}$$

Corporate IFRS-GAAP (B/S-I/S), ISBN-13: **978-1720792789**, ISBN-10: **172079278X**

<u>Law-7532</u>:

If both (**q**= [C-P]/X= Quick or Acid Test Ratio= <u>1.01</u> times), (**U**= Utilized or Starting Capital= <u>64,720</u> millions USD), (**⊩**= L/E= Leverage or Gearing Ratio= <u>99.00%</u>), (**Q**= Quoted Longterm Liabilities= <u>35,351</u> millions USD), (**F**= Fixed Cost= <u>29,740</u> millions USD), (**v**= V/S= Variable Portion= <u>19.00%</u>), (**p**= 360 P/V= Procured Inventory Days= <u>240</u> Days), (**$'**= Sales of Past Year= <u>48,851</u> millions USD), (**s**= [S/S']-1= Sales Growth= <u>5.00%</u>), (**T**= Tax Paid= <u>3,385</u> millions USD), (**I**= Interest Expense= <u>513</u> millions USD), (**M**= Margin of Contribution= <u>41,548</u> millions USD), and (**c**= C/X= Current Ratio= <u>1.20</u> times), then it's (**d**= Dividend Portion or Payout planned), is:

$$d= 1-[(Q+\$'vp[1+s]/\{360[c-q]\})/I -U]$$
$$/[M-F-I-T]$$

$$= 1- [(35,351+48,851*0.1900*240$$
$$[1+0.0500]/\{360[1.2000$$
$$-1.0100]\})/0.9900-64,720]$$
$$/[41,548-29,750-513-3,385]$$

$$= \underline{30.00\%}$$

Corporate IFRS-GAAP (B/S-I/S), ISBN-13: **978-1720792789**, ISBN-10: **172079278X**

<u>Law-7533</u>:

If both (**q**= [C-P]/X= Quick or Acid Test Ratio= <u>1.01</u> times), (**U**= Utilized or Starting Capital= <u>64,720</u> millions USD), (**l**= L/E= Leverage or Gearing Ratio= <u>99.00%</u>), (**Q**= Quoted Longterm Liabilities= <u>35,351</u> millions USD), (**F**= Fixed Cost= <u>29,740</u> millions USD), (**v**= V/S= Variable Portion= <u>19.00%</u>), (**p**= 360 P/V= Procured Inventory Days= <u>240</u> Days), (**d**= D/A= Dividend Portion or Payout= <u>30.00%</u>), (**s**= [S/S']-1= Sales Growth= <u>5.00%</u>), (**T**= Tax Paid= <u>3,385</u> millions USD), (**I**= Interest Expense= <u>513</u> millions USD), (**M**= Margin of Contribution= <u>41,548</u> millions USD), and (**c**= C/X= Current Ratio= <u>1.20</u> times), then it's (**S'**= Sales Past), must be:

$$S' = 360[c\text{-}q](l \, \{U+[M\text{-}F\text{-}I\text{-}T][1\text{-}d]\}\text{-}Q)$$
$$/\{vp[1+s]\}$$
$$= 360[1.2000\text{-}1.0100](0.9900$$
$$\{64,720+[41,548\text{-}29,750$$
$$-513\text{-}3,385][1\text{-}0.3000]\}$$
$$-35,351)/\{0.1900*240$$
$$[1+0.0500]\}$$
$$= \underline{48,851} \text{ millions USD}$$

Corporate IFRS-GAAP (B/S-I/S), ISBN-13: **978-1720792789**, ISBN-10: **172079278X**

Law-7534:

If both (**q**= [C-P]/X= Quick or Acid Test Ratio= 1.01 times), (**U**= Utilized or Starting Capital= 64,720 millions USD), (**I**= L/E= Leverage or Gearing Ratio= 99.00%), (**Q**= Quoted Longterm Liabilities= 35,351 millions USD), (**F**= Fixed Cost= 29,740 millions USD), (**S'**= Sales of Past Year= 48,851 millions USD), (**p**= 360 P/V= Procured Inventory Days= 240 Days), (**d**= D/A= Dividend Portion or Payout= 30.00%), (**s**= [S/S']-1= Sales Growth= 5.00%), (**T**= Tax Paid= 3,385 millions USD), (**I**= Interest Expense= 513 millions USD), (**M**= Margin of Contribution= 41,548 millions USD), and (**c**= C/X= Current Ratio= 1.20 times), then it's (**v**= Variable Portion or Payout planned), is:

$$v= 360[\mathbf{c}\text{-}\mathbf{q}](\mathbf{I} \{\mathbf{U}+[\mathbf{M}\text{-}\mathbf{F}\text{-}\mathbf{I}\text{-}\mathbf{T}][1\text{-}\mathbf{d}]\}\text{-}\mathbf{Q})$$
$$/\{\mathbf{S'}\mathbf{p}[1+\mathbf{s}]\}$$
$$= 360[1.2000\text{-}1.0100](0.9900$$
$$\{64,720+[41,548\text{-}29,750$$
$$-513\text{-}3,385][1\text{-}0.3000]\}$$
$$-35,351)/\{48,851$$
$$*240[1+0.0500]\}$$
$$= \underline{19.00\%}$$

Corporate IFRS-GAAP (B/S-I/S), ISBN-13: **978-1720792789**, ISBN-10: **172079278X**

<u>Law-7535</u>:

If both (**q**= [C-P]/X= Quick or Acid Test Ratio= <u>1.01</u> times), (**U**= Utilized or Starting Capital= <u>64,720</u> millions USD), (**I**= L/E= Leverage or Gearing Ratio= <u>99.00%</u>), (**Q**= Quoted Longterm Liabilities= <u>35,351</u> millions USD), (**F**= Fixed Cost= <u>29,740</u> millions USD), (**S'**= Sales of Past Year= <u>48,851</u> millions USD), (**v**= V/S= Variable Portion= <u>19.00%</u>), (**d**= D/A= Dividend Portion or Payout= <u>30.00%</u>), (**s**= [S/S']-1= Sales Growth= <u>5.00%</u>), (**T**= Tax Paid= <u>3,385</u> millions USD), (**I**= Interest Expense= <u>513</u> millions USD), (**M**= Margin of Contribution= <u>41,548</u> millions USD), and (**c**= C/X= Current Ratio= <u>1.20</u> times), then it's (**p**= Procured Inventory Days planned), is:

$$p= 360[c-q](I \{U+[M-F-I-T][1-d]\}-Q) /\{S'v[1+s]\}$$

$$= 360[1.2000-1.0100](0.9900$$
$$\{64,720+[41,548-29,750$$
$$-513-3,385][1-0.3000]\}$$
$$-35,351)/\{48,851*0.1900$$
$$[1+0.0500]\}$$

$$= \underline{240} \text{ days}$$

Corporate IFRS-GAAP (B/S-I/S), ISBN-13: **978-1720792789**, ISBN-10: **172079278X**

Law-7536:

If both (**q**= [C-P]/X= Quick or Acid Test Ratio= 1.01 times), (**U**= Utilized or Starting Capital= 64,720 millions USD), (**⌿**= L/E= Leverage or Gearing Ratio= 99.00%), (**Q**= Quoted Longterm Liabilities= 35,351 millions USD), (**F**= Fixed Cost= 29,740 millions USD), (**S'**= Sales of Past Year= 48,851 millions USD), (**v**= V/S= Variable Portion= 19.00%), (**d**= D/A= Dividend Portion or Payout= 30.00%), (**p**= 360 P/V= Procured Inventory Days= 240 Days), (**T**= Tax Paid= 3,385 millions USD), (**I**= Interest Expense= 513 millions USD), (**M**= Margin of Contribution= 41,548 millions USD), and (**c**= C/X= Current Ratio= 1.20 times), then it's (**s**= Sales Growth planned), is:

$$s= 360[\textbf{c-q}](\textbf{⌿}\{\textbf{U}+[\textbf{M-F-I-T}][1\textbf{-d}]\}\textbf{-Q})$$
$$/[\textbf{S'vp}]-1$$

$$= 360[1.2000-1.0100](0.9900\{64,720$$
$$+[41,548-29,750-513-3,385]$$
$$[1-0.3000]\}-35,351)$$
$$/[48,851*0.1900*240]-1$$

$$= 5.00\%$$

Corporate IFRS-GAAP (B/S-I/S), ISBN-13: **978-1720792789**, ISBN-10: **172079278X**

Law-7537:

If both (**q**= [C-P]/X= Quick or Acid Test Ratio= 1.01 times), (**U**= Utilized or Starting Capital= 64,720 millions USD), (**I**= L/E= Leverage or Gearing Ratio= 99.00%), (**Q**= Quoted Longterm Liabilities= 35,351 millions USD), (**F**= Fixed Cost= 29,740 millions USD), (**S'**= Sales of Past Year= 48,851 millions USD), (**v**= V/S= Variable Portion= 19.00%), (**d**= D/A= Dividend Portion or Payout= 30.00%), (**p**= 360 P/V= Procured Inventory Days= 240 Days), (**T**= Tax Paid= 3,385 millions USD), (**I**= Interest Expense= 513 millions USD), (**M**= Margin of Contribution= 41,548 millions USD), and (**s**= [S/S']-1= Sales Growth= 5.00%), then it's (**c**= Current Ratio planned), is:

$$c= q+S'vp[1+s]/[360(I \{U+[M-F-I-T]$$
$$[1-d]\}-Q)]/\{S'vp[1+s]\}$$
$$= 1.0100+48.851*0.1900*240$$
$$[1+0.0500]/[360(0.9900$$
$$\{64,720+[41,548-29,750$$
$$-513-3,385][1-0.3000]\}$$
$$-35,351)]/\{48,851*0.1900$$
$$*240[1+0.0500]\}$$
$$= \underline{1.20} \text{ times}$$

Corporate IFRS-GAAP (B/S-I/S), ISBN-13: **978-1720792789**, ISBN-10: **172079278X**

Law-7538:

If both (c= C/X= Current Ratio= 1.20 times), (U= Utilized or Starting Capital= 64,720 millions USD), (I= L/E= Leverage or Gearing Ratio= 99.00%), (Q= Quoted Longterm Liabilities= 35,351 millions USD), (F= Fixed Cost= 29,740 millions USD), (S'= Sales of Past Year= 48,851 millions USD), (v= V/S= Variable Portion= 19.00%), (d= D/A= Dividend Portion or Payout= 30.00%), (p= 360P/V= Procured Inventory Days= 240 Days), (T= Tax Paid= 3,385 millions USD), (I= Interest Expense= 513 millions USD), (M= Margin of Contribution= 41,548 millions USD), and (s= [S/S']-1= Sales Growth= 5.00%), then it's (q= Quick or Acid Test Ratio planned), is:

$$q= c\text{-}S'vp[1+s]/[360(I\ \{U+[M\text{-}F\text{-}I\text{-}T]$$
$$[1\text{-}d]\}\text{-}Q)]/\{S'vp[1+s]\}$$
$$= 1.2000\text{-}48{,}851*0.1900*240$$
$$[1+0.0500]/[360(0.9900$$
$$\{64{,}720+[41{,}548\text{-}29{,}750$$
$$\text{-}513\text{-}3{,}385][1\text{-}0.3000]\}$$
$$\text{-}35{,}351)]/\{48{,}851*0.1900$$
$$*240[1+0.0500]\}$$
$$= 1.01 \text{ times}$$

Law-7539:

If both (**U**= Utilized or Starting Capital= 64,720 millions USD), (**I**= L/E= Leverage or Gearing Ratio= 99.00%), (**F**= Fixed Cost= 29,740 millions USD), (**A**= After Tax Income= 7,899 millions USD), (**d**= D/A= Dividend Portion or Payout= 30.00%), (**T**= Tax Paid= 3,385 millions USD), (**I**= Interest Expense= 513 millions USD), (**M**= Margin of Contribution= 41,548 millions USD), and (**X**= Xpress or Current Debt= 34,196 millions USD), then it's (**Q**= Quoted Longterm Liabilities planned), is:

$$Q = I\,[U+M-F-I-T-Ad]-X$$
$$= 0.9900[64,720+41,548-29,750$$
$$-513-3,385-7,899*0.3000]$$
$$-34,196$$
$$= 35,351 \text{ millions USD}$$

Corporate IFRS-GAAP (B/S-I/S), ISBN-13: **978-1720792789**, ISBN-10: **172079278X**

Law-7540:

If both (**U**= Utilized or Starting Capital= 64,720 millions USD), (**Q**= Quoted Longterm Liabilities= 35,351 millions USD), (**F**= Fixed Cost= 29,740 millions USD), (**A**= After Tax Income= 7,899 millions USD), (**d**= D/A= Dividend Portion or Payout= 30.00%), (**T**= Tax Paid= 3,385 millions USD), (**I**= Interest Expense= 513 millions USD), (**M**= Margin of Contribution= 41,548 millions USD), and (**X**= Xpress or Current Debt= 34,196 millions USD), then it's (**I** = Quoted Longterm Liabilities planned), is:

$$I = [Q+X]/[U+M-F-I-T-Ad]$$
$$= [35,351+34,196]/[64,720+41,548$$
$$-29,750-513-3,385$$
$$-7,899*0.3000]$$
$$= 99.00\%$$

Corporate IFRS-GAAP (B/S-I/S), ISBN-13: **978-1720792789**, ISBN-10: **172079278X**

Law-7541:

If both (**I**= L/E= Leverage or Gearing Ratio= 99.00%), (**Q**= Quoted Longterm Liabilities= 35,351 millions USD), (**F**= Fixed Cost= 29,740 millions USD), (**A**= After Tax Income= 7,899 millions USD), (**d**= D/A= Dividend Portion or Payout= 30.00%), (**T**= Tax Paid= 3,385 millions USD), (**I**= Interest Expense= 513 millions USD), (**M**= Margin of Contribution= 41,548 millions USD), and (**X**= Xpress or Current Debt= 34,196 millions USD), then it's (**U**= Utilized or Starting Capital planned), is:

$$U= [Q+X]/I -[M-F-I-T-Ad]$$

$$= [35,351+34,196]/0.9900-[41,548$$
$$-29,750-513-3,385$$
$$-7,899*0.3000]$$

$$= \underline{64,720} \text{ millions USD}$$

Corporate IFRS-GAAP (B/S-I/S), ISBN-13: **978-1720792789**, ISBN-10: **172079278X**

Law-7542:

If both (I= L/E= Leverage or Gearing Ratio= 99.00%), (Q= Quoted Longterm Liabilities= 35,351 millions USD), (F= Fixed Cost= 29,740 millions USD), (A= After Tax Income= 7,899 millions USD), (d= D/A= Dividend Portion or Payout= 30.00%), (T= Tax Paid= 3,385 millions USD), (I= Interest Expense= 513 millions USD), (U= Utilized or Starting Capital= 64,720 millions USD), and (X= Xpress or Current Debt= 34,196 millions USD), then it's (M= Margin of Contribution planned), is:

$$M= F+I+T+Ad-U+[Q+X]/I$$

$$= 29,750+513+3,385+7,899*0.3000$$
$$-64,720+[35,351+34,196]$$
$$/0.9900$$

$$= 41,548 \text{ millions USD}$$

Corporate IFRS-GAAP (B/S-I/S), ISBN-13: **978-1720792789**, ISBN-10: **172079278X**

Law-7543:

If both (**I**= L/E= Leverage or Gearing Ratio= 99.00%), (**Q**= Quoted Longterm Liabilities= 35,351 millions USD), (**M**= Margin of Contribution= 41,548 millions USD), (**A**= After Tax Income= 7,899 millions USD), (**d**= D/A= Dividend Portion or Payout= 30.00%), (**T**= Tax Paid= 3,385 millions USD), (**I**= Interest Expense= 513 millions USD), (**U**= Utilized or Starting Capital= 64,720 millions USD), and (**X**= Xpress or Current Debt= 34,196 millions USD), then it's (**F**= Fixed Cost planned), is:

$$F= U+M-I-T-Ad-[Q+X]/I$$

$$= 64,720+41,548-513-3,385-7,899$$
$$*0.3000-[35,351+34,196]$$
$$/0.9900$$

$$= 29,750 \text{ millions USD}$$

Corporate IFRS-GAAP (B/S-I/S), ISBN-13: **978-1720792789**, ISBN-10: **172079278X**

Law-7544:

If both (I= L/E= Leverage or Gearing Ratio= 99.00%), (Q= Quoted Longterm Liabilities= 35,351 millions USD), (M= Margin of Contribution= 41,548 millions USD), (A= After Tax Income= 7,899 millions USD), (d= D/A= Dividend Portion or Payout= 30.00%), (T= Tax Paid= 3,385 millions USD), (F= Fixed Cost= 29,740 millions USD), (U= Utilized or Starting Capital= 64,720 millions USD), and (X= Xpress or Current Debt= 34,196 millions USD), then it's (I= Interest Expense planned), is:

$$I= U+M-F-T-Ad-[Q+X]/I$$
$$= 64{,}720+41{,}548-29{,}750-3{,}385$$
$$-7{,}899*0.3000-[35{,}351$$
$$+34{,}196]/0.9900$$
$$= \underline{513} \text{ millions USD}$$

Corporate IFRS-GAAP (B/S-I/S), ISBN-13: **978-1720792789**, ISBN-10: **172079278X**

Law-7545:

If both (**I**= L/E= Leverage or Gearing Ratio= 99.00%), (**Q**= Quoted Longterm Liabilities= 35,351 millions USD), (**M**= Margin of Contribution= 41,548 millions USD), (**A**= After Tax Income= 7,899 millions USD), (**d**= D/A= Dividend Portion or Payout= 30.00%), (**I**= Interest Expense= 513 millions USD), (**F**= Fixed Cost= 29,740 millions USD), (**U**= Utilized or Starting Capital= 64,720 millions USD), and (**X**= Xpress or Current Debt= 34,196 millions USD), then it's (**T**= Tax Paid planned), is:

$$T= U+M-F-I-Ad-[Q+X]/I$$
$$= 64,720+41,548-29,750-513$$
$$-7,899*0.3000-[35,351$$
$$+34,196]/0.9900$$
$$= \underline{3,385} \text{ millions USD}$$

Corporate IFRS-GAAP (B/S-I/S), ISBN-13: **978-1720792789**, ISBN-10: **172079278X**

Law-7546:

If both (**I**= L/E= Leverage or Gearing Ratio= 99.00%), (**Q**= Quoted Longterm Liabilities= 35,351 millions USD), (**M**= Margin of Contribution= 41,548 millions USD), (**T**= Tax Paid= 3,385 millions USD), (**d**= D/A= Dividend Portion or Payout= 30.00%), (**I**= Interest Expense= 513 millions USD), (**F**= Fixed Cost= 29,740 millions USD), (**U**= Utilized or Starting Capital= 64,720 millions USD), and (**X**= Xpress or Current Debt= 34,196 millions USD), then it's (**A**= After Tax Income planned), is:

$$A = \{U+M-F-I-T-[Q+X]/I\}/d$$
$$= \{64,720+41,548-29,750-513-3,385$$
$$-[35,351+34,196]/0.9900\}$$
$$/0.3000$$
$$= \underline{7,899} \text{ millions USD}$$

Corporate IFRS-GAAP (B/S-I/S), ISBN-13: **978-1720792789**, ISBN-10: **172079278X**

Law-7547:

If both (**I**= L/E= Leverage or Gearing Ratio= 99.00%), (**Q**= Quoted Longterm Liabilities= 35,351 millions USD), (**M**= Margin of Contribution= 41,548 millions USD), (**T**= Tax Paid= 3,385 millions USD), (**A**= After Tax Income= 7,899 millions USD), (**I**= Interest Expense= 513 millions USD), (**F**= Fixed Cost= 29,740 millions USD), (**U**= Utilized or Starting Capital= 64,720 millions USD), and (**X**= Xpress or Current Debt= 34,196 millions USD), then it's (**d**= Dividend Portion or Payout planned), is:

$$d= \{U+M-F-I-T-[Q+X]/I\}/A$$
$$= \{64,720+41,548-29,750-513-3,385$$
$$-[35,351+34,196]/0.9900\}$$
$$/7,899$$
$$= 30.00\%$$

Corporate IFRS-GAAP (B/S-I/S), ISBN-13: **978-1720792789**, ISBN-10: **172079278X**

Law-7548:

If both (**I**= L/E= Leverage or Gearing Ratio=
99.00%), (**Q**= Quoted Longterm Liabilities= 35,351
millions USD), (**M**= Margin of Contribution= 41,548
millions USD), (**T**= Tax Paid= 3,385 millions USD),
(**A**= After Tax Income= 7,899 millions USD), (**I**=
Interest Expense= 513 millions USD), (**F**= Fixed
Cost= 29,740 millions USD), (**U**= Utilized or Starting
Capital= 64,720 millions USD), and (**d**= D/A=
Dividend Portion or Payout= 30.00%), then it's (**X**=
Xpress or Current Debt planned), is:

$$X= I[U+M-F-I-T-Ad]-Q$$
$$= 0.9900[64,720+41,548-29,750-513$$
$$-3,385-7,899*0.3000]-35,351$$
$$= 34,196 \text{ millions USD}$$

Corporate IFRS-GAAP (B/S-I/S), ISBN-13: **978-1720792789**, ISBN-10: **172079278X**

Law-7549:

If both (**I**= L/E= Leverage or Gearing Ratio=
99.00%), (**c**= C/X= Current Ratio= 1.20 times), (**q**=
[C-P]/X= Quick or Acid Test Ratio= 1.01 times), (**P**=
Procured Inventories= 6,497 millions USD), (**M**=
Margin of Contribution= 41,548 millions USD), (**T**=
Tax Paid= 3,385 millions USD), (**A**= After Tax
Income= 7,899 millions USD), (**I**= Interest Expense=
513 millions USD), (**F**= Fixed Cost= 29,740 millions
USD), (**U**= Utilized or Starting Capital= 64,720
millions USD), and (**d**= D/A= Dividend Portion or
Payout= 30.00%), then it's (**Q**= Quoted Longterm
Liabilities planned), is:

$$Q= I[U+M-F-I-T-Ad]-P/[c-q]$$
$$= 0.9900[64,720+41,548-29,750$$
$$-513-3,385-7,899*0.3000]$$
$$-8,497/[1.2000-1.0100]$$
$$= 35,351 \text{ millions USD}$$

Corporate IFRS-GAAP (B/S-I/S), ISBN-13: **978-1720792789**, ISBN-10: **172079278X**

<u>Law-7550</u>:

If both (**Q**= Quoted Longterm Liabilities= <u>35,351</u> millions USD), (**c**= C/X= Current Ratio= <u>1.20</u> times), (**q**= [C-P]/X= Quick or Acid Test Ratio= <u>1.01</u> times), (**P**= Procured Inventories= <u>6,497</u> millions USD), (**M**= Margin of Contribution= <u>41,548</u> millions USD), (**T**= Tax Paid= <u>3,385</u> millions USD), (**A**= After Tax Income= <u>7,899</u> millions USD), (**I**= Interest Expense= <u>513</u> millions USD), (**F**= Fixed Cost= <u>29,740</u> millions USD), (**U**= Utilized or Starting Capital= <u>64,720</u> millions USD), and (**d**= D/A= Dividend Portion or Payout= <u>30.00%</u>), then it's (**I** = Leverage or Gearing Ratio planned), is:

$$I = \{Q+P/[c\text{-}q]\}/[U+M\text{-}F\text{-}I\text{-}T\text{-}Ad]$$
$$= \{35,351+8,497/[1.2000\text{-}1.0100]\}$$
$$/[64,720+41,548\text{-}29,750$$
$$-513\text{-}3,385\text{-}7,899*0.3000]$$
$$= \underline{99.00\%}$$

Law-7551:

If both (Q= Quoted Longterm Liabilities= 35,351 millions USD), (c= C/X= Current Ratio= 1.20 times), (q= [C-P]/X= Quick or Acid Test Ratio= 1.01 times), (P= Procured Inventories= 6,497 millions USD), (M= Margin of Contribution= 41,548 millions USD), (T= Tax Paid= 3,385 millions USD), (A= After Tax Income= 7,899 millions USD), (I= Interest Expense= 513 millions USD), (F= Fixed Cost= 29,740 millions USD), (I= L/E= Leverage or Gearing Ratio= 99.00%), and (d= D/A= Dividend Portion or Payout= 30.00%), then it's (U= Utilized or Starting Capital planned), is:

$$U= \{Q+P/[c\text{-}q]\}/I\text{-}[M\text{-}F\text{-}I\text{-}T\text{-}Ad]$$
$$= \{35,351+8,497/[1.2000\text{-}1.0100]\}$$
$$/0.9900\text{-}[41,548\text{-}29,750$$
$$\text{-}513\text{-}3,385\text{-}7,899*0.3000]$$
$$= \underline{64,720} \text{ millions USD}$$

Corporate IFRS-GAAP (B/S-I/S), ISBN-13: **978-1720792789**, ISBN-10: **172079278X**

Law-7552:

If both (**Q**= Quoted Longterm Liabilities= 35,351 millions USD), (**c**= C/X= Current Ratio= 1.20 times), (**q**= [C-P]/X= Quick or Acid Test Ratio= 1.01 times), (**P**= Procured Inventories= 6,497 millions USD), (**U**= Utilized or Starting Capital= 64,720 millions USD), (**T**= Tax Paid= 3,385 millions USD), (**A**= After Tax Income= 7,899 millions USD), (**I**= Interest Expense= 513 millions USD), (**F**= Fixed Cost= 29,740 millions USD), (**I**= L/E= Leverage or Gearing Ratio= 99.00%), and (**d**= D/A= Dividend Portion or Payout= 30.00%), then it's (**M**= Margin of Contribution planned), is:

$$M= F+I+T+Ad+\{Q+P/[c-q]\}/I$$
$$= 29,750+513+3,385+7,899*0.3000$$
$$+\{35,351+8,497/[1.2000$$
$$-1.0100]\}/0.9900$$
$$= \underline{41,548} \text{ millions USD}$$

Corporate IFRS-GAAP (B/S-I/S), ISBN-13: **978-1720792789**, ISBN-10: **172079278X**

<u>Law-7553</u>:

If both (**Q**= Quoted Longterm Liabilities= <u>35,351</u> millions USD), (**c**= C/X= Current Ratio= <u>1.20</u> times), (**q**= [C-P]/X= Quick or Acid Test Ratio= <u>1.01</u> times), (**P**= Procured Inventories= <u>6,497</u> millions USD), (**U**= Utilized or Starting Capital= <u>64,720</u> millions USD), (**T**= Tax Paid= <u>3,385</u> millions USD), (**A**= After Tax Income= <u>7,899</u> millions USD), (**I**= Interest Expense= <u>513</u> millions USD), (**M**= Margin of Contribution= <u>41,548</u> millions USD), (**I**= L/E= Leverage or Gearing Ratio= <u>99.00%</u>), and (**d**= D/A= Dividend Portion or Payout= <u>30.00%</u>), then it's (**F**= Fixed Cost planned), is:

$$F= U+M-I-T-Ad-\{Q+P/[c-q]\}/I$$
$$= 64,720+41,548-513-3,385-7,899$$
$$*0.3000-\{35,351+8,497$$
$$/[1.2000-1.0100]\}/0.9900$$
$$= \underline{29,750} \text{ millions USD}$$

Corporate IFRS-GAAP (B/S-I/S), ISBN-13: **978-1720792789**, ISBN-10: **172079278X**

Law-7554:

If both (**Q**= Quoted Longterm Liabilities= 35,351 millions USD), (**c**= C/X= Current Ratio= 1.20 times), (**q**= [C-P]/X= Quick or Acid Test Ratio= 1.01 times), (**P**= Procured Inventories= 6,497 millions USD), (**U**= Utilized or Starting Capital= 64,720 millions USD), (**T**= Tax Paid= 3,385 millions USD), (**A**= After Tax Income= 7,899 millions USD), (**F**= Fixed Cost= 29,740 millions USD), (**M**= Margin of Contribution= 41,548 millions USD), (**I**= L/E= Leverage or Gearing Ratio= 99.00%), and (**d**= D/A=Dividend Portion or Payout= 30.00%), then it's (**I**= Interest Expense planned), is:

$$I= U+M-F-T-Ad-\{Q+P/[c-q]\}/I$$
$$= 64{,}720+41{,}548-29{,}750-3{,}385$$
$$-7{,}899*0.3000-\{35{,}351$$
$$+8{,}497/[1.2000-1.0100]\}$$
$$/0.9900$$
$$= \underline{513} \text{ millions USD}$$

Corporate IFRS-GAAP (B/S-I/S), ISBN-13: **978-1720792789**, ISBN-10: **172079278X**

Law-7555:

If both (**Q**= Quoted Longterm Liabilities= 35,351 millions USD), (**c**= C/X= Current Ratio= 1.20 times), (**q**= [C-P]/X= Quick or Acid Test Ratio= 1.01 times), (**P**= Procured Inventories= 6,497 millions USD), (**U**= Utilized or Starting Capital= 64,720 millions USD), (**I**= Interest Expense= 513 millions USD), (**A**= After Tax Income= 7,899 millions USD), (**F**= Fixed Cost= 29,740 millions USD), (**M**= Margin of Contribution= 41,548 millions USD), (**I**= L/E= Leverage or Gearing Ratio= 99.00%), and (**d**= D/A= Dividend Portion or Payout= 30.00%), then it's (**T**= Tax Paid planned), is:

$$T= U+M-F-I-Ad-\{Q+P/[c-q]\}/I$$
$$= 64,720+41,548-29,750-513-7,899$$
$$*0.3000-\{35,351+8,497$$
$$/[1.2000-1.0100]\}/0.9900$$
$$= 3,385 \text{ millions USD}$$

Corporate IFRS-GAAP (B/S-I/S), ISBN-13: **978-1720792789**, ISBN-10: **172079278X**

<u>Law-7556</u>:

If both (**Q**= Quoted Longterm Liabilities= <u>35,351</u> millions USD), (**c**= C/X= Current Ratio= <u>1.20</u> times), (**q**= [C-P]/X= Quick or Acid Test Ratio= <u>1.01</u> times), (**P**= Procured Inventories= <u>6,497</u> millions USD), (**U**= Utilized or Starting Capital= <u>64,720</u> millions USD), (**I**= Interest Expense= <u>513</u> millions USD), (**T**= Tax Paid= <u>3,385</u> millions USD), (**F**= Fixed Cost= <u>29,740</u> millions USD), (**M**= Margin of Contribution= <u>41,548</u> millions USD), (**I**= L/E= Leverage or Gearing Ratio= <u>99.00%</u>), and (**d**= D/A=Dividend Portion or Payout= <u>30.00%</u>), then it's (**A**= After Tax Income planned), is:

$$A= (U+M-F-I-T-\{Q+P/[c-q]\}/I)/d$$
$$= (64,720+41,548-29,750-513-3,385$$
$$-\{35,351+8,497/[1.2000$$
$$-1.0100]\}/0.9900)/0.3000$$
$$= \underline{7,899} \text{ millions USD}$$

Corporate IFRS-GAAP (B/S-I/S), ISBN-13: **978-1720792789**, ISBN-10: **172079278X**

<u>Law-7557</u>:

If both (**Q**= Quoted Longterm Liabilities= <u>35,351</u> millions USD), (**c**= C/X= Current Ratio= <u>1.20</u> times), (**q**= [C-P]/X= Quick or Acid Test Ratio= <u>1.01</u> times), (**P**= Procured Inventories= <u>6,497</u> millions USD), (**U**= Utilized or Starting Capital= <u>64,720</u> millions USD), (**I**= Interest Expense= <u>513</u> millions USD), (**T**= Tax Paid= <u>3,385</u> millions USD), (**F**= Fixed Cost= <u>29,740</u> millions USD), (**M**= Margin of Contribution= <u>41,548</u> millions USD), (**I**= L/E= Leverage or Gearing Ratio= <u>99.00%</u>), and (**A**= After Tax Income= <u>7,899</u> millions USD), then it's (**d**= Dividend Portion or Payout planned), is:

$$d= (U+M\text{-}F\text{-}I\text{-}T\text{-}\{Q+P/[c\text{-}q]\}/I)/A$$

$$= (64{,}720+41{,}548\text{-}29{,}750\text{-}513\text{-}3{,}385$$
$$-\{35{,}351+8{,}497/[1.2000$$
$$-1.0100]\}/0.9900)/7{,}899$$

$$= \underline{30.00\%}$$

Corporate IFRS-GAAP (B/S-I/S), ISBN-13: **978-1720792789**, ISBN-10: **172079278X**

Law-7558:

If both (**Q**= Quoted Longterm Liabilities= 35,351 millions USD), (**c**= C/X= Current Ratio= 1.20 times), (**q**= [C-P]/X= Quick or Acid Test Ratio= 1.01 times), (**d**= D/A=Dividend Portion or Payout= 30.00%), (**U**= Utilized or Starting Capital= 64,720 millions USD), (**I**= Interest Expense= 513 millions USD), (**T**= Tax Paid= 3,385 millions USD), (**F**= Fixed Cost= 29,740 millions USD), (**M**= Margin of Contribution= 41,548 millions USD), (**I**= L/E= Leverage or Gearing Ratio= 99.00%), and (**A**= After Tax Income= 7,899 millions USD), then it's (**P**= Procured Inventories planned), is:

$$P= [c-q]\{I[U+M-F-I-T-Ad]-Q\}$$
$$= [1.2000-1.0100]\{0.9900)[64,720$$
$$+41,548-29,750-513-3,385$$
$$-7,899*0.3000]-35,351\}$$
$$= 8,497 \text{ millions USD}$$

Corporate IFRS-GAAP (B/S-I/S), ISBN-13: **978-1720792789**, ISBN-10: **172079278X**

Law-7559:

If both (**Q**= Quoted Longterm Liabilities= 35,351 millions USD), (**P**= Procured Inventories= 6,497 millions USD), (**q**= [C-P]/X= Quick or Acid Test Ratio= 1.01 times), (**d**= D/A=Dividend Portion or Payout= 30.00%), (**U**= Utilized or Starting Capital= 64,720 millions USD), (**I**= Interest Expense= 513 millions USD), (**T**= Tax Paid= 3,385 millions USD), (**F**= Fixed Cost= 29,740 millions USD), (**M**= Margin of Contribution= 41,548 millions USD), (**I**= L/E= Leverage or Gearing Ratio= 99.00%), and (**A**= After Tax Income= 7,899 millions USD), then it's (**c**= Current Ratio planned), is:

$$c= q+P/\{I\,[U+M-F-I-T-Ad]-Q\}$$
$$= 1.0100+6,497/\{0.9900)[64,720$$
$$+41,548-29,750-513-3,385$$
$$-7,899*0.3000]-35,351\}$$
$$= 1.20 \text{ times}$$

Corporate IFRS-GAAP (B/S-I/S), ISBN-13: **978-1720792789**, ISBN-10: **172079278X**

Law-7560:

If both (**Q**= Quoted Longterm Liabilities= 35,351 millions USD), (**P**= Procured Inventories= 6,497 millions USD), (**c**= C/X= Current Ratio= 1.20 times), (**d**= D/A= Dividend Portion or Payout= 30.00%), (**U**= Utilized or Starting Capital= 64,720 millions USD), (**I**= Interest Expense= 513 millions USD), (**T**= Tax Paid= 3,385 millions USD), (**F**= Fixed Cost= 29,740 millions USD), (**M**= Margin of Contribution= 41,548 millions USD), (**I**= L/E= Leverage or Gearing Ratio= 99.00%), and (**A**= After Tax Income= 7,899 millions USD), then it's (**q**= Quick or Acid Test Ratio planned), is:

$$q= c\text{-}P/\{I\,[U+M\text{-}F\text{-}I\text{-}T\text{-}Ad]\text{-}Q\}$$
$$= 1.2000\text{-}6,497/\{0.9900)[64,720$$
$$+41,548\text{-}29,750\text{-}513\text{-}3,385$$
$$-7,899*0.3000]\text{-}35,351\}$$
$$= 1.01 \text{ times}$$

Corporate IFRS-GAAP (B/S-I/S), ISBN-13: **978-1720792789**, ISBN-10: **172079278X**

Law-7561:

If both (**q**= [C-P]/X= Quick or Acid Test Ratio= 1.01 times), (**V**= Variable Cost= 9,746 millions USD), (**p**= 360 P/V= Procured Inventory Days= 240 Days), (**c**= C/X= Current Ratio= 1.20 times), (**d**= D/A= Dividend Portion or Payout= 30.00%), (**U**= Utilized or Starting Capital= 64,720 millions USD), (**I**= Interest Expense= 513 millions USD), (**T**= Tax Paid= 3,385 millions USD), (**F**= Fixed Cost= 29,740 millions USD), (**M**= Margin of Contribution= 41,548 millions USD), (**I**= L/E= Leverage or Gearing Ratio= 99.00%), and (**A**= After Tax Income= 7,899 millions USD), then it's (**Q**= Quoted Longterm Liabilities planned), is:

$$Q = I[U+M-F-I-T-Ad]-Vp/\{360[c-q]\}$$
$$= 0.9900[64,720+41,548-29,750-513$$
$$-3,385-7,899*0.3000]-9,746$$
$$*240/\{360[1.2000-1.0100]\}$$
$$= 35,351 \text{ millions USD}$$

Corporate IFRS-GAAP (B/S-I/S), ISBN-13: **978-1720792789**, ISBN-10: **172079278X**

Law-7562:

If both (q= [C-P]/X= Quick or Acid Test Ratio= 1.01 times), (V= Variable Cost= 9,746 millions USD), (p= 360 P/V= Procured Inventory Days= 240 Days), (c= C/X= Current Ratio= 1.20 times), (d= D/A= Dividend Portion or Payout= 30.00%), (U= Utilized or Starting Capital= 64,720 millions USD), (I= Interest Expense= 513 millions USD), (T= Tax Paid= 3,385 millions USD), (F= Fixed Cost= 29,740 millions USD), (M= Margin of Contribution= 41,548 millions USD), (Q= Quoted Longterm Liabilities= 35,351 millions USD), and (A= After Tax Income= 7,899 millions USD), then it's (I = Leverage or Gearing Ratio planned), is:

$$I = (Q+Vp/\{360[c-q]\})/[U+M-F-I-T-Ad]$$
$$= (35,351+9,746*240/\{360[1.2000$$
$$-1.0100]\})/[64,720+41,548$$
$$-29,750-513-3,385-7,899$$
$$*0.3000]$$
$$= 99.00\%$$

Corporate IFRS-GAAP (B/S-I/S), ISBN-13: **978-1720792789**, ISBN-10: **172079278X**

Law-7563:

If both (**q**= [C-P]/X= Quick or Acid Test Ratio= 1.01 times), (**V**= Variable Cost= 9,746 millions USD), (**p**= 360 P/V= Procured Inventory Days= 240 Days), (**c**= C/X= Current Ratio= 1.20 times), (**d**= D/A=Dividend Portion or Payout= 30.00%), (**f**= L/E= Leverage or Gearing Ratio= 99.00%), (**I**= Interest Expense= 513 millions USD), (**T**= Tax Paid= 3,385 millions USD), (**F**= Fixed Cost= 29,740 millions USD), (**M**= Margin of Contribution= 41,548 millions USD), (**Q**= Quoted Longterm Liabilities= 35,351 millions USD), and (**A**= After Tax Income= 7,899 millions USD),then it's (**U**= Utilized or Starting Capital planned), is:

$$U= (Q+Vp/\{360[c-q]\})/f-[M-F-I-T-Ad]$$
$$= (35,351+9,746*240/\{360[1.2000$$
$$-1.0100]\})/0.9900-[41,548$$
$$-29,750-513-3,385-7,899$$
$$*0.3000]$$
$$= \underline{64,720} \text{ millions USD}$$

Corporate IFRS-GAAP (B/S-I/S), ISBN-13: **978-1720792789**, ISBN-10: **172079278X**

Law-7564:

If both (**q**= [C-P]/X= Quick or Acid Test Ratio= 1.01 times), (**V**= Variable Cost= 9,746 millions USD), (**p**= 360 P/V= Procured Inventory Days= 240 Days), (**c**= C/X= Current Ratio= 1.20 times), (**d**= D/A= Dividend Portion or Payout= 30.00%), (**l**= L/E= Leverage or Gearing Ratio= 99.00%), (**I**= Interest Expense= 513 millions USD), (**T**= Tax Paid= 3,385 millions USD), (**F**= Fixed Cost= 29,740 millions USD), (**U**= Utilized or Starting Capital= 64,720 millions USD), (**Q**= Quoted Longterm Liabilities= 35,351 millions USD), and (**A**= After Tax Income= 7,899 millions USD), then it's (**M**= Margin of Contribution planned), is:

$$M= F+I+T+Ad-U+(Q+Vp/\{360[c-q]\})/l$$
$$= 29,750+513+3,385+7,899*0.3000$$
$$-64,720+(35,351+9,746$$
$$*240/\{360[1.2000$$
$$-1.0100]\})/0.9900$$
$$= 41,548 \text{ millions USD}$$

Corporate IFRS-GAAP (B/S-I/S), ISBN-13: **978-1720792789**, ISBN-10: **172079278X**

<u>Law-7565</u>:

If both (**q**= [C-P]/X= Quick or Acid Test Ratio= <u>1.01</u> times), (**V**= Variable Cost= <u>9,746</u> millions USD), (**p**= 360 P/V= Procured Inventory Days= <u>240</u> Days), (**c**= C/X= Current Ratio= <u>1.20</u> times), (**d**= D/A= Dividend Portion or Payout= <u>30.00%</u>), (**ℓ**= L/E= Leverage or Gearing Ratio= <u>99.00%</u>), (**I**= Interest Expense= <u>513</u> millions USD), (**T**= Tax Paid= <u>3,385</u> millions USD), (**M**= Margin of Contribution= <u>41,548</u> millions USD), (**U**= Utilized or Starting Capital= <u>64,720</u> millions USD), (**Q**= Quoted Longterm Liabilities= <u>35,351</u> millions USD), and (**A**= After Tax Income= <u>7,899</u> millions USD), then it's (**F**= Fixed Cost planned), is:

$$F= U+M-I-T-Ad-(Q+Vp/\{360[c-q]\})/\ell$$

$$= 64,720+41,548-513-3,385-7,899$$
$$*0.3000- (35,351+9,746$$
$$*240/\{360[1.2000-1.0100]\})$$
$$/0.9900$$

$$= \underline{29,750} \text{ millions USD}$$

Corporate IFRS-GAAP (B/S-I/S), ISBN-13: **978-1720792789**, ISBN-10: **172079278X**

<u>Law-7566</u>:

If both (**q**= [C-P]/X= Quick or Acid Test Ratio= <u>1.01</u> times), (**V**= Variable Cost= <u>9,746</u> millions USD), (**p**= 360 P/V= Procured Inventory Days= <u>240</u> Days), (**c**= C/X= Current Ratio= <u>1.20</u> times), (**d**= D/A= Dividend Portion or Payout= <u>30.00%</u>), (**I**= L/E= Leverage or Gearing Ratio= <u>99.00%</u>), (**F**= Fixed Cost= <u>29,740</u> millions USD), (**T**= Tax Paid= <u>3,385</u> millions USD), (**M**= Margin of Contribution= <u>41,548</u> millions USD), (**U**= Utilized or Starting Capital= <u>64,720</u> millions USD), (**Q**= Quoted Longterm Liabilities= <u>35,351</u> millions USD), and (**A**= After Tax Income= <u>7,899</u> millions USD), then it's (**I**= Interest Expense planned), is:

$$I= U+M-F-T-Ad-(Q+Vp/\{360[c-q]\})/I$$

$$= 64,720+41,548-29,750-3,385-7,899$$
$$*0.3000- (35,351+9,746*240$$
$$/\{360[1.2000-1.0100]\})$$
$$/0.9900$$

$$= \underline{513} \text{ millions USD}$$

Corporate IFRS-GAAP (B/S-I/S), ISBN-13: **978-1720792789**, ISBN-10: **172079278X**

Law-7567:

If both (**q**= [C-P]/X= Quick or Acid Test Ratio= 1.01
times), (**V**= Variable Cost= 9,746 millions USD), (**p**=
360 P/V= Procured Inventory Days= 240 Days), (**c**=
C/X= Current Ratio= 1.20 times), (**d**= D/A=
Dividend Portion or Payout= 30.00%), (**I**= L/E=
Leverage or Gearing Ratio= 99.00%), (**F**= Fixed
Cost= 29,740 millions USD), (**I**= Interest Expense=
513 millions USD), (**M**= Margin of Contribution=
41,548 millions USD), (**U**= Utilized or Starting
Capital= 64,720 millions USD), (**Q**= Quoted
Longterm Liabilities= 35,351 millions USD), and
(**A**= After Tax Income= 7,899 millions USD), then
it's (**T**= Tax Paid planned), is:

$$T= U+M-F-I-Ad-(Q+Vp/\{360[c-q]\})/I$$
$$= 64,720+41,548-29,750-513-7,899$$
$$*0.3000- (35,351+9,746*240$$
$$/\{360[1.2000-1.0100]\})$$
$$/0.9900$$
$$= \underline{3,385} \text{ millions USD}$$

Corporate IFRS-GAAP (B/S-I/S), ISBN-13: **978-1720792789**, ISBN-10: **172079278X**

Law-7568:

If both (**q**= [C-P]/X= Quick or Acid Test Ratio= 1.01 times), (**V**= Variable Cost= 9,746 millions USD), (**p**= 360 P/V= Procured Inventory Days= 240 Days), (**c**= C/X= Current Ratio= 1.20 times), (**d**= D/A= Dividend Portion or Payout= 30.00%), (**l**= L/E= Leverage or Gearing Ratio= 99.00%), (**F**= Fixed Cost= 29,740 millions USD), (**I**= Interest Expense= 513 millions USD), (**M**= Margin of Contribution= 41,548 millions USD), (**U**= Utilized or Starting Capital= 64,720 millions USD), (**Q**= Quoted Longterm Liabilities= 35,351 millions USD), and (**T**= Tax Paid= 3,385 millions USD), then it's (**A**= After Tax Income planned), is:

$$A= [U+M-F-I-T-(Q+Vp/\{360[c-q]\})/l]/d$$
$$= [64,720+41,548-29,750-513-3,385$$
$$- (35,351+9,746*240/\{360$$
$$[1.2000-1.0100]\})/0.9900]$$
$$/0.3000$$
$$= 7,899 \text{ millions USD}$$

Corporate IFRS-GAAP (B/S-I/S), ISBN-13: **978-1720792789**, ISBN-10: **172079278X**

<u>Law-7569</u>:

If both (**q**= [C-P]/X= Quick or Acid Test Ratio= <u>1.01</u> times), (**V**= Variable Cost= <u>9,746</u> millions USD), (**p**= 360 P/V= Procured Inventory Days= <u>240</u> Days), (**c**= C/X= Current Ratio= <u>1.20</u> times), (**A**= After Tax Income= <u>7,899</u> millions USD), (**I**= L/E= Leverage or Gearing Ratio= <u>99.00%</u>), (**F**= Fixed Cost= <u>29,740</u> millions USD), (**I**= Interest Expense= <u>513</u> millions USD), (**M**= Margin of Contribution= <u>41,548</u> millions USD), (**U**= Utilized or Starting Capital= <u>64,720</u> millions USD), (**Q**= Quoted Longterm Liabilities= <u>35,351</u> millions USD), and (**T**= Tax Paid= <u>3,385</u> millions USD), then it's (**d**= Dividend Portion or Payout planned), is:

$$d= [U+M-F-I-T-(Q+Vp/\{360[c-q]\})/I]/A$$
$$= [64,720+41,548-29,750-513-3,385$$
$$- (35,351+9,746*240/\{360$$
$$[1.2000-1.0100]\})/0.9900]$$
$$/7,899$$
$$= \underline{30.00\%}$$

Corporate IFRS-GAAP (B/S-I/S), ISBN-13: **978-1720792789**, ISBN-10: **172079278X**

Law-7570:

If both (q= [C-P]/X= Quick or Acid Test Ratio= 1.01 times), (d= D/A= Dividend Portion or Payout= 30.00%), (p= 360 P/V= Procured Inventory Days= 240 Days), (c= C/X= Current Ratio= 1.20 times), (A= After Tax Income= 7,899 millions USD), (I= L/E= Leverage or Gearing Ratio= 99.00%), (F= Fixed Cost= 29,740 millions USD), (I= Interest Expense= 513 millions USD), (M= Margin of Contribution= 41,548 millions USD), (U= Utilized or Starting Capital= 64,720 millions USD), (Q= Quoted Longterm Liabilities= 35,351 millions USD), and (T= Tax Paid= 3,385 millions USD), then it's (V= Variable Cost planned), is:

V= 360[**c-q**]{**l** [**U+M-F-I-T-Ad**]-**Q**}/**p**

$\quad$ = 360[1.2000-1.0100]{0.9900

$\qquad$ [64,720+41,548-29,750-513

$\qquad$ -3,385-7,899*0.3000]

$\qquad$ -35,351}/240

$\quad$ = 9,746 millions USD

Corporate IFRS-GAAP (B/S-I/S), ISBN-13: **978-1720792789**, ISBN-10: **172079278X**

Law-7571:

If both (**q**= [C-P]/X= Quick or Acid Test Ratio= 1.01 times), (**d**= D/A= Dividend Portion or Payout= 30.00%), (**V**= Variable Cost= 9,746 millions USD), (**c**= C/X= Current Ratio= 1.20 times), (**A**= After Tax Income= 7,899 millions USD), (**I**= L/E= Leverage or Gearing Ratio= 99.00%), (**F**= Fixed Cost= 29,740 millions USD), (**I**= Interest Expense= 513 millions USD), (**M**= Margin of Contribution= 41,548 millions USD), (**U**= Utilized or Starting Capital= 64,720 millions USD), (**Q**= Quoted Longterm Liabilities= 35,351 millions USD), and (**T**= Tax Paid= 3,385 millions USD), then it's (**p**= Procured Inventory Days planned), is:

$$\mathbf{p}= 360[\mathbf{c}\text{-}\mathbf{q}]\{\mathbf{I}\,[\mathbf{U}\text{+}\mathbf{M}\text{-}\mathbf{F}\text{-}\mathbf{I}\text{-}\mathbf{T}\text{-}\mathbf{Ad}]\text{-}\mathbf{Q}\}/\mathbf{V}$$

$$= 360[1.2000\text{-}1.0100]\{0.9900$$
$$[64,720+41,548\text{-}29,750\text{-}513$$
$$-3,385\text{-}7,899*0.3000]$$
$$-35,351\}/9,746$$

$$= \underline{240}\ \text{days}$$

Corporate IFRS-GAAP (B/S-I/S), ISBN-13: **978-1720792789**, ISBN-10: **172079278X**

Law-7572:

If both (q= [C-P]/X= Quick or Acid Test Ratio= 1.01 times), (d= D/A= Dividend Portion or Payout= 30.00%), (V= Variable Cost= 9,746 millions USD), (p= 360 P/V= Procured Inventory Days= 240 Days), (A= After Tax Income= 7,899 millions USD), (l= L/E= Leverage or Gearing Ratio= 99.00%), (F= Fixed Cost= 29,740 millions USD), (I= Interest Expense= 513 millions USD), (M= Margin of Contribution= 41,548 millions USD), (U= Utilized or Starting Capital= 64,720 millions USD), (Q= Quoted Longterm Liabilities= 35,351 millions USD), and (T= Tax Paid= 3,385 millions USD), then it's (c= Current Ratio planned), is:

$$c= q+Vp/(360\{I\,[U+M-F-I-T-Ad]-Q\})$$
$$= 1.0100+9,746*240/(360\{0.9900$$
$$[64,720+41,548-29,750-513$$
$$-3,385-7,899*0.3000]$$
$$-35,351\})$$
$$= 1.20 \text{ times}$$

Corporate IFRS-GAAP (B/S-I/S), ISBN-13: **978-1720792789**, ISBN-10: **172079278X**

Law-7573:

If both (**c**= C/X= Current Ratio= 1.20 times), (**d**=
D/A =Dividend Portion or Payout= 30.00%), (**V**=
Variable Cost= 9,746 millions USD), (**p**= 360 P/V=
Procured Inventory Days= 240 Days), (**A**= After Tax
Income= 7,899 millions USD), (**I**= L/E= Leverage or
Gearing Ratio= 99.00%), (**F**= Fixed Cost= 29,740
millions USD), (**I**= Interest Expense= 513 millions
USD), (**M**= Margin of Contribution= 41,548 millions
USD), (**U**= Utilized or Starting Capital= 64,720
millions USD), (**Q**= Quoted Longterm Liabilities=
35,351 millions USD), and (**T**= Tax Paid= 3,385
millions USD), then it's (**q**= Quick or Acid Test
Ratio planned), is:

$$q= c\text{-}Vp/(360\{I\,[U\text{+}M\text{-}F\text{-}I\text{-}T\text{-}Ad]\text{-}Q\})$$
$$= 1.2000\text{-}9,746*240/(360\{0.9900$$
$$[64,720+41,548\text{-}29,750\text{-}513$$
$$\text{-}3,385\text{-}7,899*0.3000]$$
$$\text{-}35,351\})$$
$$= 1.01 \text{ times}$$

Corporate IFRS-GAAP (B/S-I/S), ISBN-13: **978-1720792789**, ISBN-10: **172079278X**

<u>Law-7574</u>:

If both (**c**= C/X= Current Ratio= <u>1.20</u> times), (**d**= D/A= Dividend Portion or Payout= <u>30.00%</u>), (**$**= Sales or Revenues= <u>51,294</u> millions USD), (**v**= V/S= Variable Portion= <u>19.00%</u>), (**p**= 360 P/V= Procured Inventory Days= <u>240</u> Days), (**A**= After Tax Income= <u>7,899</u> millions USD), (**$\digamma$**= L/E= Leverage or Gearing Ratio= <u>99.00%</u>), (**F**= Fixed Cost= <u>29,740</u> millions USD), (**I**= Interest Expense= <u>513</u> millions USD), (**M**= Margin of Contribution= <u>41,548</u> millions USD), (**U**= Utilized or Starting Capital= <u>64,720</u> millions USD), (**q**= [C-P]/X= Quick or Acid Test Ratio= <u>1.01</u> times), and (**T**= Tax Paid= <u>3,385</u> millions USD), then it's (**Q**= Quoted Longterm Liabilities planned), is:

$$Q= \digamma\,[U+M-F-I-T-Ad]-$vp/\{360[c-q]\}$$
$$= 0.9900\ [64,720+41,548-29,750-513$$
$$-3,385-7,899*0.3000]-51,294$$
$$*0.1900*240/\{360[1.2000$$
$$-1.0100]\}$$
$$= \underline{35,351}\ \text{millions USD}$$

Corporate IFRS-GAAP (B/S-I/S), ISBN-13: **978-1720792789**, ISBN-10: **172079278X**

Law-7575:

If both (**c**= C/X= Current Ratio= 1.20 times), (**d**= D/A= Dividend Portion or Payout= 30.00%), (**S**= Sales or Revenues= 51,294 millions USD), (**v**= V/S= Variable Portion= 19.00%), (**p**= 360 P/V= Procured Inventory Days= 240 Days), (**A**= After Tax Income= 7,899 millions USD), (**Q**= Quoted Longterm Liabilities= 35,351 millions USD), (**F**= Fixed Cost= 29,740 millions USD), (**I**= Interest Expense= 513 millions USD), (**M**= Margin of Contribution= 41,548 millions USD), (**U**= Utilized or Starting Capital= 64,720 millions USD), (**q**= [C-P]/X= Quick or Acid Test Ratio= 1.01 times), and (**T**= Tax Paid= 3,385 millions USD), then it's (**l** = Leverage or Gearing Ratio planned), is:

$$l = (Q+Svp/\{360[c-q]\})/[U+M-F-I-T-Ad]$$
$$= (35,351+51,294*0.1900*240/\{360[1.2000-1.0100]\})/[64,720+41,548-29,750-513-3,385-7,899*0.3000]$$
$$= 99.00\%$$

Corporate IFRS-GAAP (B/S-I/S), ISBN-13: **978-1720792789**, ISBN-10: **172079278X**

<u>Law-7576</u>:

If both (**c**= C/X= Current Ratio= <u>1.20</u> times), (**d**= D/A= Dividend Portion or Payout= <u>30.00%</u>), (**S**= Sales or Revenues= <u>51,294</u> millions USD), (**v**= V/S= Variable Portion= <u>19.00%</u>), (**p**= 360 P/V= Procured Inventory Days= <u>240</u> Days), (**A**= After Tax Income= <u>7,899</u> millions USD), (**Q**= Quoted Longterm Liabilities= <u>35,351</u> millions USD), (**F**= Fixed Cost= <u>29,740</u> millions USD), (**I**= Interest Expense= <u>513</u> millions USD), (**M**= Margin of Contribution= <u>41,548</u> millions USD), (**I**= L/E= Leverage or Gearing Ratio= <u>99.00%</u>), (**q**= [C-P]/X= Quick or Acid Test Ratio= <u>1.01</u> times), and (**T**= Tax Paid= <u>3,385</u> millions USD), then it's (**U**= Utilized or Starting Capital planned), is:

$$U= (Q+Svp/\{360[c-q]\})/I -[M-F-I-T-Ad]$$
$$= (35,351+51,294*0.1900*240/\{360$$
$$[1.2000-1.0100]\})/0.9900$$
$$-[41,548-29,750-513-3,385$$
$$-7,899*0.3000]$$
$$= \underline{64,720} \text{ millions USD}$$

Corporate IFRS-GAAP (B/S-I/S), ISBN-13: **978-1720792789**, ISBN-10: **172079278X**

<u>Law-7577</u>:

If both (**c**= C/X= Current Ratio= <u>1.20</u> times), (**d**= D/A= Dividend Portion or Payout= <u>30.00%</u>), (**$**= Sales or Revenues= <u>51,294</u> millions USD), (**u**= V/S= Variable Portion= <u>19.00%</u>), (**p**= 360 P/V= Procured Inventory Days= <u>240</u> Days), (**A**= After Tax Income= <u>7,899</u> millions USD), (**Q**= Quoted Longterm Liabilities= <u>35,351</u> millions USD), (**F**= Fixed Cost= <u>29,740</u> millions USD), (**I**= Interest Expense= <u>513</u> millions USD), (**U**= Utilized or Starting Capital= <u>64,720</u> millions USD), (**⌐**= L/E= Leverage or Gearing Ratio= <u>99.00%</u>), (**q**= [C-P]/X= Quick or Acid Test Ratio= <u>1.01</u> times), and (**T**= Tax Paid= <u>3,385</u> millions USD), then it's (**M**= Margin of Contribution planned), is:

$$M= F+I+T+Ad-U+(Q+$up/\{360[c-q]\})/I$$
$$= 29,750+513+3,385+7,899*0.3000$$
$$-64,720+(35,351+51,294$$
$$*0.1900*240/\{360[1.2000$$
$$-1.0100]\})/0.9900$$
$$= \underline{41,548} \text{ millions USD}$$

Corporate IFRS-GAAP (B/S-I/S), ISBN-13: **978-1720792789**, ISBN-10: **172079278X**

Law-7578:

If both (**c**= C/X= Current Ratio= 1.20 times), (**d**= D/A= Dividend Portion or Payout= 30.00%), (**S**= Sales or Revenues= 51,294 millions USD), (**v**= V/S= Variable Portion= 19.00%), (**p**= 360 P/V= Procured Inventory Days= 240 Days), (**A**= After Tax Income= 7,899 millions USD), (**Q**= Quoted Longterm Liabilities= 35,351 millions USD), (**M**= Margin of Contribution= 41,548 millions USD), (**I**= Interest Expense= 513 millions USD), (**U**= Utilized or Starting Capital= 64,720 millions USD), (**I**= L/E= Leverage or Gearing Ratio= 99.00%), (**q**= [C-P]/X= Quick or Acid Test Ratio= 1.01 times), and (**T**= Tax Paid= 3,385 millions USD), then it's (**F**= Fixed Cost planned), is:

$$F= U+M-I-T-Ad-(Q+Sup/\{360[c-q]\})/I$$
$$= 64,720+41,548-513-3,385-7,899$$
$$*0.3000- (35,351+51,294$$
$$*0.1900*240/\{360[1.2000$$
$$-1.0100]\})/0.9900$$
$$= \underline{29,750} \text{ millions USD}$$

Corporate IFRS-GAAP (B/S-I/S), ISBN-13: **978-1720792789**, ISBN-10: **172079278X**

Law-7579:

If both (**c**= C/X= Current Ratio= 1.20 times), (**d**= D/A= Dividend Portion or Payout= 30.00%), (**$**= Sales or Revenues= 51,294 millions USD), (**υ**= V/S= Variable Portion= 19.00%), (**p**= 360 P/V= Procured Inventory Days= 240 Days), (**A**= After Tax Income= 7,899 millions USD), (**Q**= Quoted Longterm Liabilities= 35,351 millions USD), (**M**= Margin of Contribution= 41,548 millions USD), (**F**= Fixed Cost= 29,740 millions USD), (**U**= Utilized or Starting Capital= 64,720 millions USD), (**I**= L/E= Leverage or Gearing Ratio= 99.00%), (**q**= [C-P]/X= Quick or Acid Test Ratio= 1.01 times), and (**T**= Tax Paid= 3,385 millions USD), then it's (**I**= Interest Expense planned), is:

$$I= U+M-F-T-Ad-(Q+\$up/\{360[c-q]\})/I$$
$$= 64,720+41,548-29,750-3,385$$
$$-7,899*0.3000-(35,351$$
$$+51,294*0.1900*240$$
$$/\{360[1.2000-1.0100]\})$$
$$/0.9900$$
$$= \underline{513} \text{ millions USD}$$

Corporate IFRS-GAAP (B/S-I/S), ISBN-13: **978-1720792789**, ISBN-10: **172079278X**

<u>Law-7580</u>:

If both (**c**= C/X= Current Ratio= <u>1.20</u> times), (**d**= D/A= Dividend Portion or Payout= <u>30.00%</u>), (**S**= Sales or Revenues= <u>51,294</u> millions USD), (**v**= V/S= Variable Portion= <u>19.00%</u>), (**p**= 360 P/V= Procured Inventory Days= <u>240</u> Days), (**A**= After Tax Income= <u>7,899</u> millions USD), (**Q**= Quoted Longterm Liabilities= <u>35,351</u> millions USD), (**M**= Margin of Contribution= <u>41,548</u> millions USD), (**F**= Fixed Cost= <u>29,740</u> millions USD), (**U**= Utilized or Starting Capital= <u>64,720</u> millions USD), (**I**= L/E= Leverage or Gearing Ratio= <u>99.00%</u>), (**q**= [C-P]/X= Quick or Acid Test Ratio= <u>1.01</u> times), and (**I**= Interest Expense= <u>513</u> millions USD), then it's (**T**= Tax Paid planned), is:

$$T= U+M-F-I-Ad-(Q+Svp/\{360[c-q]\})/I$$
$$= 64,720+41,548-29,750-513-7,899$$
$$*0.3000-(35,351+51,294$$
$$*0.1900*240/\{360[1.2000$$
$$-1.0100]\})/0.9900$$
$$= \underline{3,385} \text{ millions USD}$$

Corporate IFRS-GAAP (B/S-I/S), ISBN-13: **978-1720792789**, ISBN-10: **172079278X**

<u>Law-7581</u>:

If both (**c**= C/X= Current Ratio= <u>1.20</u> times), (**d**= D/A= Dividend Portion or Payout= <u>30.00%</u>), (**$**= Sales or Revenues= <u>51,294</u> millions USD), (**v**= V/S= Variable Portion= <u>19.00%</u>), (**p**= 360 P/V= Procured Inventory Days= <u>240</u> Days), (**T**= Tax Paid= <u>3,385</u> millions USD), (**Q**= Quoted Longterm Liabilities= <u>35,351</u> millions USD), (**M**= Margin of Contribution= <u>41,548</u> millions USD), (**F**= Fixed Cost= <u>29,740</u> millions USD), (**U**= Utilized or Starting Capital= <u>64,720</u> millions USD), (**$\digamma$**= L/E= Leverage or Gearing Ratio= <u>99.00%</u>), (**q**= [C-P]/X= Quick or Acid Test Ratio= <u>1.01</u> times), and (**I**= Interest Expense= <u>513</u> millions USD), then it's (**A**= After Tax Income planned), is:

$$A= [U+M-F-I-T-(Q+$vp/\{360[c-q]\})/\digamma]/d$$
$$= [64,720+41,548-29,750-513-3,385$$
$$- (35,351+51,294*0.1900$$
$$*240/\{360[1.2000-1.0100]\})$$
$$/0.9900]/0.3000$$
$$= \underline{7,899} \text{ millions USD}$$

Corporate IFRS-GAAP (B/S-I/S), ISBN-13: **978-1720792789**, ISBN-10: **172079278X**

<u>Law-7582</u>:

If both (**c**= C/X= Current Ratio= <u>1.20</u> times), (**A**= After Tax Income= <u>7,899</u> millions USD), (**S**= Sales or Revenues= <u>51,294</u> millions USD), (**v**= V/S= Variable Portion= <u>19.00%</u>), (**p**= 360 P/V= Procured Inventory Days= <u>240</u> Days), (**T**= Tax Paid= <u>3,385</u> millions USD), (**Q**= Quoted Longterm Liabilities= <u>35,351</u> millions USD), (**M**= Margin of Contribution= <u>41,548</u> millions USD), (**F**= Fixed Cost= <u>29,740</u> millions USD), (**U**= Utilized or Starting Capital= <u>64,720</u> millions USD), (**I**= L/E= Leverage or Gearing Ratio= <u>99.00%</u>), (**q**= [C-P]/X= Quick or Acid Test Ratio= <u>1.01</u> times), and (**I**= Interest Expense= <u>513</u> millions USD), then it's (**d**= Dividend Portion or Payout planned), is:

$$d = [U+M-F-I-T-(Q+Svp/\{360[c-q]\})/I]/A$$

$$= [64,720+41,548-29,750-513-3,385$$
$$- (35,351+51,294*0.1900$$
$$*240/\{360[1.2000-1.0100]\})$$
$$/0.9900]/7,899$$

$$= \underline{30.00\%}$$

Corporate IFRS-GAAP (B/S-I/S), ISBN-13: **978-1720792789**, ISBN-10: **172079278X**

Law-7583:

If both (**c**= C/X= Current Ratio= 1.20 times), (**A**= After Tax Income= 7,899 millions USD), (**d**= D/A= Dividend Portion or Payout= 30.00%), (**v**= V/S= Variable Portion= 19.00%), (**p**= 360 P/V= Procured Inventory Days= 240 Days), (**T**= Tax Paid= 3,385 millions USD), (**Q**= Quoted Longterm Liabilities= 35,351 millions USD), (**M**= Margin of Contribution= 41,548 millions USD), (**F**= Fixed Cost= 29,740 millions USD), (**U**= Utilized or Starting Capital= 64,720 millions USD), (**l**= L/E= Leverage or Gearing Ratio= 99.00%), (**q**= [C-P]/X= Quick or Acid Test Ratio= 1.01 times), and (**I**= Interest Expense= 513 millions USD), then it's (**S**= Sales or Revenues planned), is:

$$S= 360[c\text{-}q]\{l\,[U+M\text{-}F\text{-}I\text{-}T\text{-}Ad]\text{-}Q\}/[vp]$$

$$= 360[1.2000\text{-}1.0100]\{0.9900$$
$$[64,720+41,548\text{-}29,750\text{-}513$$
$$\text{-}3,385\text{-}7,899*0.3000]\text{-}35,351\}$$
$$/[0.1900*240]$$

$$= 51,294 \text{ millions USD}$$

Corporate IFRS-GAAP (B/S-I/S), ISBN-13: **978-1720792789**, ISBN-10: **172079278X**

Law-7584:

If both (**c**= C/X= Current Ratio= 1.20 times), (**A**= After Tax Income= 7,899 millions USD), (**d**= D/A= Dividend Portion or Payout= 30.00%), (**S**= Sales or Revenues= 51,294 millions USD), (**p**= 360 P/V= Procured Inventory Days= 240 Days), (**T**= Tax Paid= 3,385 millions USD), (**Q**= Quoted Longterm Liabilities= 35,351 millions USD), (**M**= Margin of Contribution= 41,548 millions USD), (**F**= Fixed Cost= 29,740 millions USD), (**U**= Utilized or Starting Capital= 64,720 millions USD), (**I**= L/E= Leverage or Gearing Ratio= 99.00%), (**q**= [C-P]/X= Quick or Acid Test Ratio= 1.01 times), and (**I**= Interest Expense= 513 millions USD), then it's (**v**= Variable Portion planned), is:

$$v= 360[c-q]\{I\ [U+M-F-I-T-Ad]-Q\}/[Sp]$$
$$= 360[1.2000-1.0100]\{0.9900[64.720$$
$$+41,548-29,750-513-3,385$$
$$-7,899*0.3000]-35,351\}$$
$$/[51,294*240]$$
$$= 19.00\%$$

Law-7585:

If both (**c**= C/X= Current Ratio= 1.20 times), (**A**= After Tax Income= 7,899 millions USD), (**d**= D/A= Dividend Portion or Payout= 30.00%), (**υ**= V/S= Variable Portion= 19.00%), (**S**= Sales or Revenues= 51,294 millions USD), (**T**= Tax Paid= 3,385 millions USD), (**Q**= Quoted Longterm Liabilities= 35,351 millions USD), (**M**= Margin of Contribution= 41,548 millions USD), (**F**= Fixed Cost= 29,740 millions USD), (**U**= Utilized or Starting Capital= 64,720 millions USD), (**l**= L/E= Leverage or Gearing Ratio= 99.00%), (**q**= [C-P]/X= Quick or Acid Test Ratio= 1.01 times), and (**I**= Interest Expense= 513 millions USD), then it's (**p**= Procured Inventory Days planned), is:

$$\mathbf{p}= 360[\mathbf{c}\text{-}\mathbf{q}]\{\mathbf{l}\,[\mathbf{U}+\mathbf{M}\text{-}\mathbf{F}\text{-}\mathbf{I}\text{-}\mathbf{T}\text{-}\mathbf{Ad}]\text{-}\mathbf{Q}\}/[\mathbf{Sυ}]$$

$$= 360[1.2000\text{-}1.0100]\{0.9900[64,720$$
$$+41,548\text{-}29,750\text{-}513\text{-}3,385$$
$$-7,899*0.3000]\text{-}35,351\}$$
$$/[51,294*0.1900]$$

$$= \underline{240} \text{ days}$$

Corporate IFRS-GAAP (B/S-I/S), ISBN-13: **978-1720792789**, ISBN-10: **172079278X**

<u>Law-7586</u>:

If both (**p**= 360 P/V= Procured Inventory Days= <u>240</u> Days), (**A**= After Tax Income= <u>7,899</u> millions USD), (**d**= D/A= Dividend Portion or Payout= <u>30.00%</u>), (**v**= V/S= Variable Portion= <u>19.00%</u>), (**S**= Sales or Revenues= <u>51,294</u> millions USD), (**T**= Tax Paid= <u>3,385</u> millions USD), (**Q**= Quoted Longterm Liabilities= <u>35,351</u> millions USD), (**M**= Margin of Contribution= <u>41,548</u> millions USD), (**F**= Fixed Cost= <u>29,740</u> millions USD), (**U**= Utilized or Starting Capital= <u>64,720</u> millions USD), (**I**= L/E= Leverage or Gearing Ratio= <u>99.00%</u>), (**q**= [C-P]/X= Quick or Acid Test Ratio= <u>1.01</u> times), and (**I**= Interest Expense= <u>513</u> millions USD), then it's (**c**= Current Ratio planned), is:

$$c= q+Svp/(360\{I[U+M-F-I-T-Ad]-Q\})$$
$$= 1.0100+51,294*0.1900*240/(360$$
$$\{0.9900[64,720+41,548$$
$$-29,750-513-3,385-7,899$$
$$*0.3000]-35,351\})$$
$$= \underline{1.20} \text{ times}$$

Corporate IFRS-GAAP (B/S-I/S), ISBN-13: **978-1720792789**, ISBN-10: **172079278X**

Law-7587:

If both (**p**= 360 P/V= Procured Inventory Days= 240 Days), (**A**= After Tax Income= 7,899 millions USD), (**d**= D/A= Dividend Portion or Payout= 30.00%), (**v**= V/S= Variable Portion= 19.00%), (**S**= Sales or Revenues= 51,294 millions USD), (**T**= Tax Paid= 3,385 millions USD), (**Q**= Quoted Longterm Liabilities= 35,351 millions USD), (**M**= Margin of Contribution= 41,548 millions USD), (**F**= Fixed Cost= 29,740 millions USD), (**U**= Utilized or Starting Capital= 64,720 millions USD), (**l**= L/E= Leverage or Gearing Ratio= 99.00%), (**c**= C/X= Current Ratio= 1.20 times), and (**I**= Interest Expense= 513 millions USD), then it's (**q**= Quick or Acid Test Ratio planned), is:

$$q= c\text{-}Svp/(360\{l\,[U+M\text{-}F\text{-}I\text{-}T\text{-}Ad]\text{-}Q\})$$
$$= 1.2000\text{-}51,294*0.1900*240/(360$$
$$\{0.9900[64,720+41,548$$
$$-29,750\text{-}513\text{-}3,385\text{-}7,899$$
$$*0.3000]\text{-}35,351\})$$
$$= 1.01 \text{ times}$$

Corporate IFRS-GAAP (B/S-I/S), ISBN-13: **978-1720792789**, ISBN-10: **172079278X**

Law-7588:

If both (**p**= 360 P/V= Procured Inventory Days= 240 Days), (**A**= After Tax Income= 7,899 millions USD), (**d**= D/A= Dividend Portion or Payout= 30.00%), (**v**= V/S= Variable Portion= 19.00%), (**S'**= Sales of Past Year= 48,851 millions USD), (**s**= [S/S']-1= Sales Growth= 5.00%), (**T**= Tax Paid= 3,385 millions USD), (**q**= [C-P]/X= Quick or Acid Test Ratio= 1.01 times), (**M**= Margin of Contribution= 41,548 millions USD), (**F**= Fixed Cost= 29,740 millions USD), (**U**= Utilized or Starting Capital= 64,720 millions USD), (**I**= L/E= Leverage or Gearing Ratio= 99.00%), (**c**= C/X= Current Ratio= 1.20 times), and (**I**= Interest Expense= 513 millions USD), then it's (**Q**= Quoted Longterm Liabilities planned), is:

$$Q = I[U+M-F-I-T-Ad]-S'vp[1+s]$$
$$/\{360[c-q]\}$$
$$= 0.9900[64,720+41,548-29,750-513$$
$$-3,385-7,899*0.3000]-48,851$$
$$*0.1900*240[1+0.0500]$$
$$/\{360[1.2000-1.0100]\}$$
$$= 35,351 \text{ millions USD}$$

Law-7589:

If both (**p**= 360 P/V= Procured Inventory Days= 240 Days), (**A**= After Tax Income= 7,899 millions USD), (**d**= D/A= Dividend Portion or Payout= 30.00%), (**v**= V/S= Variable Portion= 19.00%), (**S'**= Sales of Past Year= 48,851 millions USD), (**s**= [S/S']-1= Sales Growth= 5.00%), (**T**= Tax Paid= 3,385 millions USD), (**q**= [C-P]/X= Quick or Acid Test Ratio= 1.01 times), (**M**= Margin of Contribution= 41,548 millions USD), (**F**= Fixed Cost= 29,740 millions USD), (**U**= Utilized or Starting Capital= 64,720 millions USD), (**Q**= Quoted Longterm Liabilities= 35,351 millions USD), (**c**= C/X= Current Ratio= 1.20 times), and (**I**= Interest Expense= 513 millions USD), then it's (**I** = Leverage or Gearing Ratio planned), is:

$$\mathbf{I} = (\mathbf{Q} + \mathbf{S'vp}[1+\mathbf{s}]/\{360[\mathbf{c}\text{-}\mathbf{q}]\})$$
$$/[\mathbf{U} + \mathbf{M} \text{-} \mathbf{F} \text{-} \mathbf{I} \text{-} \mathbf{T} \text{-} \mathbf{Ad}]$$
$$= (35,351 + 48,851 * 0.1900 * 240$$
$$[1 + 0.0500]/\{360[1.2000$$
$$-1.0100]\})/[64,720 + 41,548$$
$$-29,750 - 513 - 3,385 - 7,899$$
$$*0.3000]$$
$$= 99.00\%$$

Corporate IFRS-GAAP (B/S-I/S), ISBN-13: **978-1720792789**, ISBN-10: **172079278X**

<u>Law-7590</u>:

If both (**p**= 360 P/V= Procured Inventory Days= <u>240</u>
Days), (**A**= After Tax Income= <u>7,899</u> millions USD),
(**d**= D/A= Dividend Portion or Payout= <u>30.00%</u>), (**v**=
V/S= Variable Portion= <u>19.00%</u>), (**$'**= Sales of Past
Year= <u>48,851</u> millions USD), (**s**= [S/S']-1= Sales
Growth= <u>5.00%</u>), (**T**= Tax Paid= <u>3,385</u> millions
USD), (**q**= [C-P]/X= Quick or Acid Test Ratio= <u>1.01</u>
times), (**M**= Margin of Contribution= <u>41,548</u> millions
USD), (**F**= Fixed Cost= <u>29,740</u> millions USD), (**l**=
L/E= Leverage or Gearing Ratio= <u>99.00%</u>), (**Q**=
Quoted Longterm Liabilities= <u>35,351</u> millions USD),
(**c**= C/X= Current Ratio= <u>1.20</u> times), and (**I**= Interest
Expense= <u>513</u> millions USD), then it's (**U**= Utilized
or Starting Capital planned), is:

$$\mathbf{U}= (\mathbf{Q}+\mathbf{\$'vp}[1+\mathbf{s}]/\{360[\mathbf{c}-\mathbf{q}]\})/\mathbf{l}$$
$$-[\mathbf{M}-\mathbf{F}-\mathbf{I}-\mathbf{T}-\mathbf{Ad}]$$
$$= (35,351+48,851*0.1900*240$$
$$[1+0.0500]/\{360[1.2000$$
$$-1.0100]\})/0.9900-[41,548$$
$$-29,750-513-3,385-7,899$$
$$*0.3000]$$
$$= \underline{64,720} \text{ millions USD}$$

Corporate IFRS-GAAP (B/S-I/S), ISBN-13: **978-1720792789**, ISBN-10: **172079278X**

Law-7591:

If both (**p**= 360 P/V= Procured Inventory Days= 240 Days), (**A**= After Tax Income= 7,899 millions USD), (**d**= D/A= Dividend Portion or Payout= 30.00%), (**v**= V/S= Variable Portion= 19.00%), (**S'**= Sales of Past Year= 48,851 millions USD), (**s**= [S/S']-1= Sales Growth= 5.00%), (**T**= Tax Paid= 3,385 millions USD), (**q**= [C-P]/X= Quick or Acid Test Ratio= 1.01 times), (**U**= Utilized or Starting Capital= 64,720 millions USD), (**F**= Fixed Cost= 29,740 millions USD), (**I**= L/E= Leverage or Gearing Ratio= 99.00%), (**Q**= Quoted Longterm Liabilities= 35,351 millions USD), (**c**= C/X= Current Ratio= 1.20 times), and (**I**= Interest Expense= 513 millions USD), then it's (**M**= Margin of Contribution planned), is:

$$M = F+I+T+Ad-U+(Q+S'vp[1+s]$$
$$/\{360[c-q]\})/I$$
$$= 29,750+513+3,385+7,899*0.3000$$
$$-64,720+(35,351+48,851$$
$$*0.1900*240[1+0.0500]$$
$$/\{360[1.2000-1.0100]\})$$
$$/0.9900$$
$$= 41,548 \text{ millions USD}$$

Corporate IFRS-GAAP (B/S-I/S), ISBN-13: **978-1720792789**, ISBN-10: **172079278X**

<u>Law-7592</u>:

If both (**p**= 360 P/V= Procured Inventory Days= <u>240</u> Days), (**A**= After Tax Income= <u>7,899</u> millions USD), (**d**= D/A= Dividend Portion or Payout= <u>30.00%</u>), (**v**= V/S= Variable Portion= <u>19.00%</u>), (**S'**= Sales of Past Year= <u>48,851</u> millions USD), (**s**= [S/S']-1= Sales Growth= <u>5.00%</u>), (**T**= Tax Paid= <u>3,385</u> millions USD), (**q**= [C-P]/X= Quick or Acid Test Ratio= <u>1.01</u> times), (**U**= Utilized or Starting Capital= <u>64,720</u> millions USD), (**M**= Margin of Contribution= <u>41,548</u> millions USD), (**I**= L/E= Leverage or Gearing Ratio= <u>99.00%</u>), (**Q**= Quoted Longterm Liabilities= <u>35,351</u> millions USD), (**c**= C/X= Current Ratio= <u>1.20</u> times), and (**I**= Interest Expense= <u>513</u> millions USD), then it's (**F**= Fixed Cost planned), is:

$$F= U+M-I-T-Ad-(Q+S'vp[1+s]$$
$$/\{360[c-q]\})/I$$
$$= 64,720+41,548-513-3,385-7,899$$
$$*0.3000- (35,351+48,851$$
$$*0.1900*240[1+0.0500]$$
$$/\{360[1.2000-1.0100]\})$$
$$/0.9900$$
$$= \underline{29,750} \text{ millions USD}$$

Corporate IFRS-GAAP (B/S-I/S), ISBN-13: **978-1720792789**, ISBN-10: **172079278X**

Law-7593:

If both (**p**= 360 P/V= Procured Inventory Days= 240
Days), (**A**= After Tax Income= 7,899 millions USD),
(**d**= D/A= Dividend Portion or Payout= 30.00%), (**u**=
V/S= Variable Portion= 19.00%), (**S'**= Sales of Past
Year= 48,851 millions USD), (**s**= [S/S']-1= Sales
Growth= 5.00%), (**T**= Tax Paid= 3,385 millions
USD), (**q**= [C-P]/X= Quick or Acid Test Ratio= 1.01
times), (**U**= Utilized or Starting Capital= 64,720
millions USD), (**M**= Margin of Contribution= 41,548
millions USD), (**I**= L/E= Leverage or Gearing Ratio=
99.00%), (**Q**= Quoted Longterm Liabilities= 35,351
millions USD), (**c**= C/X= Current Ratio= 1.20 times),
and (**F**= Fixed Cost= 29,740 millions USD), then it's
(**I**= Interest Expense planned), is:

$$I = U+M-F-T-Ad-(Q+S'up[1+s]$$
$$/\{360[c-q]\})/I$$
$$= 64,720+41,548-29,750-3,385-7,899$$
$$*0.3000-(35,351+48,851$$
$$*0.1900*240[1+0.0500]$$
$$/\{360[1.2000-1.0100]\})$$
$$/0.9900$$
$$= 513 \text{ millions USD}$$

Corporate IFRS-GAAP (B/S-I/S), ISBN-13: **978-1720792789**, ISBN-10: **172079278X**

Law-7594:

If both (**p**= 360 P/V= Procured Inventory Days= <u>240</u>
Days), (**A**= After Tax Income= <u>7,899</u> millions USD),
(**d**= D/A= Dividend Portion or Payout= <u>30.00%</u>), (**v**=
V/S= Variable Portion= <u>19.00%</u>), (**S'**= Sales of Past
Year= <u>48,851</u> millions USD), (**s**= [S/S']-1= Sales
Growth= <u>5.00%</u>), (**I**= Interest Expense= <u>513</u> millions
USD), (**q**= [C-P]/X= Quick or Acid Test Ratio= <u>1.01</u>
times), (**U**= Utilized or Starting Capital= <u>64,720</u>
millions USD), (**M**= Margin of Contribution= <u>41,548</u>
millions USD), (**I**= L/E= Leverage or Gearing Ratio=
<u>99.00%</u>), (**Q**= Quoted Longterm Liabilities= <u>35,351</u>
millions USD), (**c**= C/X= Current Ratio= <u>1.20</u> times),
and (**F**= Fixed Cost= <u>29,740</u> millions USD), then it's
(**T**= Tax Paid planned), is:

$$T= U+M-F-I-Ad-(Q+S'vp[1+s]$$
$$/\{360[c-q]\})/I$$
$$= 64,720+41,548-29,750-513-7,899$$
$$*0.3000-(35,351+48,851$$
$$*0.1900*240[1+0.0500]$$
$$/\{360[1.2000-1.0100]\})$$
$$/0.9900$$

$$= 3,385 \text{ millions USD}$$

Corporate IFRS-GAAP (B/S-I/S), ISBN-13: **978-1720792789**, ISBN-10: **172079278X**

<u>Law-7595</u>:

If both (**p**= 360 P/V= Procured Inventory Days= <u>240</u> Days), (**T**= Tax Paid= <u>3,385</u> millions USD), (**d**= D/A= Dividend Portion or Payout= <u>30.00%</u>), (**v**= V/S= Variable Portion= <u>19.00%</u>), (**S'**= Sales of Past Year= <u>48,851</u> millions USD), (**s**= [S/S']-1= Sales Growth= <u>5.00%</u>), (**I**= Interest Expense= <u>513</u> millions USD), (**q**= [C-P]/X= Quick or Acid Test Ratio= <u>1.01</u> times), (**U**= Utilized or Starting Capital= <u>64,720</u> millions USD), (**M**= Margin of Contribution= <u>41,548</u> millions USD), (**I**= L/E= Leverage or Gearing Ratio= <u>99.00%</u>), (**Q**= Quoted Longterm Liabilities= <u>35,351</u> millions USD), (**c**= C/X= Current Ratio= <u>1.20</u> times), and (**F**= Fixed Cost= <u>29,740</u> millions USD), then it's (**A**= After Tax Income planned), is:

$$A = [U+M-F-I-T-(Q+S'vp[1+s]/\{360[c-q]\})/I]/d$$

$$= [64,720+41,548-29,750-513-3,385$$
$$- (35,351+48,851*0.1900$$
$$*240[1+0.0500]/\{360[1.2000$$
$$-1.0100]\})/0.9900]/0.3000$$

$$= \underline{7,899} \text{ millions USD}$$

Corporate IFRS-GAAP (B/S-I/S), ISBN-13: **978-1720792789**, ISBN-10: **172079278X**

Law-7596:

If both (**p**= 360 P/V= Procured Inventory Days= <u>240</u>
Days), (**T**= Tax Paid= <u>3,385</u> millions USD), (**A**=
After Tax Income= <u>7,899</u> millions USD), (**v**= V/S=
Variable Portion= <u>19.00%</u>), (**S'**= Sales of Past Year=
<u>48,851</u> millions USD), (**s**= [S/S']-1= Sales Growth=
<u>5.00%</u>), (**I**= Interest Expense= <u>513</u> millions USD),
(**q**= [C-P]/X= Quick or Acid Test Ratio= <u>1.01</u> times),
(**U**= Utilized or Starting Capital= <u>64,720</u> millions
USD), (**M**= Margin of Contribution= <u>41,548</u> millions
USD), (**l**= L/E= Leverage or Gearing Ratio=
<u>99.00%</u>), (**Q**= Quoted Longterm Liabilities= <u>35,351</u>
millions USD), (**c**= C/X= Current Ratio= <u>1.20</u> times),
and (**F**= Fixed Cost= <u>29,740</u> millions USD), then it's
(**d**= Dividend Portion or Payout planned), is:

$$d= [U+M-F-I-T-(Q+S'vp[1+s]$$
$$/\{360[c-q]\})/l\ ]/A$$
$$= [64,720+41,548-29,750-513-3,385$$
$$- (35,351+48,851*0.1900$$
$$*240[1+0.0500]/\{360[1.2000$$
$$-1.0100]\})/0.9900]/7,899$$
$$= 30.00\%$$

Corporate IFRS-GAAP (B/S-I/S), ISBN-13: **978-1720792789**, ISBN-10: **172079278X**

Law-7597:

If both (**p**= 360 P/V= Procured Inventory Days= 240
Days), (**T**= Tax Paid= 3,385 millions USD), (**A**=
After Tax Income= 7,899 millions USD), (**v**= V/S=
Variable Portion= 19.00%), (**d**= D/A= Dividend
Portion or Payout= 30.00%), (**s**= [S/S']-1= Sales
Growth= 5.00%), (**I**= Interest Expense= 513 millions
USD), (**q**= [C-P]/X= Quick or Acid Test Ratio= 1.01
times), (**U**= Utilized or Starting Capital= 64,720
millions USD), (**M**= Margin of Contribution= 41,548
millions USD), (**I**= L/E= Leverage or Gearing Ratio=
99.00%), (**Q**= Quoted Longterm Liabilities= 35,351
millions USD), (**c**= C/X= Current Ratio= 1.20 times),
and (**F**= Fixed Cost= 29,740 millions USD), then it's
(**S'**= Sales Past), must be:

$$S' = 360[c\text{-}q]\{I[U+M\text{-}F\text{-}I\text{-}T\text{-}Ad]\text{-}Q\}$$
$$/\{vp[1+s]$$
$$= 360[1.2000\text{-}1.0100]\{0.9900$$
$$[64,720+41,548\text{-}29,750$$
$$-513\text{-}3,385\text{-}7,899*0.3000]$$
$$-35,351\}/\{0.1900*240$$
$$[1+0.0500]\}$$
$$= 48,851 \text{ millions USD}$$

Corporate IFRS-GAAP (B/S-I/S), ISBN-13: **978-1720792789**, ISBN-10: **172079278X**

<u>Law-7598</u>:

If both (**p**= 360 P/V= Procured Inventory Days= <u>240</u> Days), (**T**= Tax Paid= <u>3,385</u> millions USD), (**A**= After Tax Income= <u>7,899</u> millions USD), (**d**= D/A= Dividend Portion or Payout= <u>30.00%</u>), (**S'**= Sales of Past Year= <u>48,851</u> millions USD), (**s**= [S/S']-1= Sales Growth= <u>5.00%</u>), (**I**= Interest Expense= <u>513</u> millions USD), (**q**= [C-P]/X= Quick or Acid Test Ratio= <u>1.01</u> times), (**U**= Utilized or Starting Capital= <u>64,720</u> millions USD), (**M**= Margin of Contribution= <u>41,548</u> millions USD), (**I**= L/E= Leverage or Gearing Ratio= <u>99.00%</u>), (**Q**= Quoted Longterm Liabilities= <u>35,351</u> millions USD), (**c**= C/X= Current Ratio= <u>1.20</u> times), and (**F**= Fixed Cost= <u>29,740</u> millions USD), then it's (**v**= Variable Portion planned), is:

$$\mathbf{v}= 360[\mathbf{c}\text{-}\mathbf{q}]\{\mathbf{I}\,[\mathbf{U}\text{+}\mathbf{M}\text{-}\mathbf{F}\text{-}\mathbf{I}\text{-}\mathbf{T}\text{-}\mathbf{Ad}]\text{-}\mathbf{Q}\}$$
$$/\{\mathbf{S'}\mathbf{p}[1\text{+}\mathbf{s}]$$
$$= 360[1.2000\text{-}1.0100]\{0.9900[64,720$$
$$+41,548\text{-}29,750\text{-}513\text{-}3,385$$
$$-7,899*0.3000]\text{-}35,351\}$$
$$/\{48,851*240[1\text{+}0.0500]\}$$
$$= \underline{19.00\%}$$

Corporate IFRS-GAAP (B/S-I/S), ISBN-13: **978-1720792789**, ISBN-10: **172079278X**

<u>Law-7599</u>:

If both (**v**= V/S= Variable Portion= <u>19.00%</u>), (**T**= Tax Paid= <u>3,385</u> millions USD), (**A**= After Tax Income= <u>7,899</u> millions USD), (**d**= D/A= Dividend Portion or Payout= <u>30.00%</u>), (**$'**= Sales of Past Year= <u>48,851</u> millions USD), (**s**= [S/S']-1= Sales Growth= <u>5.00%</u>), (**I**= Interest Expense= <u>513</u> millions USD), (**q**= Quick or Acid Test Ratio= [C-P]/X= <u>1.01</u> times), (**U**= Utilized or Starting Capital= <u>64,720</u> millions USD), (**M**= Margin of Contribution= <u>41,548</u> millions USD), (**f**= L/E= Leverage or Gearing Ratio= <u>99.00%</u>), (**Q**= Quoted Longterm Liabilities= <u>35,351</u> millions USD), (**c**= C/X= Current Ratio= <u>1.20</u> times), and (**F**= Fixed Cost= <u>29,740</u> millions USD), then it's (**p**= Procured Inventory Days planned), is:

$$p= 360[c\text{-}q]\{f\,[U+M\text{-}F\text{-}I\text{-}T\text{-}Ad]\text{-}Q\}$$
$$/\{\$'v[1+s]$$
$$= 360[1.2000\text{-}1.0100]\{0.9900$$
$$[64,720+41,548\text{-}29,750$$
$$-513\text{-}3,385\text{-}7,899*0.3000]$$
$$-35,351\}/\{48,851*0.1900$$
$$[1+0.0500]\}$$
$$= \underline{240} \text{ days}$$

Corporate IFRS-GAAP (B/S-I/S), ISBN-13: **978-1720792789**, ISBN-10: **172079278X**

Law-7600:

If both (v= V/S= Variable Portion= 19.00%), (T= Tax
Paid= 3,385 millions USD), (A= After Tax Income=
7,899 millions USD), (d= D/A= Dividend Portion or
Payout= 30.00%), (S'= Sales of Past Year= 48,851
millions USD), (p= 360 P/V= Procured Inventory
Days= 240 Days), (I= Interest Expense= 513 millions
USD), (q= [C-P]/X= Quick or Acid Test Ratio= 1.01
times), (U= Utilized or Starting Capital= 64,720
millions USD), (M= Margin of Contribution= 41,548
millions USD), (I= L/E= Leverage or Gearing Ratio=
99.00%), (Q= Quoted Longterm Liabilities= 35,351
millions USD), (c= C/X= Current Ratio= 1.20 times),
and (F= Fixed Cost= 29,740 millions USD), then it's
(s= Sales Growth planned), is:

$$s= 360[\mathbf{c\text{-}q}]\{\mathbf{/}[\mathbf{U+M\text{-}F\text{-}I\text{-}T\text{-}Ad}]\text{-}\mathbf{Q}\}$$
$$/[\mathbf{S'vp}]\text{-}1$$
$$= 360[1.2000\text{-}1.0100]\{0.9900$$
$$[64,720+41,548\text{-}29,750$$
$$\text{-}513\text{-}3,385\text{-}7.899*0.3000]$$
$$\text{-}35,351\}/[48,851*0.1900$$
$$*240]\text{-}1$$
$$= 5.00\%$$

Corporate IFRS-GAAP (B/S-I/S), ISBN-13: **978-1720792789**, ISBN-10: **172079278X**

<u>Law-7601</u>:

If both (**v**= V/S= Variable Portion= <u>19.00%</u>), (**T**= Tax Paid= <u>3,385</u> millions USD), (**A**= After Tax Income= <u>7,899</u> millions USD), (**d**= D/A= Dividend Portion or Payout= <u>30.00%</u>), (**S'**= Sales of Past Year= <u>48,851</u> millions USD), (**p**= 360 P/V= Procured Inventory Days= <u>240</u> Days), (**I**= Interest Expense= <u>513</u> millions USD), (**q**= [C-P]/X= Quick or Acid Test Ratio= <u>1.01</u> times), (**U**= Utilized or Starting Capital= <u>64,720</u> millions USD), (**M**= Margin of Contribution= <u>41,548</u> millions USD), (**I**= L/E= Leverage or Gearing Ratio= <u>99.00%</u>), (**Q**= Quoted Longterm Liabilities= <u>35,351</u> millions USD), (**s**= [S/S']-1= Sales Growth= <u>5.00%</u>), and (**F**= Fixed Cost= <u>29,740</u> millions USD), then it's (**c**= Current Ratio planned), is:

$$c= q+S'vp[1+s]/(360\{I\,[U+M-F-I$$
$$-T-Ad]-Q\})$$
$$= 1.0100+48,851*0.1900*240$$
$$[1+0.0500]/(360\{0.9900$$
$$[64,720+41,548-29,750-513$$
$$-3,385-7,899*0.3000]$$
$$-35,351\})$$
$$= \underline{1.20} \text{ times}$$

Corporate IFRS-GAAP (B/S-I/S), ISBN-13: **978-1720792789**, ISBN-10: **172079278X**

Law-7602:

If both (v= V/S= Variable Portion= 19.00%), (**T**= Tax Paid= 3,385 millions USD), (**A**= After Tax Income= 7,899 millions USD), (**d**= D/A= Dividend Portion or Payout= 30.00%), (**$**= Sales of Past Year= 48,851 millions USD), (**p**= 360 P/V= Procured Inventory Days= 240 Days), (**I**= Interest Expense= 513 millions USD), (**c**= C/X= Current Ratio= 1.20 times),(**U**= Utilized or Starting Capital= 64,720 millions USD), (**M**= Margin of Contribution= 41,548 millions USD), (**F**= L/E= Leverage or Gearing Ratio= 99.00%), (**Q**= Quoted Longterm Liabilities= 35,351 millions USD), (**s**= [S/S']-1= Sales Growth= 5.00%), and (**F**= Fixed Cost= 29,740 millions USD), then it's (**q**= Quick or Acid Test Ratio planned), is:

$$q= c\text{-}\mathbf{\$}vp[1+s]/(360\{\boldsymbol{I}[U+M\text{-}F\text{-}I \\ \text{-}T\text{-}Ad]\text{-}Q\})$$
$$= 1.2000\text{-}48,851*0.1900*240 \\ [1+0.0500]/(360\{0.9900 \\ [64,720+41,548\text{-}29,750 \\ \text{-}513\text{-}3,385\text{-}7,899*0.3000] \\ \text{-}35,351\})$$
$$= 1.01 \text{ times}$$

Corporate IFRS-GAAP (B/S-I/S), ISBN-13: **978-1720792789**, ISBN-10: **172079278X**

<u>Law-7603</u>:

If both (**I**= Interest Expense= <u>513</u> millions USD), (**t**= T/B= Tax Rate= <u>30.00%</u>), (**D**= Dividend Paid= <u>2,370</u> millions USD), (**U**= Utilized or Starting Capital= <u>64,720</u> millions USD), (**M**= Margin of Contribution= <u>41,548</u> millions USD), (**I**= L/E= Leverage or Gearing Ratio= <u>99.00%</u>), (**X**= Xpress or Current Debt= <u>34,196</u> millions USD), and (**F**= Fixed Cost= <u>29,740</u> millions USD), then it's (**Q**= Quoted Longterm Liabilities planned), is:

$$Q = I\{U+[M-F-I][1-t]-D]\}-X$$
$$= 0.9900\{64,720+[41,548-29,750$$
$$-513][1-0.3000]-2,370]$$
$$-34,196$$
$$= \underline{35,351} \text{ millions USD}$$

Corporate IFRS-GAAP (B/S-I/S), ISBN-13: **978-1720792789**, ISBN-10: **172079278X**

Law-7604:

If both (**I**= Interest Expense= <u>513</u> millions USD), (**t**= T/B= Tax Rate= <u>30.00%</u>), (**D**= Dividend Paid= <u>2,370</u> millions USD), (**U**= Utilized or Starting Capital= <u>64,720</u> millions USD), (**M**= Margin of Contribution= <u>41,548</u> millions USD), (**Q**= Quoted Longterm Liabilities= <u>35,351</u> millions USD), (**X**= Xpress or Current Debt= <u>34,196</u> millions USD), and (**F**= Fixed Cost= <u>29,740</u> millions USD), then it's (**I**= L/E= Leverage or Gearing Ratio planned), is:

$$I = [Q+X]/\{U+[M-F-I][1-t]-D]\}$$
$$= [35,351+34,196]/\{64,720+[41,548$$
$$-29,750-513][1-0.3000]$$
$$-2,370]$$
$$= 99.00\%$$

Corporate IFRS-GAAP (B/S-I/S), ISBN-13: **978-1720792789**, ISBN-10: **172079278X**

Law-7605:

If both (**I**= Interest Expense= <u>513</u> millions USD), (**t**= T/B= Tax Rate= <u>30.00%</u>), (**D**= Dividend Paid= <u>2,370</u> millions USD), (**I**= L/E= Leverage or Gearing Ratio= <u>99.00%</u>), (**M**= Margin of Contribution= <u>41,548</u> millions USD), (**Q**= Quoted Longterm Liabilities= <u>35,351</u> millions USD), (**X**= Xpress or Current Debt= <u>34,196</u> millions USD), and (**F**= Fixed Cost= <u>29,740</u> millions USD), then it's (**U**= Utilized or Starting Capital planned), is:

$$U = D - [M-F-I][1-t] + [Q+X]/I$$
$$= 2,370 - [41,548-29,750-513]$$
$$[1-0.3000] + [35,351$$
$$+34,196]/0.9900$$
$$= \underline{64,720} \text{ millions USD}$$

Corporate IFRS-GAAP (B/S-I/S), ISBN-13: **978-1720792789**, ISBN-10: **172079278X**

Law-7606:

If both (**I**= Interest Expense= 513 millions USD), (**t**= T/B= Tax Rate= 30.00%), (**D**= Dividend Paid= 2,370 millions USD), (**I**= L/E= Leverage or Gearing Ratio= 99.00%), (**U**= Utilized or Starting Capital= 64,720 millions USD), (**Q**= Quoted Longterm Liabilities= 35,351 millions USD), (**X**= Xpress or Current Debt= 34,196 millions USD), and (**F**= Fixed Cost= 29,740 millions USD), then it's (**M**= Margin of Contribution planned), is:

$$M= F+I+\{D-U+[Q+X]/I\}/[1-t]$$
$$= 29{,}750+513+\{2{,}370-64{,}720$$
$$+[35{,}351+34{,}196]/0.9900\}$$
$$/[1-0.3000]$$
$$= \underline{41{,}548} \text{ millions USD}$$

Corporate IFRS-GAAP (B/S-I/S), ISBN-13: **978-1720792789**, ISBN-10: **172079278X**

<u>Law-7607</u>:

If both (**I**= Interest Expense= <u>513</u> millions USD), (**t**= T/B= Tax Rate= <u>30.00%</u>), (**D**= Dividend Paid= <u>2,370</u> millions USD), (**I**= L/E= Leverage or Gearing Ratio= <u>99.00%</u>), (**U**= Utilized or Starting Capital= <u>64,720</u> millions USD), (**Q**= Quoted Longterm Liabilities= <u>35,351</u> millions USD), (**X**= Xpress or Current Debt= <u>34,196</u> millions USD), and (**M**= Margin of Contribution= <u>41,548</u> millions USD), then it's (**F**= Fixed Cost planned), is:

$$F= M\text{-}I\text{-}\{D\text{-}U+[Q+X]/I\}/[1\text{-}t]$$
$$= 41,548\text{-}513\text{-}\{2,370\text{-}64,720$$
$$+[35,351+34,196]/0.9900\}$$
$$/[1\text{-}0.3000]$$
$$= \underline{29,750} \text{ millions USD}$$

Corporate IFRS-GAAP (B/S-I/S), ISBN-13: **978-1720792789**, ISBN-10: **172079278X**

Law-7608:

If both (**F**= Fixed Cost= 29,740 millions USD), (**t**= T/B= Tax Rate= 30.00%), (**D**= Dividend Paid= 2,370 millions USD), (**I**= L/E= Leverage or Gearing Ratio= 99.00%), (**U**= Utilized or Starting Capital= 64,720 millions USD), (**Q**= Quoted Longterm Liabilities= 35,351 millions USD), (**X**= Xpress or Current Debt= 34,196 millions USD), and (**M**= Margin of Contribution= 41,548 millions USD), then it's (**I**= Interest Expense planned), is:

$$I = M-F-\{D-U+[Q+X]/I\}/[1-t]$$
$$= 41,548-29,750-\{2,370-64,720$$
$$+[35,351+34,196]/0.9900\}$$
$$/[1-0.3000]$$
$$= \underline{513} \text{ millions USD}$$

Corporate IFRS-GAAP (B/S-I/S), ISBN-13: **978-1720792789**, ISBN-10: **172079278X**

Law-7609:

If both (**F**= Fixed Cost= <u>29,740</u> millions USD), (**I**= Interest Expense= <u>513</u> millions USD), (**D**= Dividend Paid= <u>2,370</u> millions USD), (**I**= L/E= Leverage or Gearing Ratio= <u>99.00%</u>), (**U**= Utilized or Starting Capital= <u>64,720</u> millions USD), (**Q**= Quoted Longterm Liabilities= <u>35,351</u> millions USD), (**X**= Xpress or Current Debt= <u>34,196</u> millions USD), and (**M**= Margin of Contribution= <u>41,548</u> millions USD), then it's (**t**= Tax Rate planned), is:

$$\textbf{t} = 1 - \{\textbf{D-U}+[\textbf{Q+X}]/\textbf{I}\}/[\textbf{M-F-I}]$$
$$= 1 - \{2{,}370 - 64{,}720 + [35{,}351 + 34{,}196]/0.9900\}/[41{,}548 - 29{,}750 - 513]$$
$$= \underline{30.00\%}$$

Corporate IFRS-GAAP (B/S-I/S), ISBN-13: **978-1720792789**, ISBN-10: **172079278X**

Law-7610:

If both (**F**= Fixed Cost= 29,740 millions USD), (**I**= Interest Expense= 513 millions USD), (**t**= T/B= Tax Rate= 30.00%), (**I**= L/E= Leverage or Gearing Ratio= 99.00%), (**U**= Utilized or Starting Capital= 64,720 millions USD), (**Q**= Quoted Longterm Liabilities= 35,351 millions USD), (**X**= Xpress or Current Debt= 34,196 millions USD), and (**M**= Margin of Contribution= 41,548 millions USD), then it's (**D**= Dividend Paid planned), is:

$$D= U+[M-F-I][1-t]-[Q+X]/I$$
$$= 64,720+[41,548-29,750-513]$$
$$[1-0.3000]-[35,351$$
$$+34,196]/0.9900$$
$$= \underline{2,370} \text{ millions USD}$$

Corporate IFRS-GAAP (B/S-I/S), ISBN-13: **978-1720792789**, ISBN-10: **172079278X**

Law-7611:

If both (**F**= Fixed Cost= 29,740 millions USD), (**I**= Interest Expense= 513 millions USD), (**t**= T/B= Tax Rate= 30.00%), (**I**= L/E= Leverage or Gearing Ratio= 99.00%), (**U**= Utilized or Starting Capital= 64,720 millions USD), (**Q**= Quoted Longterm Liabilities= 35,351 millions USD), (**D**= Dividend Paid= 2,370 millions USD), and (**M**= Margin of Contribution= 41,548 millions USD), then it's (**X**= Xpress or Current Debt planned), is:

$$X = I\{U+[M-F-I][1-t]-D\}-Q$$
$$= 0.9900\{64,720+[41,548-29,750$$
$$-513][1-0.3000]-2,370\}$$
$$-35,351$$
$$= 34,196 \text{ millions USD}$$

Corporate IFRS-GAAP (B/S-I/S), ISBN-13: **978-1720792789**, ISBN-10: **172079278X**

<u>Law-7612</u>:

If both (**F**= Fixed Cost= <u>29,740</u> millions USD), (**I**= Interest Expense= <u>513</u> millions USD), (**t**= T/B= Tax Rate= <u>30.00%</u>), (**I**= L/E= Leverage or Gearing Ratio= <u>99.00%</u>), (**U**= Utilized or Starting Capital= <u>64,720</u> millions USD), (**c**= C/X= Current Ratio= <u>1.20</u> times), (**q**= [C-P]/X= Quick or Acid Test Ratio= <u>1.01</u> times), (**P**= Procured Inventories= <u>6,497</u> millions USD), (**D**= Dividend Paid= <u>2,370</u> millions USD), and (**M**= Margin of Contribution= <u>41,548</u> millions USD), then it's (**Q**= Quoted Longterm Liabilities planned), is:

$$Q= I\{U+[M-F-I][1-t]-D\}-P/[c-q]$$
$$= 0.9900\{64,720+[41,548-29,750$$
$$-513][1-0.3000]-2,370\}$$
$$-6,497/[1.2000-1.0100]$$
$$= \underline{35,351} \text{ millions USD}$$

Corporate IFRS-GAAP (B/S-I/S), ISBN-13: **978-1720792789**, ISBN-10: **172079278X**

Law-7613:

If both (**F**= Fixed Cost= 29,740 millions USD), (**I**= Interest Expense= 513 millions USD), (**t**= T/B= Tax Rate= 30.00%), (**Q**= Quoted Longterm Liabilities= 35,351 millions USD), (**U**= Utilized or Starting Capital= 64,720 millions USD), (**c**= C/X= Current Ratio= 1.20 times), (**q**= [C-P]/X= Quick or Acid Test Ratio= 1.01 times), (**P**= Procured Inventories= 6,497 millions USD), (**D**= Dividend Paid= 2,370 millions USD), and (**M**= Margin of Contribution= 41,548 millions USD), then it's (**I**= L/E= Leverage or Gearing Ratio planned), is:

$$I = \{Q+P/[c-q]\}/\{U+[M-F-I][1-t]-D\}$$
$$= \{35,351+6,497/[1.2000-1.0100]\}$$
$$/\{64,720+[41,548-29,750$$
$$-513][1-0.3000]-2,370\}$$
$$= 99.00\%$$

Corporate IFRS-GAAP (B/S-I/S), ISBN-13: **978-1720792789**, ISBN-10: **172079278X**

Law-7614:

If both (**F**= Fixed Cost= 29,740 millions USD), (**I**= Interest Expense= 513 millions USD), (**t**= T/B= Tax Rate= 30.00%), (**Q**= Quoted Longterm Liabilities= 35,351 millions USD), (**I**= L/E= Leverage or Gearing Ratio= 99.00%), (**c**= C/X= Current Ratio= 1.20 times), (**q**= [C-P]/X= Quick or Acid Test Ratio= 1.01 times), (**P**= Procured Inventories= 6,497 millions USD), (**D**= Dividend Paid= 2,370 millions USD), and (**M**= Margin of Contribution= 41,548 millions USD), then it's (**U**= Utilized or Starting Capital planned), is:

$$\mathbf{U}= \{\mathbf{Q}+\mathbf{P}/[\mathbf{c}-\mathbf{q}]\}/\mathbf{I} - \{[\mathbf{M}-\mathbf{F}-\mathbf{I}][1-\mathbf{t}]-\mathbf{D}\}$$
$$= \{35,351+6,497/[1.2000-1.0100]\}$$
$$/0.9900-\{41,548-29,750$$
$$-513][1-0.3000]-2,370\}$$
$$= 64,720 \text{ millions USD}$$

Corporate IFRS-GAAP (B/S-I/S), ISBN-13: **978-1720792789**, ISBN-10: **172079278X**

<u>Law-7615</u>:

If both (**F**= Fixed Cost= <u>29,740</u> millions USD), (**I**= Interest Expense= <u>513</u> millions USD), (**t**= T/B= Tax Rate= <u>30.00%</u>), (**Q**= Quoted Longterm Liabilities= <u>35,351</u> millions USD), (**I**= L/E= Leverage or Gearing Ratio= <u>99.00%</u>), (**c**= C/X= Current Ratio= <u>1.20</u> times), (**q**= [C-P]/X= Quick or Acid Test Ratio= <u>1.01</u> times), (**P**= Procured Inventories= <u>6,497</u> millions USD), (**D**= Dividend Paid= <u>2,370</u> millions USD), and (**U**= Utilized or Starting Capital= <u>64,720</u> millions USD), then it's (**M**= Margin of Contribution planned), is:

$$M= F+I+(D-U+\{Q+P/[c-q]\}/I)/[1-t]$$
$$= 29,750+513+(2,370-64,720+\{35,351$$
$$+6,497/[1.2000-1.0100]\}$$
$$/0.9900)/[1-0.3000]$$
$$= \underline{41,548} \text{ millions USD}$$

Corporate IFRS-GAAP (B/S-I/S), ISBN-13: **978-1720792789**, ISBN-10: **172079278X**

Law-7616:

If both (**M**= Margin of Contribution= 41,548 millions USD), (**I**= Interest Expense= 513 millions USD), (**t**= T/B= Tax Rate= 30.00%), (**Q**= Quoted Longterm Liabilities= 35,351 millions USD), (**I**= L/E= Leverage or Gearing Ratio= 99.00%), (**c**= C/X= Current Ratio= 1.20 times), (**q**= [C-P]/X= Quick or Acid Test Ratio= 1.01 times), (**P**= Procured Inventories= 6,497 millions USD), (**D**= Dividend Paid= 2,370 millions USD), and (**U**= Utilized or Starting Capital= 64,720 millions USD), then it's (**F**= Fixed Cost planned), is:

$$F= M\text{-}I\text{-}(D\text{-}U+\{Q+P/[c\text{-}q]\}/I\,)/[1\text{-}t]$$

$$= 41,548\text{-}513\text{-}(2,370\text{-}64,720+\{35,351$$
$$+6,497/[1.2000\text{-}1.0100]\}$$
$$/0.9900)/[1\text{-}0.3000]$$

$$= 29,750 \text{ millions USD}$$

Corporate IFRS-GAAP (B/S-I/S), ISBN-13: **978-1720792789**, ISBN-10: **172079278X**

Law-7617:

If both (**M**= Margin of Contribution= 41,548 millions
USD), (**F**= Fixed Cost= 29,740 millions USD), (**t**=
T/B= Tax Rate= 30.00%), (**Q**= Quoted Longterm
Liabilities= 35,351 millions USD), (**I**= L/E=
Leverage or Gearing Ratio= 99.00%), (**c**= C/X=
Current Ratio= 1.20 times), (**q**= [C-P]/X= Quick or
Acid Test Ratio= 1.01 times), (**P**= Procured
Inventories= 6,497 millions USD), (**D**= Dividend
Paid= 2,370 millions USD), and (**U**= Utilized or
Starting Capital= 64,720 millions USD), then it's (**I**=
Interest Expense planned), is:

$$I= M-F-(D-U+\{Q+P/[c-q]\}/I)/[1-t]$$
$$= 41,548-29,750-(2,370-64,720$$
$$+\{35,351+6,497/[1.2000$$
$$-1.0100]\}/0.9900)$$
$$/[1-0.3000]$$
$$= \underline{513} \text{ millions USD}$$

Corporate IFRS-GAAP (B/S-I/S), ISBN-13: **978-1720792789**, ISBN-10: **172079278X**

<u>Law-7618</u>:

If both (**M**= Margin of Contribution= <u>41,548</u> millions USD), (**F**= Fixed Cost= <u>29,740</u> millions USD), (**I**= Interest Expense= <u>513</u> millions USD), (**Q**= Quoted Longterm Liabilities= <u>35,351</u> millions USD), (**I**= L/E= Leverage or Gearing Ratio= <u>99.00%</u>), (**c**= C/X= Current Ratio= <u>1.20</u> times), (**q**= [C-P]/X= Quick or Acid Test Ratio= <u>1.01</u> times), (**P**= Procured Inventories= <u>6,497</u> millions USD), (**D**= Dividend Paid= <u>2,370</u> millions USD), and (**U**= Utilized or Starting Capital= <u>64,720</u> millions USD), then it's (**t**= Tax Rate planned), is:

$$\begin{aligned}
\mathbf{t} &= 1\text{-}(\mathbf{D\text{-}U}+\{\mathbf{Q}+\mathbf{P}/[\mathbf{c\text{-}q}]\}/\boldsymbol{I}\,)/[\mathbf{M\text{-}F\text{-}I}] \\
&= 1\text{-}(2,370\text{-}64,720+\{35,351+6,497 \\
&\quad /[1.2000\text{-}1.0100]\}/0.9900) \\
&\quad /[41,548\text{-}29,750\text{-}513] \\
&= 30.00\%
\end{aligned}$$

Corporate IFRS-GAAP (B/S-I/S), ISBN-13: **978-1720792789**, ISBN-10: **172079278X**

Law-7619:

If both (**M**= Margin of Contribution= 41,548 millions USD), (**F**= Fixed Cost= 29,740 millions USD), (**I**= Interest Expense= 513 millions USD), (**Q**= Quoted Longterm Liabilities= 35,351 millions USD), (**I**= L/E= Leverage or Gearing Ratio= 99.00%), (**c**= C/X= Current Ratio= 1.20 times), (**q**= [C-P]/X= Quick or Acid Test Ratio= 1.01 times), (**P**= Procured Inventories= 6,497 millions USD), (**t**= T/B= Tax Rate= 30.00%), and (**U**= Utilized or Starting Capital= 64,720 millions USD), then it's (**D**= Dividend Paid planned), is:

$$D= U+[M-F-I][1-t]-\{Q+P/[c-q]\}/I$$
$$= 64,720+[41,548-29,750-513]$$
$$[1-0.3000]-\{35,351+6,497$$
$$/[1.2000-1.0100]\}/0.9900$$
$$= 2,370 \text{ millions USD}$$

Corporate IFRS-GAAP (B/S-I/S), ISBN-13: **978-1720792789**, ISBN-10: **172079278X**

Law-7620:

If both (**M**= Margin of Contribution= 41,548 millions USD), (**F**= Fixed Cost= 29,740 millions USD), (**I**= Interest Expense= 513 millions USD), (**Q**= Quoted Longterm Liabilities= 35,351 millions USD), (**l**= L/E= Leverage or Gearing Ratio= 99.00%), (**c**= C/X= Current Ratio= 1.20 times), (**q**= [C-P]/X= Quick or Acid Test Ratio= 1.01 times), (**D**= Dividend Paid= 2,370 millions USD), (**t**= T/B= Tax Rate= 30.00%), and (**U**= Utilized or Starting Capital= 64,720 millions USD), then it's (**P**= Procured Inventories planned), is:

$$P= [c-q](l \{U+[M-F-I][1-t]-D\}-Q)$$
$$= [1.2000-1.0100](0.9900\{64,720$$
$$+[41,548-29,750-513]$$
$$[1-0.3000]-2,370\}-35,351)$$
$$= 6,497 \text{ millions USD}$$

Corporate IFRS-GAAP (B/S-I/S), ISBN-13: **978-1720792789**, ISBN-10: **172079278X**

Law-7621:

If both (**M**= Margin of Contribution= 41,548 millions USD), (**F**= Fixed Cost= 29,740 millions USD), (**I**= Interest Expense= 513 millions USD), (**Q**= Quoted Longterm Liabilities= 35,351 millions USD), (**l**= L/E= Leverage or Gearing Ratio= 99.00%), (**P**= Procured Inventories= 6,497 millions USD), (**q**= Quick or Acid Test Ratio= [C-P]/X= 1.01 times), (**D**= Dividend Paid= 2,370 millions USD), (**t**= T/B= Tax Rate= 30.00%), and (**U**= Utilized or Starting Capital= 64,720 millions USD), then it's (**c**= Current Ratio planned), is:

$$c= q+P/(l \{U+[M-F-I][1-t]-D\}-Q)$$
$$= 1.0100+6,497/(0.9900\{64,720$$
$$+[41,548-29,750-513]$$
$$[1-0.3000]-2,370\}-35,351)$$
$$= \underline{1.20} \text{ times}$$

Corporate IFRS-GAAP (B/S-I/S), ISBN-13: **978-1720792789**, ISBN-10: **172079278X**

Law-7622:

If both (**M**= Margin of Contribution= 41,548 millions USD), (**F**= Fixed Cost= 29,740 millions USD), (**I**= Interest Expense= 513 millions USD), (**Q**= Quoted Longterm Liabilities= 35,351 millions USD), (**I**= L/E= Leverage or Gearing Ratio= 99.00%), (**P**= Procured Inventories= 6,497 millions USD), (**c**= C/X= Current Ratio= 1.20 times), (**D**= Dividend Paid= 2,370 millions USD), (**t**= T/B= Tax Rate= 30.00%), and (**U**= Utilized or Starting Capital= 64,720 millions USD), then it's (**q**= Quick or Acid Test Ratio planned), is:

$$q= c\text{-}P/(I\{U+[M\text{-}F\text{-}I][1\text{-}t]\text{-}D\}\text{-}Q)$$
$$= 1.2000\text{-}6,497/(0.9900\{64,720$$
$$+[41,548\text{-}29,750\text{-}513]$$
$$[1\text{-}0.3000]\text{-}2,370\}\text{-}35,351)$$
$$= 1.01 \text{ times}$$

Corporate IFRS-GAAP (B/S-I/S), ISBN-13: **978-1720792789**, ISBN-10: **172079278X**

Law-7623:

If both (**M**= Margin of Contribution= 41,548 millions USD), (**F**= Fixed Cost= 29,740 millions USD), (**I**= Interest Expense= 513 millions USD), (**q**= [C-P]/X= Quick or Acid Test Ratio= 1.01 times), (**l**= L/E= Leverage or Gearing Ratio= 99.00%), (**p**= 360 P/V= Procured Inventory Days= 240 Days), (**V**= Variable Cost= 9,746 millions USD), (**c**= C/X= Current Ratio= 1.20 times), (**D**= Dividend Paid= 2,370 millions USD), (**t**= T/B= Tax Rate= 30.00%), and (**U**= Utilized or Starting Capital= 64,720 millions USD), then it's (**Q**= Quoted Longterm Liabilities planned), is:

$$Q = l\{U+[M-F-I][1-t]-D\}-Vp/\{360[c-q]\}$$
$$= 0.9900\{64,720+[41,548-29,750$$
$$-513][1-0.3000]-2,370\}$$
$$-9,746*240/\{360[1.2000$$
$$-1.0100]\}$$
$$= 35,351 \text{ millions USD}$$

Law-7624:

If both (**M**= Margin of Contribution= <u>41,548</u> millions USD), (**F**= Fixed Cost= <u>29,740</u> millions USD), (**I**= Interest Expense= <u>513</u> millions USD), (**q**= [C-P]/X= Quick or Acid Test Ratio= <u>1.01</u> times), (**Q**= Quoted Longterm Liabilities= <u>35,351</u> millions USD), (**p**= 360 P/V= Procured Inventory Days= <u>240</u> Days), (**V**= Variable Cost= <u>9,746</u> millions USD), (**c**= C/X= Current Ratio= <u>1.20</u> times), (**D**= Dividend Paid= <u>2,370</u> millions USD), (**t**= T/B= Tax Rate= <u>30.00%</u>), and (**U**= Utilized or Starting Capital= <u>64,720</u> millions USD), then it's (**I** = Quoted Longterm Liabilities planned), is:

$$I = (\mathbf{Q} + \mathbf{Vp}/\{360[\mathbf{c}\text{-}\mathbf{q}]\})/\{\mathbf{U} + [\mathbf{M}\text{-}\mathbf{F}\text{-}\mathbf{I}][1\text{-}\mathbf{t}]\text{-}\mathbf{D}\}$$

$$= (35,351 + 9,746*240/\{360[1.2000 \\ -1.0100]\})/\{64,720 + [41,548 \\ -29,750 - 513][1 - 0.3000] \\ -2,370\}$$

$$= \underline{99.00\%}$$

Corporate IFRS-GAAP (B/S-I/S), ISBN-13: **978-1720792789**, ISBN-10: **172079278X**

<u>Law-7625</u>:

If both (**M**= Margin of Contribution= <u>41,548</u> millions USD), (**F**= Fixed Cost= <u>29,740</u> millions USD), (**I**= Interest Expense= <u>513</u> millions USD), (**q**= [C-P]/X= Quick or Acid Test Ratio= <u>1.01</u> times), (**Q**= Quoted Longterm Liabilities= <u>35,351</u> millions USD), (**p**= 360 P/V= Procured Inventory Days= <u>240</u> Days), (**V**= Variable Cost= <u>9,746</u> millions USD), (**c**= C/X= Current Ratio= <u>1.20</u> times), (**D**= Dividend Paid= <u>2,370</u> millions USD), (**t**= T/B= Tax Rate= <u>30.00%</u>), and (**I**= L/E= Leverage or Gearing Ratio= <u>99.00%</u>), then it's (**U**= Utilized or Starting Capital planned), is:

$$U= (Q+Vp/\{360[c\text{-}q]\})/I - \{[M\text{-}F\text{-}I][1\text{-}t]\text{-}D\}$$

$$= (35,351+9,746*240/\{360[1.2000$$
$$-1.0100]\})/0.9900 - \{[41,548$$
$$-29,750\text{-}513][1\text{-}0.3000]$$
$$-2,370\}$$

$$= \underline{64,720} \text{ millions USD}$$

Corporate IFRS-GAAP (B/S-I/S), ISBN-13: **978-1720792789**, ISBN-10: **172079278X**

Law-7626:

If both (**U**= Utilized or Starting Capital= 64,720 millions USD), (**F**= Fixed Cost= 29,740 millions USD), (**I**= Interest Expense= 513 millions USD), (**q**= [C-P]/X= Quick or Acid Test Ratio= 1.01 times), (**Q**= Quoted Longterm Liabilities= 35,351 millions USD), (**p**= 360 P/V= Procured Inventory Days= 240 Days), (**V**= Variable Cost= 9,746 millions USD), (**c**= C/X= Current Ratio= 1.20 times), (**D**= Dividend Paid= 2,370 millions USD), (**t**= T/B= Tax Rate= 30.00%), and (**I**= L/E= Leverage or Gearing Ratio= 99.00%), then it's (**M**= Margin of Contribution planned), is:

$$M = F+I+[D-U+(Q+Vp/\{360[c-q]\})/I]/[1-t]$$

$$= 29,750+513+[2,370-64,720$$
$$+(35,351+9,746*240/\{360$$
$$[1.2000-1.0100]\})/0.9900]$$
$$/[1-0.3000]$$

$$= 41,548 \text{ millions USD}$$

Corporate IFRS-GAAP (B/S-I/S), ISBN-13: **978-1720792789**, ISBN-10: **172079278X**

Law-7627:

If both (**U**= Utilized or Starting Capital= 64,720 millions USD), (**M**= Margin of Contribution= 41,548 millions USD), (**I**= Interest Expense= 513 millions USD), (**q**= [C-P]/X= Quick or Acid Test Ratio= 1.01 times), (**Q**= Quoted Longterm Liabilities= 35,351 millions USD), (**p**= 360 P/V= Procured Inventory Days= 240 Days), (**V**= Variable Cost= 9,746 millions USD), (**c**= C/X= Current Ratio= 1.20 times), (**D**= Dividend Paid= 2,370 millions USD), (**t**= T/B= Tax Rate= 30.00%), and (**I**= L/E= Leverage or Gearing Ratio= 99.00%), then it's (**F**= Fixed Cost planned), is:

$$F= M\text{-}I\text{-}[D\text{-}U+(Q+Vp/\{360[c\text{-}q]\})/I]/[1\text{-}t]$$
$$= 41{,}548\text{-}513\text{-}[2{,}370\text{-}64{,}720$$
$$+(35{,}351+9{,}746*240/\{360$$
$$[1.2000\text{-}1.0100]\})/0.9900]$$
$$/[1\text{-}0.3000]$$
$$= 29{,}750 \text{ millions USD}$$

Corporate IFRS-GAAP (B/S-I/S), ISBN-13: **978-1720792789**, ISBN-10: **172079278X**

<u>Law-7628</u>:

If both (**U**= Utilized or Starting Capital= <u>64,720</u> millions USD), (**F**= Fixed Cost= <u>29,740</u> millions USD), (**M**= Margin of Contribution= <u>41,548</u> millions USD), (**q**= [C-P]/X= Quick or Acid Test Ratio= <u>1.01</u> times), (**Q**= Quoted Longterm Liabilities= <u>35,351</u> millions USD), (**p**= 360 P/V= Procured Inventory Days= <u>240</u> Days), (**V**= Variable Cost= <u>9,746</u> millions USD), (**c**= C/X= Current Ratio= <u>1.20</u> times), (**D**= Dividend Paid= <u>2,370</u> millions USD), (**t**= T/B= Tax Rate= <u>30.00%</u>), and (**/**= L/E= Leverage or Gearing Ratio= <u>99.00%</u>), then it's (**I**= Interest Expense planned), is:

$$I= M-F-[D-U+(Q+Vp/\{360[c-q]\})//]/[1-t]$$
$$= 41,548-29,750-[2,370-64,720$$
$$+(35,351+9,746*240/\{360$$
$$[1.2000-1.0100]\})/0.9900]$$
$$/[1-0.3000]$$
$$= \underline{513} \text{ millions USD}$$

Corporate IFRS-GAAP (B/S-I/S), ISBN-13: **978-1720792789**, ISBN-10: **172079278X**

<u>Law-7629</u>:

If both (**U**= Utilized or Starting Capital= <u>64,720</u> millions USD), (**F**= Fixed Cost= <u>29,740</u> millions USD), (**M**= Margin of Contribution= <u>41,548</u> millions USD), (**q**= [C-P]/X= Quick or Acid Test Ratio= <u>1.01</u> times), (**Q**= Quoted Longterm Liabilities= <u>35,351</u> millions USD), (**p**= 360 P/V= Procured Inventory Days= <u>240</u> Days), (**V**= Variable Cost= <u>9,746</u> millions USD), (**c**= C/X= Current Ratio= <u>1.20</u> times), (**D**= Dividend Paid= <u>2,370</u> millions USD), (**I**= Interest Expense= <u>513</u> millions USD), and (**l**= L/E= Leverage or Gearing Ratio= <u>99.00%</u>), then it's (**t**= Tax Rate planned), is:

$$\mathbf{t}= 1-[\mathbf{D}-\mathbf{U}+(\mathbf{Q}+\mathbf{Vp}/\{360[\mathbf{c}-\mathbf{q}]\})/\mathbf{l}\ ]/[\mathbf{M}-\mathbf{F}-\mathbf{I}]$$

$$= 1-[2,370-64,720+(35,351+9,746$$
$$*240/\{360[1.2000-1.0100]\})$$
$$/0.9900]/[41,548-29,750-513]$$

$$= \underline{30.00\%}$$

Corporate IFRS-GAAP (B/S-I/S), ISBN-13: **978-1720792789**, ISBN-10: **172079278X**

<u>Law-7630</u>:

If both (**U**= Utilized or Starting Capital= <u>64,720</u> millions USD), (**F**= Fixed Cost= <u>29,740</u> millions USD), (**M**= Margin of Contribution= <u>41,548</u> millions USD), (**q**= [C-P]/X= Quick or Acid Test Ratio= <u>1.01</u> times), (**Q**= Quoted Longterm Liabilities= <u>35,351</u> millions USD), (**p**= 360 P/V= Procured Inventory Days= <u>240</u> Days), (**V**= Variable Cost= <u>9,746</u> millions USD), (**c**= C/X= Current Ratio= <u>1.20</u> times), (**t**= T/B= Tax Rate= <u>30.00%</u>), (**I**= Interest Expense= <u>513</u> millions USD), and (**I**= L/E= Leverage or Gearing Ratio= <u>99.00%</u>), then it's (**D**= Dividend Paid planned), is:

$$D= U+[M-F-I][1-t]-(Q+Vp/\{360[c-q]\})/I$$

$$\begin{aligned} &= 64{,}720+[41{,}548{-}29{,}750{-}513] \\ &\quad [1{-}0.3000]{-}(35{,}351{+}9{,}746 \\ &\quad *240/\{360[1.2000{-}1.0100]\}) \\ &\quad /0.9900 \\ &= \underline{2{,}370} \text{ millions USD} \end{aligned}$$

Corporate IFRS-GAAP (B/S-I/S), ISBN-13: **978-1720792789**, ISBN-10: **172079278X**

<u>Law-7631</u>:

If both (**U**= Utilized or Starting Capital= <u>64,720</u> millions USD), (**F**= Fixed Cost= <u>29,740</u> millions USD), (**M**= Margin of Contribution= <u>41,548</u> millions USD), (**q**= [C-P]/X= Quick or Acid Test Ratio= <u>1.01</u> times), (**Q**= Quoted Longterm Liabilities= <u>35,351</u> millions USD), (**p**= 360 P/V= Procured Inventory Days= <u>240</u> Days), (**D**= Dividend Paid= <u>2,370</u> millions USD), (**c**= C/X= Current Ratio= <u>1.20</u> times), (**t**= T/B= Tax Rate= <u>30.00%</u>), (**I**= Interest Expense= <u>513</u> millions USD), and (**I**= L/E= Leverage or Gearing Ratio= <u>99.00%</u>), then it's (**V**= Variable Cost planned), is:

$$V = 360[c\text{-}q](I\{U+[M\text{-}F\text{-}I][1\text{-}t]\}\text{-}Q)/p$$
$$= 360[1.2000\text{-}1.0100](0.9900\{64,720$$
$$+[41,548\text{-}29,750\text{-}513]$$
$$[1\text{-}0.3000]\}\text{-}35,351)/240$$
$$= \underline{9,746} \text{ millions USD}$$

Corporate IFRS-GAAP (B/S-I/S), ISBN-13: **978-1720792789**, ISBN-10: **172079278X**

Law-7632:

If both (**U**= Utilized or Starting Capital= 64,720 millions USD), (**F**= Fixed Cost= 29,740 millions USD), (**M**= Margin of Contribution= 41,548 millions USD), (**q**= [C-P]/X= Quick or Acid Test Ratio= 1.01 times), (**Q**= Quoted Longterm Liabilities= 35,351 millions USD), (**V**= Variable Cost= 9,746 millions USD), (**D**= Dividend Paid= 2,370 millions USD), (**c**= C/X= Current Ratio= 1.20 times), (**t**= T/B= Tax Rate= 30.00%), (**I**= Interest Expense= 513 millions USD), and (**I**= L/E= Leverage or Gearing Ratio= 99.00%), then it's (**p**= Procured Inventory Days planned), is:

$$\mathbf{p}= 360[\mathbf{c}-\mathbf{q}](\mathbf{I}\ \{\mathbf{U}+[\mathbf{M}-\mathbf{F}-\mathbf{I}][1-\mathbf{t}]\}-\mathbf{Q})/\mathbf{V}$$
$$= 360[1.2000-1.0100](0.9900$$
$$\{64,720+[41,548-29,750$$
$$-513][1-0.3000]\}-35,351)$$
$$/9,746$$

$$= \underline{240}\ \text{days}$$

Corporate IFRS-GAAP (B/S-I/S), ISBN-13: **978-1720792789**, ISBN-10: **172079278X**

Law-7633:

If both (**U**= Utilized or Starting Capital= 64,720 millions USD), (**F**= Fixed Cost= 29,740 millions USD), (**M**= Margin of Contribution= 41,548 millions USD), (**q**= [C-P]/X= Quick or Acid Test Ratio= 1.01 times), (**Q**= Quoted Longterm Liabilities= 35,351 millions USD), (**V**= Variable Cost= 9,746 millions USD), (**D**= Dividend Paid= 2,370 millions USD), (**p**= 360 P/V= Procured Inventory Days= 240 Days), (**t**= T/B= Tax Rate= 30.00%), (**I**= Interest Expense= 513 millions USD), and (**I**= L/E= Leverage or Gearing Ratio= 99.00%), then it's (**c**= Current Ratio planned), is:

$$c = q + Vp/[360(I\{U+[M-F-I][1-t]\}-Q)]$$
$$= 1.0100+9,746*240/[360(0.9900$$
$$\{64,720+[41,548-29,750$$
$$-513][1-0.3000]\}- 35,351)]$$
$$= 1.20 \text{ times}$$

<u>Law-7634</u>:

If both (**U**= Utilized or Starting Capital= <u>64,720</u> millions USD), (**F**= Fixed Cost= <u>29,740</u> millions USD), (**M**= Margin of Contribution= <u>41,548</u> millions USD), (**c**= C/X= Current Ratio= <u>1.20</u> times), (**Q**= Quoted Longterm Liabilities= <u>35,351</u> millions USD), (**V**= Variable Cost= <u>9,746</u> millions USD), (**D**= Dividend Paid= <u>2,370</u> millions USD), (**p**= 360 P/V= Procured Inventory Days= <u>240</u> Days), (**t**= T/B= Tax Rate= <u>30.00%</u>), (**I**= Interest Expense= <u>513</u> millions USD), and (**l**= L/E= Leverage or Gearing Ratio= <u>99.00%</u>), then it's (**q**= Quick Acid Test Ratio planned), is:

$$q= c\text{-}Vp/[360(I\,\{U+[M\text{-}F\text{-}I][1\text{-}t]\}\text{-}Q)]$$
$$= 1.2000\text{-}9{,}746*240/[360(0.9900$$
$$\{64{,}720+[41{,}548\text{-}29{,}750$$
$$-513][1\text{-}0.3000]\}\text{-}35{,}351)]$$
$$= \underline{1.01}\text{ times}$$

Corporate IFRS-GAAP (B/S-I/S), ISBN-13: **978-1720792789**, ISBN-10: **172079278X**

Law-7635:

If both (**U**= Utilized or Starting Capital= 64,720 millions USD), (**F**= Fixed Cost= 29,740 millions USD), (**M**= Margin of Contribution= 41,548 millions USD), (**c**= C/X= Current Ratio= 1.20 times), (**q**= Quick or Acid Test Ratio= [C-P]/X= 1.01 times), (**v**= V/S= Variable Portion= 19.00%), (**S**= Sales or Revenues= 51,294 millions USD), (**D**= Dividend Paid= 2,370 millions USD), (**p**= 360 P/V= Procured Inventory Days= 240 Days), (**t**= T/B= Tax Rate= 30.00%), (**I**= Interest Expense= 513 millions USD), and (**I**= L/E= Leverage or Gearing Ratio= 99.00%), then it's (**Q**= Quoted Longterm Liabilities planned), is:

$$Q = I\{U+[M-F-I][1-t]-D\}-Svp/\{360[c-q]\}$$
$$= 0.9900\{64,720+[41,548-29,750$$
$$-513][1-0.3000]-2,370\}$$
$$-51,294*0.1900*240$$
$$/\{360[1.2000-1.0100]\}$$
$$= 35,351 \text{ millions USD}$$

Corporate IFRS-GAAP (B/S-I/S), ISBN-13: **978-1720792789**, ISBN-10: **172079278X**

Law-7636:

If both (**U**= Utilized or Starting Capital= 64,720 millions USD), (**F**= Fixed Cost= 29,740 millions USD), (**M**= Margin of Contribution= 41,548 millions USD), (**c**= C/X= Current Ratio= 1.20 times), (**q**= Quick or Acid Test Ratio= [C-P]/X= 1.01 times), (**v**= V/S= Variable Portion= 19.00%), (**S**= Sales or Revenues= 51,294 millions USD), (**D**= Dividend Paid= 2,370 millions USD), (**p**= 360 P/V= Procured Inventory Days= 240 Days), (**t**= T/B= Tax Rate= 30.00%), (**I**= Interest Expense= 513 millions USD), and (**Q**= Quoted Longterm Liabilities= 35,351 millions USD), then it's (**l** = Leverage or Gearing Ratio planned), is:

$$l = (\mathbf{Q} + \mathbf{Svp} / \{360[\mathbf{c} - \mathbf{q}]\})$$
$$/ \{\mathbf{U} + [\mathbf{M} - \mathbf{F} - \mathbf{I}][1 - \mathbf{t}] - \mathbf{D}\}$$
$$= (35,351 + 51,294 * 0.1900 * 240$$
$$/ \{360[1.2000 - 1.0100]\})$$
$$/ \{64,720 + [41,548 - 29,750$$
$$-513][1 - 0.3000] - 2,370\}$$
$$= 99.00\%$$

Corporate IFRS-GAAP (B/S-I/S), ISBN-13: **978-1720792789**, ISBN-10: **172079278X**

<u>Law-7637</u>:

If both (**Q**= Quoted Longterm Liabilities= <u>35,351</u> millions USD), (**F**= Fixed Cost= <u>29,740</u> millions USD), (**M**= Margin of Contribution= <u>41,548</u> millions USD), (**c**= C/X= Current Ratio= <u>1.20</u> times), (**q**= Quick or Acid Test Ratio= [C-P]/X= <u>1.01</u> times), (**v**= V/S= Variable Portion= <u>19.00%</u>), (**S**= Sales or Revenues= <u>51,294</u> millions USD), (**D**= Dividend Paid= <u>2,370</u> millions USD), (**p**= 360 P/V= Procured Inventory Days= <u>240</u> Days), (**t**= T/B= Tax Rate= <u>30.00%</u>), (**I**= Interest Expense= <u>513</u> millions USD), and (**I**= L/E= Leverage or Gearing Ratio= <u>99.00%</u>), then it's (**U**= Utilized or Starting Capital planned), is:

$$U = (Q+Svp/\{360[c-q]\})/I - \{[M-F-I][1-t]-D\}$$
$$= (35,351+51,294*0.1900*240$$
$$/\{360[1.2000-1.0100]\})$$
$$/0.9900-\{[41,548-29,750$$
$$-513][1-0.3000]-2,370\}$$
$$= \underline{64,720} \text{ millions USD}$$

Corporate IFRS-GAAP (B/S-I/S), ISBN-13: **978-1720792789**, ISBN-10: **172079278X**

Law-7638:

If both (**Q**= Quoted Longterm Liabilities= 35,351 millions USD), (**F**= Fixed Cost= 29,740 millions USD), (**U**= Utilized or Starting Capital= 64,720 millions USD), (**c**= C/X= Current Ratio= 1.20 times), (**q**= [C-P]/X= Quick or Acid Test Ratio= 1.01 times), (**v**= V/S= Variable Portion= 19.00%), (**S**= Sales or Revenues= 51,294 millions USD), (**D**= Dividend Paid= 2,370 millions USD), (**p**= 360 P/V= Procured Inventory Days= 240 Days), (**t**= T/B= Tax Rate= 30.00%), (**I**= Interest Expense= 513 millions USD), and (**l**= L/E= Leverage or Gearing Ratio= 99.00%), then it's (**M**= Margin of Contribution planned), is:

$$M= F+I+[D-U+(Q+Svp/\{360[c-q]\})/l\,]/[1-t]$$
$$= 29,750+513+[2,370-64,720$$
$$+(35,351+51,294*0.1900$$
$$*240/\{360[1.2000$$
$$-1.0100]\})/0.9900]$$
$$/[1-0.3000]$$
$$= \underline{41,548} \text{ millions USD}$$

Corporate IFRS-GAAP (B/S-I/S), ISBN-13: **978-1720792789**, ISBN-10: **172079278X**

<u>Law-7639</u>:

If both (**Q**= Quoted Longterm Liabilities= <u>35,351</u> millions USD), (**M**= Margin of Contribution= <u>41,548</u> millions USD), (**U**= Utilized or Starting Capital= <u>64,720</u> millions USD), (**c**= C/X= Current Ratio= <u>1.20</u> times), (**q**= [C-P]/X= Quick or Acid Test Ratio= <u>1.01</u> times), (**v**= V/S= Variable Portion= <u>19.00%</u>), (**$**= Sales or Revenues= <u>51,294</u> millions USD), (**D**= Dividend Paid= <u>2,370</u> millions USD), (**p**= 360 P/V= Procured Inventory Days= <u>240</u> Days), (**t**= T/B= Tax Rate= <u>30.00%</u>), (**I**= Interest Expense= <u>513</u> millions USD), and (**F**= L/E= Leverage or Gearing Ratio= <u>99.00%</u>), then it's (**F**= Fixed Cost planned), is:

$$F= M-I-[D-U+(Q+\$vp/\{360[c-q]\})/F]/[1-t]$$
$$= 41,548-513-[2,370-64,720$$
$$+(35,351+51,294*0.1900$$
$$*240/\{360[1.2000$$
$$-1.0100]\})/0.9900]$$
$$/[1-0.3000]$$
$$= \underline{29,750} \text{ millions USD}$$

Corporate IFRS-GAAP (B/S-I/S), ISBN-13: **978-1720792789**, ISBN-10: **172079278X**

Law-7640:

If both (**Q**= Quoted Longterm Liabilities= 35,351 millions USD), (**M**= Margin of Contribution= 41,548 millions USD), (**U**= Utilized or Starting Capital= 64,720 millions USD), (**c**= C/X= Current Ratio= 1.20 times), (**q**= [C-P]/X= Quick or Acid Test Ratio= 1.01 times), (**v**= V/S= Variable Portion= 19.00%), (**S**= Sales or Revenues= 51,294 millions USD), (**D**= Dividend Paid= 2,370 millions USD), (**p**= 360 P/V= Procured Inventory Days= 240 Days), (**t**= T/B= Tax Rate= 30.00%), (**F**= Fixed Cost= 29,740 millions USD), and (**I**= L/E= Leverage or Gearing Ratio= 99.00%), then it's (**I**= Interest Expense planned), is:

$$\text{I= M-F-[D-U+(Q+Svp/\{360[c-q]\})/I]/[1-t]}$$

$$= 41,548\text{-}29,750\text{-}[2,370\text{-}64,720$$
$$+(35,351+51,294*0.1900$$
$$*240/\{360[1.2000$$
$$-1.0100]\})/0.9900]$$
$$/[1\text{-}0.3000]$$

$$= \underline{513} \text{ millions USD}$$

Corporate IFRS-GAAP (B/S-I/S), ISBN-13: **978-1720792789**, ISBN-10: **172079278X**

Law-7641:

If both (**Q**= Quoted Longterm Liabilities= 35,351 millions USD), (**M**= Margin of Contribution= 41,548 millions USD), (**U**= Utilized or Starting Capital= 64,720 millions USD), (**c**= C/X= Current Ratio= 1.20 times), (**q**= [C-P]/X= Quick or Acid Test Ratio= 1.01 times), (**v**= V/S= Variable Portion= 19.00%), (**S**= Sales or Revenues= 51,294 millions USD), (**D**= Dividend Paid= 2,370 millions USD), (**p**= 360 P/V= Procured Inventory Days= 240 Days), (**I**= Interest Expense= 513 millions USD), (**F**= Fixed Cost= 29,740 millions USD), and (**I**= L/E= Leverage or Gearing Ratio= 99.00%), then it's (**t**= Tax Rate planned), is:

$$t= 1\text{-}[D\text{-}U+(Q+Svp/\{360[c\text{-}q]\})/I]/[M\text{-}F\text{-}I]$$
$$= 1\text{-}[2,370\text{-}64,720+(35,351+51,294$$
$$*0.1900*240/\{360[1.2000$$
$$-1.0100]\})/0.9900]/[41,548$$
$$-29,750\text{-}513]$$
$$= 30.00\%$$

Corporate IFRS-GAAP (B/S-I/S), ISBN-13: **978-1720792789**, ISBN-10: **172079278X**

Law-7642:

If both (**Q**= Quoted Longterm Liabilities= 35,351 millions USD), (**M**= Margin of Contribution= 41,548 millions USD), (**U**= Utilized or Starting Capital= 64,720 millions USD), (**c**= C/X= Current Ratio= 1.20 times), (**q**= [C-P]/X= Quick or Acid Test Ratio= 1.01 times), (**v**= V/S= Variable Portion= 19.00%), (**S**= Sales or Revenues= 51,294 millions USD), (**t**= T/B= Tax Rate= 30.00%), (**p**= 360 P/V= Procured Inventory Days= 240 Days), (**I**= Interest Expense= 513 millions USD), (**F**= Fixed Cost= 29,740 millions USD), and (**l**= L/E= Leverage or Gearing Ratio= 99.00%), then it's (**D**= Dividend Paid planned), is:

$$D = U + [M\text{-}F\text{-}I][1\text{-}t] - (Q + Svp/\{360[c\text{-}q]\})/l$$
$$= 64{,}720 + [41{,}548\text{-}29{,}750\text{-}513]$$
$$[1\text{-}0.300] - (35{,}351 + 51{,}294$$
$$*0.1900*240/\{360[1.2000$$
$$-1.0100]\})/0.9900$$
$$= \underline{2{,}370} \text{ millions USD}$$

Corporate IFRS-GAAP (B/S-I/S), ISBN-13: **978-1720792789**, ISBN-10: **172079278X**

Law-7643:

If both (Q= Quoted Longterm Liabilities= 35,351 millions USD), (M= Margin of Contribution= 41,548 millions USD), (U= Utilized or Starting Capital= 64,720 millions USD), (c= C/X= Current Ratio= 1.20 times), (q= [C-P]/X= Quick or Acid Test Ratio= 1.01 times), (v= V/S= Variable Portion= 19.00%), (D= Dividend Paid= 2,370 millions USD), (t= T/B= Tax Rate= 30.00%), (p= 360 P/V= Procured Inventory Days= 240 Days), (I= Interest Expense= 513 millions USD), (F= Fixed Cost= 29,740 millions USD), and (I= L/E= Leverage or Gearing Ratio= 99.00%), then it's (S= Sales or Revenues planned), is:

$$S= 360[c\text{-}q](I\{U+[M\text{-}F\text{-}I][1\text{-}t]\text{-}D\}\text{-}Q)/[vp]$$
$$= 360[1.2000\text{-}1.0100](0.9900$$
$$\{64,720+[41,548\text{-}29,750$$
$$-513][1\text{-}0.3000]\text{-}2,370\}$$
$$-35,351)/[0.1900*240]$$
$$= 51,294 \text{ millions USD}$$

Corporate IFRS-GAAP (B/S-I/S), ISBN-13: **978-1720792789**, ISBN-10: **172079278X**

Law-7644:

If both (**Q**= Quoted Longterm Liabilities= 35,351 millions USD), (**M**= Margin of Contribution= 41,548 millions USD), (**U**= Utilized or Starting Capital= 64,720 millions USD), (**c**= C/X= Current Ratio= 1.20 times), (**q**= [C-P]/X= Quick or Acid Test Ratio= 1.01 times), (**$**= Sales or Revenues= 51,294 millions USD), (**D**= Dividend Paid= 2,370 millions USD), (**t**= T/B= Tax Rate= 30.00%), (**p**= 360 P/V= Procured Inventory Days= 240 Days), (**I**= Interest Expense= 513 millions USD), (**F**= Fixed Cost= 29,740 millions USD), and (**I**= L/E= Leverage or Gearing Ratio= 99.00%), then it's (**v**= Variable Portion planned), is:

$$v= 360[\textbf{c-q}](\textbf{\textit{I}}\{U+[\textbf{M-F-I}][1-\textbf{t}]-\textbf{D}\}-\textbf{Q})/[\textbf{\$p}]$$

$$= 360[1.2000-1.0100](0.9900$$
$$\{64,720+[41,548-29,750$$
$$-513][1-0.3000]-2,370\}$$
$$-35,351)/[51,294*240]$$

$$= \underline{19.00\%}$$

Corporate IFRS-GAAP (B/S-I/S), ISBN-13: **978-1720792789**, ISBN-10: **172079278X**

Law-7645:

If both ($\mathbf{Q}$= Quoted Longterm Liabilities= 35,351 millions USD), ($\mathbf{M}$= Margin of Contribution= 41,548 millions USD), ($\mathbf{U}$= Utilized or Starting Capital= 64,720 millions USD), ($\mathbf{c}$= C/X= Current Ratio= 1.20 times), ($\mathbf{q}$= [C-P]/X= Quick or Acid Test Ratio= 1.01 times), ($\mathbf{\$}$= Sales or Revenues= 51,294 millions USD), ($\mathbf{D}$= Dividend Paid= 2,370 millions USD), ($\mathbf{t}$= T/B= Tax Rate= 30.00%), ($\mathbf{v}$= V/S= Variable Portion= 19.00%), ($\mathbf{I}$= Interest Expense= 513 millions USD), ($\mathbf{F}$= Fixed Cost= 29,740 millions USD), and ($\mathbf{I}$= L/E= Leverage or Gearing Ratio= 99.00%), then it's ($\mathbf{p}$= Procured Inventory Days planned), is:

$$\mathbf{p}= 360[\mathbf{c\text{-}q}](\mathbf{I}\;\{\mathbf{U}+[\mathbf{M\text{-}F\text{-}I}][1\text{-}\mathbf{t}]\text{-}\mathbf{D}\}\text{-}\mathbf{Q})/[\mathbf{\$v}]$$

$$= 360[1.2000\text{-}1.0100](0.9900$$
$$\{64,720+[41,548\text{-}29,750$$
$$\text{-}513][1\text{-}0.3000]\text{-}2,370\}$$
$$\text{-}35,351)/[51,294*0.1900]$$

$$= \underline{240}\text{ days}$$

Corporate IFRS-GAAP (B/S-I/S), ISBN-13: **978-1720792789**, ISBN-10: **172079278X**

<u>Law-7646</u>:

If both (**Q**= Quoted Longterm Liabilities= <u>35,351</u> millions USD), (**M**= Margin of Contribution= <u>41,548</u> millions USD), (**U**= Utilized or Starting Capital= <u>64,720</u> millions USD), (**p**= 360 P/V= Procured Inventory Days= <u>240</u> Days), (**q**= [C-P]/X= Quick or Acid Test Ratio= <u>1.01</u> times), (**S**= Sales or Revenues= <u>51,294</u> millions USD), (**D**= Dividend Paid= <u>2,370</u> millions USD), (**t**= T/B= Tax Rate= <u>30.00%</u>), (**v**= V/S= Variable Portion= <u>19.00%</u>), (**I**= Interest Expense= <u>513</u> millions USD), (**F**= Fixed Cost= <u>29,740</u> millions USD), and (**I**= L/E= Leverage or Gearing Ratio= <u>99.00%</u>), then it's (**c**= Current Ratio planned), is:

$$c= q+Svp/[360(I\{U+[M-F-I][1-t]-D\}-Q)]$$

$$= 1.0100+51,294*0.1900*240$$
$$/[360(0.9900\{64,720$$
$$+[41,548-29,750-513]$$
$$[1-0.3000]-2,370\}$$
$$-35,351)]$$

$$= \underline{1.20} \text{ times}$$

Corporate IFRS-GAAP (B/S-I/S), ISBN-13: **978-1720792789**, ISBN-10: **172079278X**

Law-7647:

If both (**Q**= Quoted Longterm Liabilities= 35,351 millions USD), (**M**= Margin of Contribution= 41,548 millions USD), (**U**= Utilized or Starting Capital= 64,720 millions USD), (**p**= 360 P/V= Procured Inventory Days= 240 Days), (**c**= C/X= Current Ratio= 1.20 times), (**$**= Sales or Revenues= 51,294 millions USD), (**D**= Dividend Paid= 2,370 millions USD), (**t**= T/B= Tax Rate= 30.00%), (**v**= V/S= Variable Portion= 19.00%), (**I**= Interest Expense= 513 millions USD), (**F**= Fixed Cost= 29,740 millions USD), and (**I**= L/E= Leverage or Gearing Ratio= 99.00%), then it's (**q**= Quick or Rapid Test Ratio planned), is:

$$q= c-\$vp/[360(I\{U+[M-F-I][1-t]-D\}-Q)]$$
$$= 1.2000-51,294*0.1900*240/[360$$
$$(0.9900\{64,720+[41,548$$
$$-29,750-513][1-0.3000]$$
$$-2,370\}-35,351)]$$
$$= 1.01 \text{ times}$$

Corporate IFRS-GAAP (B/S-I/S), ISBN-13: **978-1720792789**, ISBN-10: **172079278X**

<u>Law-7648</u>:

If both (**q**= [C-P]/X= Quick or Acid Test Ratio= <u>1.01</u> times), (**M**= Margin of Contribution= <u>41,548</u> millions USD), (**U**= Utilized or Starting Capital= <u>64,720</u> millions USD), (**p**= 360 P/V= Procured Inventory Days= <u>240</u> Days), (**c**= C/X= Current Ratio= <u>1.20</u> times), (**$'**= Sales of Past Year= <u>48,851</u> millions USD), (**s**= [S/S']-1= Sales Growth= <u>5.00%</u>), (**D**= Dividend Paid= <u>2,370</u> millions USD), (**t**= T/B= Tax Rate= <u>30.00%</u>), (**v**= V/S= Variable Portion= <u>19.00%</u>), (**I**= Interest Expense= <u>513</u> millions USD), (**F**= Fixed Cost= <u>29,740</u> millions USD), and (**/**= L/E= Leverage or Gearing Ratio= <u>99.00%</u>), then it's (**Q**= Quoted Longterm Liabilities planned), is:

$$Q= I \{U+[M-F-I][1-t]-D\}$$
$$-\$'vp[1+s]/\{360[c-q]\}$$
$$= 0.9900\{64,720+[41,548-29,750$$
$$-513][1-0.3000]-2,370\}$$
$$-48,851*0.1900*240$$
$$[1+0.0500]/\{360\{1.2000$$
$$-1.0100]\}$$
$$= \underline{35,351} \text{ millions USD}$$

Corporate IFRS-GAAP (B/S-I/S), ISBN-13: **978-1720792789**, ISBN-10: **172079278X**

Law-7649:

If both (**q**= [C-P]/X= Quick or Acid Test Ratio= 1.01 times), (**M**= Margin of Contribution= 41,548 millions USD), (**U**= Utilized or Starting Capital= 64,720 millions USD), (**p**= 360 P/V= Procured Inventory Days= 240 Days), (**c**= C/X= Current Ratio= 1.20 times), (**S'**= Sales of Past Year= 48,851 millions USD), (**s**= [S/S']-1= Sales Growth= 5.00%), (**D**= Dividend Paid= 2,370 millions USD), (**t**= T/B= Tax Rate= 30.00%), (**v**= V/S= Variable Portion= 19.00%), (**I**= Interest Expense= 513 millions USD), (**F**= Fixed Cost= 29,740 millions USD), and (**Q**= Quoted Longterm Liabilities= 35,351 millions USD), then it's (**l** =Leverage or Gearing Ratio planned), is:

$$l = (Q+S'vp[1+s]/\{360[c-q]\})$$
$$/\{U+[M-F-I][1-t]-D\}$$
$$= (35{,}351+48{,}851*0.1900*240$$
$$[1+0.0500]/\{360\{1.2000$$
$$-1.0100]\})/\{64{,}720+[41{,}548$$
$$-29{,}750-513][1-0.3000]$$
$$-2{,}370\}$$
$$= 99.00\%$$

Corporate IFRS-GAAP (B/S-I/S), ISBN-13: **978-1720792789**, ISBN-10: **172079278X**

Law-7650:

If both (**q**= [C-P]/X= Quick or Acid Test Ratio= 1.01
times), (**M**= Margin of Contribution= 41,548 millions
USD), (**I**= L/E= Leverage or Gearing Ratio=
99.00%), (**p**= 360 P/V= Procured Inventory Days=
240 Days), (**c**= C/X= Current Ratio= 1.20 times), (**S'**=
Sales of Past Year= 48,851 millions USD), (**s**=
[S/S']-1= Sales Growth= 5.00%), (**D**= Dividend
Paid= 2,370 millions USD), (**t**= T/B= Tax Rate=
30.00%), (**v**= V/S= Variable Portion= 19.00%), (**I**=
Interest Expense= 513 millions USD), (**F**= Fixed
Cost= 29,740 millions USD), and (**Q**= Quoted
Longterm Liabilities= 35,351 millions USD), then
it's (**U**= Utilized or Starting Capital planned), is:

$$U= (Q+S'vp[1+s]/\{360[c-q]\})/I$$
$$-\{[M-F-I][1-t]-D\}$$
$$= (35,351+48,851*0.1900*240$$
$$[1+0.0500]/\{360\{1.2000$$
$$-1.0100]\})/0.9900-\{[41,548$$
$$-29,750-513][1-0.3000]$$
$$-2,370\}$$
$$= 64,720 \text{ millions USD}$$

Law-7651:

If both (**q**= [C-P]/X= Quick or Acid Test Ratio= 1.01
times), (**U**= Utilized or Starting Capital= 64,720
millions USD), (**I**= L/E= Leverage or Gearing Ratio=
99.00%), (**p**= 360 P/V= Procured Inventory Days=
240 Days), (**c**= C/X= Current Ratio= 1.20 times), (**S'**=
Sales of Past Year= 48,851 millions USD), (**s**=
[S/S']-1= Sales Growth= 5.00%), (**D**= Dividend
Paid= 2,370 millions USD), (**t**= T/B= Tax Rate=
30.00%), (**v**= V/S= Variable Portion= 19.00%), (**I**=
Interest Expense= 513 millions USD), (**F**= Fixed
Cost= 29,740 millions USD), and (**Q**= Quoted
Longterm Liabilities= 35,351 millions USD), then
it's (**M**= Margin of Contribution planned), is:

$$M= F+I+[D-U+(Q+S'vp[1+s]$$
$$/\{360[c-q]\})/I-U]/[1-t]$$
$$= 29,750+513+[2,370-64,720$$
$$+(35,351+48,851*0.1900$$
$$*240[1+0.0500]/\{360$$
$$\{1.2000-1.0100]\})/0.9900]$$
$$/[1-0.3000]$$
$$= 41,548 \text{ millions USD}$$

Corporate IFRS-GAAP (B/S-I/S), ISBN-13: **978-1720792789**, ISBN-10: **172079278X**

Law-7652:

If both (**q**= [C-P]/X= Quick or Acid Test Ratio= 1.01
times), (**U**= Utilized or Starting Capital= 64,720
millions USD), (**I**= L/E= Leverage or Gearing Ratio=
99.00%), (**p**= 360 P/V= Procured Inventory Days=
240 Days), (**c**= C/X= Current Ratio= 1.20 times), (**S'**=
Sales of Past Year= 48,851 millions USD), (**s**=
[S/S']-1= Sales Growth= 5.00%), (**D**= Dividend
Paid= 2,370 millions USD), (**t**= T/B= Tax Rate=
30.00%), (**v**= V/S= Variable Portion= 19.00%), (**I**=
Interest Expense= 513 millions USD), (**M**= Margin of
Contribution= 41,548 millions USD), and (**Q**=
Quoted Longterm Liabilities= 35,351 millions USD),
then it's (**F**= Fixed Cost planned), is:

$$F= M\text{-}I\text{-}[D\text{-}U+(Q+S'vp[1+s]/\{360[c\text{-}q]\})/I]$$
$$/[1\text{-}t]$$
$$= 41,548\text{-}513\text{-}[2,370\text{-}64,720+(35,351$$
$$+48,851*0.1900*240$$
$$[1+0.0500]/\{360\{1.2000$$
$$-1.0100]\})/0.9900]/[1\text{-}0.3000]$$
$$= 29,750 \text{ millions USD}$$

Corporate IFRS-GAAP (B/S-I/S), ISBN-13: **978-1720792789**, ISBN-10: **172079278X**

<u>Law-7653</u>:

If both (**q**= [C-P]/X= Quick or Acid Test Ratio= <u>1.01</u> times), (**U**= Utilized or Starting Capital= <u>64,720</u> millions USD), (**I**= L/E= Leverage or Gearing Ratio= <u>99.00%</u>), (**p**= 360 P/V= Procured Inventory Days= <u>240</u> Days), (**c**= C/X= Current Ratio= <u>1.20</u> times), (**S'**= Sales of Past Year= <u>48,851</u> millions USD), (**s**= [S/S']-1= Sales Growth= <u>5.00%</u>), (**D**= Dividend Paid= <u>2,370</u> millions USD), (**t**= T/B= Tax Rate= <u>30.00%</u>), (**v**= V/S= Variable Portion= <u>19.00%</u>), (**F**= Fixed Cost= <u>29,740</u> millions USD), (**M**= Margin of Contribution= <u>41,548</u> millions USD), and (**Q**= Quoted Longterm Liabilities= <u>35,351</u> millions USD), then it's (**I**= Interest Expense planned), is:

$$I= M-F-[D-U+(Q+S'vp[1+s]/\{360[c-q]\})/I]$$
$$/[1-t]$$
$$= 41,548-29,750-[2,370-64,720$$
$$+(35,351+48,851*0.1900$$
$$*240[1+0.0500]/\{360$$
$$\{1.2000-1.0100]\})/0.9900]$$
$$/[1-0.3000]$$
$$= \underline{513} \text{ millions USD}$$

Corporate IFRS-GAAP (B/S-I/S), ISBN-13: **978-1720792789**, ISBN-10: **172079278X**

Law-7654:

If both (**q**= [C-P]/X= Quick or Acid Test Ratio= 1.01
times), (**U**= Utilized or Starting Capital= 64,720
millions USD), (**I**= L/E= Leverage or Gearing Ratio=
99.00%), (**p**= 360 P/V= Procured Inventory Days=
240 Days), (**c**= C/X= Current Ratio= 1.20 times), (**S'**=
Sales of Past Year= 48,851 millions USD), (**s**=
[S/S']-1= Sales Growth= 5.00%), (**D**= Dividend
Paid= 2,370 millions USD), (**I**= Interest Expense=
513 millions USD), (**v**= V/S= Variable Portion=
19.00%), (**F**= Fixed Cost= 29,740 millions USD),
(**M**= Margin of Contribution= 41,548 millions USD),
and (**Q**= Quoted Longterm Liabilities= 35,351
millions USD), then it's (**t**= Tax Rate planned), is:

$$t= 1-[D-U+(Q+S'vp[1+s]/\{360[c-q]\})/I]$$
$$/[M-F-I]$$

$$= 1-[2,370-64,720+(35,351+48,851$$
$$*0.1900*240[1+0.0500]$$
$$/\{360\{1.2000-1.0100\}\})$$
$$/0.9900]/[41,548-29,750$$
$$-513]$$

$$= 30.00\%$$

Corporate IFRS-GAAP (B/S-I/S), ISBN-13: **978-1720792789**, ISBN-10: **172079278X**

Law-7655:

If both (**q**= [C-P]/X= Quick or Acid Test Ratio= 1.01 times), (**U**= Utilized or Starting Capital= 64,720 millions USD), (**I**= L/E= Leverage or Gearing Ratio= 99.00%), (**p**= 360 P/V= Procured Inventory Days= 240 Days), (**c**= C/X= Current Ratio= 1.20 times), (**S'**= Sales of Past Year= 48,851 millions USD), (**s**= [S/S']-1= Sales Growth= 5.00%), (**t**= T/B= Tax Rate= 30.00%), (**I**= Interest Expense= 513 millions USD), (**v**= V/S= Variable Portion= 19.00%), (**F**= Fixed Cost= 29,740 millions USD), (**M**= Margin of Contribution= 41,548 millions USD), and (**Q**= Quoted Longterm Liabilities= 35,351 millions USD), then it's (**D**= Dividend Paid planned), is:

$$D= U+[M-F-I][1-t]-(Q+S'vp[1+s]$$
$$/\{360[c-q]\})/I$$
$$= 64,720+[41,548-29,750-513]$$
$$[1-0.3000]-(35,351+48,851$$
$$*0.1900*240[1+0.0500]$$
$$/\{360[1.2000-1.0100]\})$$
$$/0.9900$$
$$= 2,370 \text{ millions USD}$$

Corporate IFRS-GAAP (B/S-I/S), ISBN-13: **978-1720792789**, ISBN-10: **172079278X**

Law-7656:

If both (**q**= [C-P]/X= Quick or Acid Test Ratio= 1.01
times), (**U**= Utilized or Starting Capital= 64,720
millions USD), (**l**= L/E= Leverage or Gearing Ratio=
99.00%), (**p**= 360 P/V= Procured Inventory Days=
240 Days), (**c**= C/X= Current Ratio= 1.20 times), (**D**=
Dividend Paid= 2,370 millions USD), (**s**= [S/S']-1=
Sales Growth= 5.00%), (**t**= T/B= Tax Rate=
30.00%), (**I**= Interest Expense= 513 millions USD),
(**v**= V/S= Variable Portion= 19.00%), (**F**= Fixed
Cost= 29,740 millions USD), (**M**= Margin of
Contribution= 41,548 millions USD), and (**Q**=
Quoted Longterm Liabilities= 35,351 millions USD),
then it's (**S'**= Sales Past), must be:

$$S' = 360[\mathbf{c}\text{-}\mathbf{q}](\mathit{l}\{\mathbf{U}+[\mathbf{M}\text{-}\mathbf{F}\text{-}\mathbf{I}][1\text{-}\mathbf{t}]\text{-}\mathbf{D}\}\text{-}\mathbf{Q})$$
$$/\{\mathbf{vp}[1+\mathbf{s}]\}$$
$$= 360[1.2000\text{-}1.0100](0.9900$$
$$\{64,720+[41,548\text{-}29,750$$
$$\text{-}513][1\text{-}0.3000]\text{-}2,370\}$$
$$\text{-}35,351)/\{0.1900*240$$
$$[1+0.0500]\}$$
$$= 48,851 \text{ millions USD}$$

Corporate IFRS-GAAP (B/S-I/S), ISBN-13: **978-1720792789**, ISBN-10: **172079278X**

<u>Law-7657</u>:

If both (**q**= [C-P]/X= Quick or Acid Test Ratio= <u>1.01</u> times), (**U**= Utilized or Starting Capital= <u>64,720</u> millions USD), (**⌐**= L/E= Leverage or Gearing Ratio= <u>99.00%</u>), (**p**= 360 P/V= Procured Inventory Days= <u>240</u> Days), (**c**= C/X= Current Ratio= <u>1.20</u> times), (**D**= Dividend Paid= <u>2,370</u> millions USD), (**s**= [S/S']-1= Sales Growth= <u>5.00%</u>), (**t**= T/B= Tax Rate= <u>30.00%</u>), (**I**= Interest Expense= <u>513</u> millions USD), (**S'**= Sales of Past Year= <u>48,851</u> millions USD), (**F**= Fixed Cost= <u>29,740</u> millions USD), (**M**= Margin of Contribution= <u>41,548</u> millions USD), and (**Q**= Quoted Longterm Liabilities= <u>35,351</u> millions USD), then it's (**v**= Variable Portion planned), is:

$$v= 360[c-q](\⌐\{U+[M-F-I][1-t]-D\}-Q)$$
$$/\{S'p[1+s]\}$$

$$= 360\{1.2000-1.0100\}(0.9900$$
$$\{64,720+[41,548-29,750$$
$$-513][1-0.3000]-2,370\}$$
$$-35,351)/\{48,851*240$$
$$[1+0.0500]\}$$

$$= \underline{19.00\%}$$

Corporate IFRS-GAAP (B/S-I/S), ISBN-13: **978-1720792789**, ISBN-10: **172079278X**

<u>Law-7658</u>:

If both (**q**= [C-P]/X= Quick or Acid Test Ratio= <u>1.01</u> times), (**U**= Utilized or Starting Capital= <u>64,720</u> millions USD), (**⌐**= L/E= Leverage or Gearing Ratio= <u>99.00%</u>), (**v**= V/S= Variable Portion= <u>19.00%</u>), (**c**= C/X= Current Ratio= <u>1.20</u> times), (**D**= Dividend Paid= <u>2,370</u> millions USD), (**s**= [S/S']-1= Sales Growth= <u>5.00%</u>), (**t**= T/B= Tax Rate= <u>30.00%</u>), (**I**= Interest Expense= <u>513</u> millions USD), (**S'**= Sales of Past Year= <u>48,851</u> millions USD), (**F**= Fixed Cost= <u>29,740</u> millions USD), (**M**= Margin of Contribution= <u>41,548</u> millions USD), and (**Q**= Quoted Longterm Liabilities= <u>35,351</u> millions USD), then it's (**p**= Procured Inventory Days planned), is:

$$p= 360[c\text{-}q](\mathit{I}\{U+[M\text{-}F\text{-}I][1\text{-}t]\text{-}D\}\text{-}Q)$$
$$/\{S'v[1+s]\}$$

$$= 360\{1.2000\text{-}1.0100](0.9900$$
$$\{64,720+[41,548\text{-}29,750$$
$$-513][1\text{-}0.3000]\text{-}2,370\}$$
$$-35,351)/\{48,851*0.1900$$
$$[1+0.0500]\}$$

$$= \underline{240} \text{ days}$$

Corporate IFRS-GAAP (B/S-I/S), ISBN-13: **978-1720792789**, ISBN-10: **172079278X**

<u>Law-7659</u>:

If both (**q**= [C-P]/X= Quick or Acid Test Ratio= <u>1.01</u> times), (**U**= Utilized or Starting Capital= <u>64,720</u> millions USD), (**I**= L/E= Leverage or Gearing Ratio= <u>99.00%</u>), (**v**= V/S= Variable Portion= <u>19.00%</u>), (**c**= C/X= Current Ratio= <u>1.20</u> times), (**D**= Dividend Paid= <u>2,370</u> millions USD), (**p**= 360 P/V= Procured Inventory Days= <u>240</u> Days), (**t**= T/B= Tax Rate= <u>30.00%</u>), (**I**= Interest Expense= <u>513</u> millions USD), (**$'**= Sales of Past Year= <u>48,851</u> millions USD), (**F**= Fixed Cost= <u>29,740</u> millions USD), (**M**= Margin of Contribution= <u>41,548</u> millions USD), and (**Q**= Quoted Longterm Liabilities= <u>35,351</u> millions USD), then it's (**s**= Sales Growth planned), is:

$$s= 360[\textbf{c-q}](\textbf{I}\{\textbf{U}+[\textbf{M-F-I}][1-\textbf{t}]-\textbf{D}\}-\textbf{Q})$$
$$/[\textbf{\$'vp}]-1$$
$$= 360\{1.2000-1.0100](0.9900$$
$$\{64,720+[41,548-29,750$$
$$-513][1-0.3000]-2,370\}$$
$$-35,351)/[48,851*0.1900$$
$$*240]-1$$
$$= \underline{5.00\%}$$

Corporate IFRS-GAAP (B/S-I/S), ISBN-13: **978-1720792789**, ISBN-10: **172079278X**

Law-7660:

If both (**q**= [C-P]/X= Quick or Acid Test Ratio= 1.01 times), (**U**= Utilized or Starting Capital= 64,720 millions USD), (**l**= L/E= Leverage or Gearing Ratio= 99.00%), (**v**= V/S= Variable Portion= 19.00%), (**s**= [S/S']-1= Sales Growth= 5.00%), (**D**= Dividend Paid= 2,370 millions USD), (**p**= 360 P/V= Procured Inventory Days= 240 Days), (**t**= T/B= Tax Rate= 30.00%), (**I**= Interest Expense= 513 millions USD), (**S'**= Sales of Past Year= 48,851 millions USD), (**F**= Fixed Cost= 29,740 millions USD), (**M**= Margin of Contribution= 41,548 millions USD), and (**Q**= Quoted Longterm Liabilities= 35,351 millions USD), then it's (**c**= Current Ratio planned), is:

$$c= q+S'vp[1+s]/[360[c-q](l\{U+[M-F-I][1-t]-D\}-Q)]$$

$$= 1.0100+48,851*0.1900*240[1+0.0500]/[360(0.9900\{64,720+[41,548-29,750-513][1-0.3000]-2,370\}-35,351)]$$

$$= 1.20 \text{ times}$$

Corporate IFRS-GAAP (B/S-I/S), ISBN-13: **978-1720792789**, ISBN-10: **172079278X**

Law-7661:

If both (**c**= C/X= Current Ratio= 1.20 times), (**U**= Utilized or Starting Capital= 64,720 millions USD), (**I**= L/E= Leverage or Gearing Ratio= 99.00%), (**v**= V/S= Variable Portion= 19.00%), (**s**= [S/S']-1= Sales Growth= 5.00%), (**D**= Dividend Paid= 2,370 millions USD), (**p**= 360 P/V= Procured Inventory Days= 240 Days), (**t**= T/B= Tax Rate= 30.00%), (**I**= Interest Expense= 513 millions USD), (**S'**= Sales of Past Year= 48,851 millions USD), (**F**= Fixed Cost= 29,740 millions USD), (**M**= Margin of Contribution= 41,548 millions USD), and (**Q**= Quoted Longterm Liabilities= 35,351 millions USD), then it's (**q**= Quick or Acid Test Ratio planned), is:

$$\text{q}= \text{c-S'vp}[1+\text{s}]/[360[\text{c-q}](\textit{I}\{\text{U}+[\text{M-F-I}]$$
$$[1\text{-t}]\text{-D}\}\text{-Q})]$$
$$= 1.2000\text{-}48,851*0.1900*240$$
$$[1+0.0500]/[360(0.9900$$
$$\{64,720+[41,548\text{-}29,750$$
$$\text{-}513][1\text{-}0.3000]\text{-}2,370\}$$
$$\text{-}35,351)]$$
$$= 1.01 \text{ times}$$

Corporate IFRS-GAAP (B/S-I/S), ISBN-13: **978-1720792789**, ISBN-10: **172079278X**

<u>Law-7662</u>:

If both (**U**= Utilized or Starting Capital= <u>64,720</u> millions USD), (**I**= L/E= Leverage or Gearing Ratio= <u>99.00%</u>), (**A**= After Tax Income= <u>7,899</u> millions USD), (**d**= D/A= Dividend Portion or Payout= <u>30.00%</u>), (**t**= T/B= Tax Rate= <u>30.00%</u>), (**I**= Interest Expense= <u>513</u> millions USD), (**F**= Fixed Cost= <u>29,740</u> millions USD), (**M**= Margin of Contribution= <u>41,548</u> millions USD), and (**X**= Xpress or Current Debt= <u>34,196</u> millions USD), then it's (**Q**= Quoted Longterm Liabilities planned), is:

$$Q= I\{U+[M\text{-}F\text{-}I][1\text{-}t]\text{-}Ad\}\text{-}X$$
$$= 0.9900\{64,720+[41,548\text{-}29,750$$
$$-513][1\text{-}0.3000]\text{-}7,899$$
$$*0.3000\}\text{-}34,196$$
$$= \underline{35,351} \text{ millions USD}$$

Corporate IFRS-GAAP (B/S-I/S), ISBN-13: **978-1720792789**, ISBN-10: **172079278X**

<u>Law-7663</u>:

If both (**U**= Utilized or Starting Capital= <u>64,720</u> millions USD), (**Q**= Quoted Longterm Liabilities= <u>35,351</u> millions USD), (**A**= After Tax Income= <u>7,899</u> millions USD), (**d**= D/A= Dividend Portion or Payout= <u>30.00%</u>), (**t**= T/B= Tax Rate= <u>30.00%</u>), (**I**= Interest Expense= <u>513</u> millions USD), (**F**= Fixed Cost= <u>29,740</u> millions USD), (**M**= Margin of Contribution= <u>41,548</u> millions USD), and (**X**= Xpress or Current Debt= <u>34,196</u> millions USD), then it's (**I** = Leverage or Gearing Ratio planned), is:

$$I = [Q+X]/\{U+[M-F-I][1-t]-Ad\}$$
$$= [35,351+34,196]/\{64,720+[41,548$$
$$-29,750-513][1-0.3000]$$
$$-7,899*0.3000\}$$
$$= \underline{99.00\%}$$

Corporate IFRS-GAAP (B/S-I/S), ISBN-13: **978-1720792789**, ISBN-10: **172079278X**

<u>Law-7664</u>:

If both (**Q**= Quoted Longterm Liabilities= <u>35,351</u> millions USD), (**I**= L/E= Leverage or Gearing Ratio= <u>99.00%</u>), (**A**= After Tax Income= <u>7,899</u> millions USD), (**d**= D/A= Dividend Portion or Payout= <u>30.00%</u>), (**t**= T/B= Tax Rate= <u>30.00%</u>), (**I**= Interest Expense= <u>513</u> millions USD), (**F**= Fixed Cost= <u>29,740</u> millions USD), (**M**= Margin of Contribution= <u>41,548</u> millions USD), and (**X**= Xpress or Current Debt= <u>34,196</u> millions USD), then it's (**U**= Utilized or Starting Capital planned), is:

$$U= [Q+X]/I-\{[M-F-I][1-t]-Ad\}$$
$$= [35,351+34,196]/0.9900-\{[41,548$$
$$-29,750-513][1-0.3000]$$
$$-7,899*0.3000\}$$
$$= \underline{64,720} \text{ millions USD}$$

Corporate IFRS-GAAP (B/S-I/S), ISBN-13: **978-1720792789**, ISBN-10: **172079278X**

Law-7665:

If both (**Q**= Quoted Longterm Liabilities= 35,351 millions USD), (**I**= L/E= Leverage or Gearing Ratio= 99.00%), (**A**= After Tax Income= 7,899 millions USD), (**d**= D/A= Dividend Portion or Payout= 30.00%), (**t**= T/B= Tax Rate= 30.00%), (**I**= Interest Expense= 513 millions USD), (**F**= Fixed Cost= 29,740 millions USD), (**U**= Utilized or Starting Capital= 64,720 millions USD), and (**X**= Xpress or Current Debt= 34,196 millions USD), then it's (**M**= Margin of Contribution planned), is:

$$M= F+I+\{Ad-U+[Q+X]/I \}/[1-t]$$
$$= 29,750+513+\{7,899*0.3000$$
$$-64,720+[35,351+34,196]$$
$$/0.9900\}/[1-0.3000]$$
$$= 41,548 \text{ millions USD}$$

Corporate IFRS-GAAP (B/S-I/S), ISBN-13: **978-1720792789**, ISBN-10: **172079278X**

Law-7666:

If both (**Q**= Quoted Longterm Liabilities= 35,351 millions USD), (**I**= L/E= Leverage or Gearing Ratio= 99.00%), (**A**= After Tax Income= 7,899 millions USD), (**d**= D/A= Dividend Portion or Payout= 30.00%), (**t**= T/B= Tax Rate= 30.00%), (**I**= Interest Expense= 513 millions USD), (**M**= Margin of Contribution= 41,548 millions USD), (**U**= Utilized or Starting Capital= 64,720 millions USD), and (**X**= Xpress or Current Debt= 34,196 millions USD), then it's (**F**= Fixed Cost planned), is:

$$F= M-I-\{Ad-U+[Q+X]/I\}/[1-t]$$
$$= 41,548-513-\{7,899*0.3000$$
$$-64,720+[35,351+34,196]$$
$$/0.9900\}/[1-0.3000]$$
$$= 29,750 \text{ millions USD}$$

Corporate IFRS-GAAP (B/S-I/S), ISBN-13: **978-1720792789**, ISBN-10: **172079278X**

Law-7667:

If both (**Q**= Quoted Longterm Liabilities= 35,351 millions USD), (**F**= L/E= Leverage or Gearing Ratio= 99.00%), (**A**= After Tax Income= 7,899 millions USD), (**d**= D/A= Dividend Portion or Payout= 30.00%), (**t**= T/B= Tax Rate= 30.00%), (**F**= Fixed Cost= 29,740 millions USD), (**M**= Margin of Contribution= 41,548 millions USD), (**U**= Utilized or Starting Capital= 64,720 millions USD), and (**X**= Xpress or Current Debt= 34,196 millions USD), then it's (**I**= Interest Expense planned), is:

$$I = M-F-\{Ad-U+[Q+X]/I\}/[1-t]$$
$$= 41,548-29,750-\{7,899*0.3000$$
$$-64,720+[35,351+34,196]$$
$$/0.9900\}/[1-0.3000]$$
$$= 513 \text{ millions USD}$$

Corporate IFRS-GAAP (B/S-I/S), ISBN-13: **978-1720792789**, ISBN-10: **172079278X**

<u>Law-7668</u>:

If both (**Q**= Quoted Longterm Liabilities= <u>35,351</u> millions USD), (**I**= L/E= Leverage or Gearing Ratio= <u>99.00%</u>), (**A**= After Tax Income= <u>7,899</u> millions USD), (**d**= D/A= Dividend Portion or Payout= <u>30.00%</u>), (**I**= Interest Expense= <u>513</u> millions USD), (**F**= Fixed Cost= <u>29,740</u> millions USD), (**M**= Margin of Contribution= <u>41,548</u> millions USD), (**U**= Utilized or Starting Capital= <u>64,720</u> millions USD), and (**X**= Xpress or Current Debt= <u>34,196</u> millions USD), then it's (**t**= Tax Rate planned), is:

$$t= 1-\{\textbf{Ad-U}+[\textbf{Q+X}]/\textbf{I}\}/[\textbf{M-F-I}]$$
$$= 1-\{7,899*0.3000-64,720+[35,351$$
$$+34,196]/0.9900\}/[41,548$$
$$-29,750-513]$$
$$= \underline{30.00\%}$$

Corporate IFRS-GAAP (B/S-I/S), ISBN-13: **978-1720792789**, ISBN-10: **172079278X**

Law-7669:

If both ($\mathbf{Q}$= Quoted Longterm Liabilities= 35,351 millions USD), ($\mathbf{I}$= L/E= Leverage or Gearing Ratio= 99.00%), ($\mathbf{t}$= T/B= Tax Rate= 30.00%), ($\mathbf{d}$= D/A= Dividend Portion or Payout= 30.00%), ($\mathbf{I}$= Interest Expense= 513 millions USD), ($\mathbf{F}$= Fixed Cost= 29,740 millions USD), ($\mathbf{M}$= Margin of Contribution= 41,548 millions USD), ($\mathbf{U}$= Utilized or Starting Capital= 64,720 millions USD), and ($\mathbf{X}$= Xpress or Current Debt= 34,196 millions USD), then it's ($\mathbf{A}$= After Tax Income planned), is:

$$\mathbf{A} = \{\mathbf{U}-[\mathbf{Q}+\mathbf{X}]/\mathbf{I}+[\mathbf{M}-\mathbf{F}-\mathbf{I}][1-\mathbf{t}]\}/\mathbf{d}$$
$$= \{64,720-[35,351+34,196]/0.9900$$
$$+[41,548-29,750-513]$$
$$[1-0.3000]\}/0.3000$$
$$= 7,899 \text{ millions USD}$$

Corporate IFRS-GAAP (B/S-I/S), ISBN-13: **978-1720792789**, ISBN-10: **172079278X**

<u>Law-7670</u>:

If both (Q= Quoted Longterm Liabilities= <u>35,351</u> millions USD), (I= L/E= Leverage or Gearing Ratio= <u>99.00%</u>), (t= T/B= Tax Rate= <u>30.00%</u>), (A= After Tax Income= <u>7,899</u> millions USD), (I= Interest Expense= <u>513</u> millions USD), (F= Fixed Cost= <u>29,740</u> millions USD), (M= Margin of Contribution= <u>41,548</u> millions USD), (U= Utilized or Starting Capital= <u>64,720</u> millions USD), and (X= Xpress or Current Debt= <u>34,196</u> millions USD), then it's (d= Dividend Portion or Payout planned), is:

$$d= \{U-[Q+X]/I+[M-F-I][1-t]\}/A$$
$$= \{64,720-[35,351+34,196]/0.9900$$
$$+[41,548-29,750-513]$$
$$[1-0.3000])/0.3000$$
$$= \underline{30.00\%}$$

Corporate IFRS-GAAP (B/S-I/S), ISBN-13: **978-1720792789**, ISBN-10: **172079278X**

Law-7671:

If both (**Q**= Quoted Longterm Liabilities= 35,351 millions USD), (**I**= L/E= Leverage or Gearing Ratio= 99.00%), (**t**= T/B= Tax Rate= 30.00%), (**A**= After Tax Income= 7,899 millions USD), (**I**= Interest Expense= 513 millions USD), (**F**= Fixed Cost= 29,740 millions USD), (**M**= Margin of Contribution= 41,548 millions USD), (**U**= Utilized or Starting Capital= 64,720 millions USD), and (**d**= D/A= Dividend Portion or Payout= 30.00%), then it's (**X**= Xpress or Current Debt planned), is:

$$X= I\{U+[M-F-I][1-t]-Ad\}-Q$$
$$= 0.9900\{64,720+[41,548-29,750$$
$$-513][1-0.3000]-7,899$$
$$*0.3000\}-35,351$$
$$= 34,196 \text{ millions USD}$$

Corporate IFRS-GAAP (B/S-I/S), ISBN-13: **978-1720792789**, ISBN-10: **172079278X**

Law-7672:

If both (**P**= Procured Inventories= 6,497 millions
USD), (**q**= [C-P]/X= Quick or Acid Test Ratio= 1.01
times), (**c**= C/X= Current Ratio= 1.20 times), (**I**=
L/E= Leverage or Gearing Ratio= 99.00%), (**t**= T/B=
Tax Rate= 30.00%), (**A**= After Tax Income= 7,899
millions USD), (**I**= Interest Expense= 513 millions
USD), (**F**= Fixed Cost= 29,740 millions USD), (**M**=
Margin of Contribution= 41,548 millions USD), (**U**=
Utilized or Starting Capital= 64,720 millions USD),
and (**d**= D/A= Dividend Portion or Payout= 30.00%),
then it's (**Q**= Quoted Longterm Liabilities planned),
is:

$$Q= I\{U+[M\text{-}F\text{-}I][1\text{-}t]\text{-}Ad\}\text{-}P/[c\text{-}q]$$
$$= 0.9900\{64,720+[41,548\text{-}29,750$$
$$\text{-}513][1\text{-}0.3000]\text{-}7,899$$
$$*0.3000\}\text{-}6,497/[1.2000$$
$$\text{-}1.0100]$$
$$= 35,351 \text{ millions USD}$$

Corporate IFRS-GAAP (B/S-I/S), ISBN-13: **978-1720792789**, ISBN-10: **172079278X**

Law-7673:

   If both (**P**= Procured Inventories= 6,497 millions
USD), (**q**= [C-P]/X= Quick or Acid Test Ratio= 1.01
times), (**c**= C/X= Current Ratio= 1.20 times), (**Q**=
Quoted Longterm Liabilities= 35,351 millions USD),
(**t**= T/B= Tax Rate= 30.00%), (**A**= After Tax
Income= 7,899 millions USD), (**I**= Interest Expense=
513 millions USD), (**F**= Fixed Cost= 29,740 millions
USD), (**M**= Margin of Contribution= 41,548 millions
USD), (**U**= Utilized or Starting Capital= 64,720
millions USD), and (**d**= D/A= Dividend Portion or
Payout= 30.00%), then it's (**I** = Leverage or Gearing
Ratio planned), is:

$$\mathbf{I} = \{\mathbf{Q}+\mathbf{P}/[\mathbf{c}\text{-}\mathbf{q}]\}/\{\mathbf{U}+[\mathbf{M}\text{-}\mathbf{F}\text{-}\mathbf{I}][1\text{-}\mathbf{t}]\text{-}\mathbf{Ad}\}$$

$$= \{35,351+6,497/[1.2000\text{-}1.0100]\}$$
$$/\{64,720+[41,548\text{-}29,750$$
$$\text{-}513][1\text{-}0.3000]\text{-}7,899$$
$$*0.3000\}$$

   $= 99.00\%$

Corporate IFRS-GAAP (B/S-I/S), ISBN-13: **978-1720792789**, ISBN-10: **172079278X**

Law-7674:

If both (**P**= Procured Inventories= 6,497 millions
USD), (**q**= [C-P]/X= Quick or Acid Test Ratio= 1.01
times), (**c**= C/X= Current Ratio= 1.20 times), (**l**=
L/E= Leverage or Gearing Ratio= 99.00%), (**Q**=
Quoted Longterm Liabilities= 35,351 millions USD),
(**t**= T/B= Tax Rate= 30.00%), (**A**= After Tax
Income= 7,899 millions USD), (**l**= Interest Expense=
513 millions USD), (**F**= Fixed Cost= 29,740 millions
USD), (**M**= Margin of Contribution= 41,548 millions
USD), and (**d**= D/A= Dividend Portion or Payout=
30.00%), then it's (**U**= Utilized or Starting Capital
planned), is:

$$U= Ad+\{Q+P/[c-q]\}/l -[M-F-l][1-t]$$
$$= 7,899*0.3000+\{35,351+6,497$$
$$/[1.2000-1.0100]\}/0.9900$$
$$-[41,548-29,750-513]$$
$$[1-0.3000]$$
$$= 64,720 \text{ millions USD}$$

Corporate IFRS-GAAP (B/S-I/S), ISBN-13: **978-1720792789**, ISBN-10: **172079278X**

Law-7675:

If both (**P**= Procured Inventories= 6,497 millions
USD), (**q**= [C-P]/X= Quick or Acid Test Ratio= 1.01
times), (**c**= C/X= Current Ratio= 1.20 times), (**l**=
L/E= Leverage or Gearing Ratio= 99.00%), (**Q**=
Quoted Longterm Liabilities= 35,351 millions USD),
(**t**= T/B= Tax Rate= 30.00%), (**A**= After Tax
Income= 7,899 millions USD), (**I**= Interest Expense=
513 millions USD), (**F**= Fixed Cost= 29,740 millions
USD), (**U**= Utilized or Starting Capital= 64,720
millions USD), and (**d**= D/A= Dividend Portion or
Payout= 30.00%), then it's (**M**= Margin of
Contribution planned), is:

$$M= F+I+(Ad-U+\{Q+P/[c-q]\}/I\ )/[1-t]$$
$$= 29,750+513+(7,899*0.3000$$
$$-64,720+\{35,351+6,497$$
$$/[1.2000-1.0100]\}/0.9900)$$
$$/[1-0.3000]$$
$$= 41,548 \text{ millions USD}$$

Corporate IFRS-GAAP (B/S-I/S), ISBN-13: **978-1720792789**, ISBN-10: **172079278X**

<u>Law-7676</u>:

If both (**P**= Procured Inventories= <u>6,497</u> millions USD), (**q**= [C-P]/X= Quick or Acid Test Ratio= <u>1.01</u> times), (**c**= C/X= Current Ratio= <u>1.20</u> times), (**l**= L/E= Leverage or Gearing Ratio= <u>99.00%</u>), (**Q**= Quoted Longterm Liabilities= <u>35,351</u> millions USD), (**t**= T/B= Tax Rate= <u>30.00%</u>), (**A**= After Tax Income= <u>7,899</u> millions USD), (**I**= Interest Expense= <u>513</u> millions USD), (**M**= Margin of Contribution= <u>41,548</u> millions USD), (**U**= Utilized or Starting Capital= <u>64,720</u> millions USD), and (**d**= D/A= Dividend Portion or Payout= <u>30.00%</u>), then it's (**F**= Fixed Cost planned), is:

$$F= M-I-(Ad-U+\{Q+P/[c-q]\}/l)/[1-t]$$
$$= 41,548-513-(7,899*0.3000$$
$$-64,720+\{35,351+6,497$$
$$/[1.2000-1.0100]\}/0.9900)$$
$$/[1-0.3000]$$
$$= \underline{29,750} \text{ millions USD}$$

Corporate IFRS-GAAP (B/S-I/S), ISBN-13: **978-1720792789**, ISBN-10: **172079278X**

<u>Law-7677</u>:

If both (**P**= Procured Inventories= <u>6,497</u> millions USD), (**q**= [C-P]/X= Quick or Acid Test Ratio= <u>1.01</u> times), (**c**= C/X= Current Ratio= <u>1.20</u> times), (**l**= L/E= Leverage or Gearing Ratio= <u>99.00%</u>), (**Q**= Quoted Longterm Liabilities= <u>35,351</u> millions USD), (**t**= T/B= Tax Rate= <u>30.00%</u>), (**A**= After Tax Income= <u>7,899</u> millions USD), (**F**= Fixed Cost= <u>29,740</u> millions USD), (**M**= Margin of Contribution= <u>41,548</u> millions USD), (**U**= Utilized or Starting Capital= <u>64,720</u> millions USD), and (**d**= D/A= Dividend Portion or Payout= <u>30.00%</u>), then it's (**I**= Interest Expense planned), is:

$$I= M\text{-}F\text{-}(Ad\text{-}U+\{Q+P/[c\text{-}q]\}/I)/[1\text{-}t]$$
$$= 41,548\text{-}29,750\text{-}(7,899*0.3000$$
$$-64,720+\{35,351+6,497$$
$$/[1.2000\text{-}1.0100]\}/0.9900)$$
$$/[1\text{-}0.3000]$$
$$= \underline{513} \text{ millions USD}$$

Corporate IFRS-GAAP (B/S-I/S), ISBN-13: **978-1720792789**, ISBN-10: **172079278X**

Law-7678:

If both (**P**= Procured Inventories= 6,497 millions USD), (**q**= [C-P]/X= Quick or Acid Test Ratio= 1.01 times), (**c**= C/X= Current Ratio= 1.20 times), (**I**= L/E= Leverage or Gearing Ratio= 99.00%), (**Q**= Quoted Longterm Liabilities= 35,351 millions USD), (**I**= Interest Expense= 513 millions USD), (**A**= After Tax Income= 7,899 millions USD), (**F**= Fixed Cost= 29,740 millions USD), (**M**= Margin of Contribution= 41,548 millions USD), (**U**= Utilized or Starting Capital= 64,720 millions USD), and (**d**= D/A= Dividend Portion or Payout= 30.00%), then it's (**t**= Tax Rate planned), is:

$$t = 1-(\mathbf{Ad}\text{-}\mathbf{U}+\{\mathbf{Q}+\mathbf{P}/[\mathbf{c}\text{-}\mathbf{q}]\}/\mathbf{I})/[\mathbf{M}\text{-}\mathbf{F}\text{-}\mathbf{I}]$$

$$= 1-(7,899*0.3000-64,720+\{35,351$$
$$+6,497/[1.2000-1.0100]\}$$
$$/0.9900)/[41,548-29,750$$
$$-513]$$

$$= 30.00\%$$

Corporate IFRS-GAAP (B/S-I/S), ISBN-13: **978-1720792789**, ISBN-10: **172079278X**

Law-7679:

If both (**P**= Procured Inventories= 6,497 millions USD), (**q**= [C-P]/X= Quick or Acid Test Ratio= 1.01 times), (**c**= C/X= Current Ratio= 1.20 times), (**l**= L/E= Leverage or Gearing Ratio= 99.00%), (**Q**= Quoted Longterm Liabilities= 35,351 millions USD), (**I**= Interest Expense= 513 millions USD), (**t**= T/B= Tax Rate= 30.00%), (**F**= Fixed Cost= 29,740 millions USD), (**M**= Margin of Contribution= 41,548 millions USD), (**U**= Utilized or Starting Capital= 64,720 millions USD), and (**d**= D/A= Dividend Portion or Payout= 30.00%), then it's (**A**= After Tax Income planned), is:

$$\mathbf{A} = (\mathbf{U} - \{\mathbf{Q} + \mathbf{P}/[\mathbf{c}\text{-}\mathbf{q}]\}/\mathbf{l} + [\mathbf{M}\text{-}\mathbf{F}\text{-}\mathbf{I}][1\text{-}\mathbf{t}])/\mathbf{d}$$

$$= (64{,}720 - \{35{,}351 + 6{,}497/[1.2000 - 1.0100]\}/0.9900 + [41{,}548 - 29{,}750 - 513][1 - 0.3000])/0.3000$$

$$= \underline{7{,}899} \text{ millions USD}$$

Corporate IFRS-GAAP (B/S-I/S), ISBN-13: **978-1720792789**, ISBN-10: **172079278X**

Law-7680:

If both (**P**= Procured Inventories= 6,497 millions USD), (**q**= [C-P]/X= Quick or Acid Test Ratio= 1.01 times), (**c**= C/X= Current Ratio= 1.20 times), (**I**= L/E= Leverage or Gearing Ratio= 99.00%), (**Q**= Quoted Longterm Liabilities= 35,351 millions USD), (**I**= Interest Expense= 513 millions USD), (**t**= T/B= Tax Rate= 30.00%), (**F**= Fixed Cost= 29,740 millions USD), (**M**= Margin of Contribution= 41,548 millions USD), (**U**= Utilized or Starting Capital= 64,720 millions USD), and (**A**= After Tax Income= 7,899 millions USD), then it's (**d**= Dividend Portion or Payout planned), is:

$$d= (U-\{Q+P/[c-q]\}/I+[M-F-I][1-t])/A$$
$$= (64,720-\{35,351+6,497/[1.2000$$
$$-1.0100]\}/0.9900+[41,548$$
$$-29,750-513][1-0.3000])$$
$$/7,899$$
$$= 30.00\%$$

Corporate IFRS-GAAP (B/S-I/S), ISBN-13: **978-1720792789**, ISBN-10: **172079278X**

Law-7681:

If both (**d**= D/A= Dividend Portion or Payout= 30.00%), (**q**= [C-P]/X= Quick or Acid Test Ratio= 1.01 times), (**c**= C/X= Current Ratio= 1.20 times), (**I**= L/E= Leverage or Gearing Ratio= 99.00%), (**Q**= Quoted Longterm Liabilities= 35,351 millions USD), (**I**= Interest Expense= 513 millions USD), (**t**= T/B= Tax Rate= 30.00%), (**F**= Fixed Cost= 29,740 millions USD), (**M**= Margin of Contribution= 41,548 millions USD), (**U**= Utilized or Starting Capital= 64,720 millions USD), and (**A**= After Tax Income= 7,899 millions USD), then it's (**P**= Procured Inventories planned), is:

$$P= [c\text{-}q](I \{U+[M\text{-}F\text{-}I][1\text{-}t]\text{-}Ad\}\text{-}Q)$$
$$= [1.2000\text{-}1.0100](0.9900\{64,720$$
$$+[41,548\text{-}29,750\text{-}513]$$
$$[1\text{-}0.3000]\text{-}7,899*0.3000\}$$
$$-35,351)$$
$$= \underline{6,497} \text{ millions USD}$$

Corporate IFRS-GAAP (B/S-I/S), ISBN-13: **978-1720792789**, ISBN-10: **172079278X**

Law-7682:

If both (**d**= D/A= Dividend Portion or Payout=
30.00%), (**q**= [C-P]/X= Quick or Acid Test Ratio=
1.01 times), (**P**= Procured Inventories= 6,497
millions USD), (**I**= L/E= Leverage or Gearing Ratio=
99.00%), (**Q**= Quoted Longterm Liabilities= 35,351
millions USD), (**I**= Interest Expense= 513 millions
USD), (**t**= T/B= Tax Rate= 30.00%), (**F**= Fixed
Cost= 29,740 millions USD), (**M**= Margin of
Contribution= 41,548 millions USD), (**U**= Utilized or
Starting Capital= 64,720 millions USD), and (**A**=
After Tax Income= 7,899 millions USD), then it's
(**c**= Current Ratio planned), is:

$$c= q+P/(I \{U+[M-F-I][1-t]-Ad\}-Q)$$
$$= 1.0100+6,497/(0.9900\{64,720$$
$$+[41,548-29,750-513]$$
$$[1-0.3000]-7,899*0.3000\}$$
$$-35,351)$$
$$= 1.20 \text{ times}$$

Corporate IFRS-GAAP (B/S-I/S), ISBN-13: **978-1720792789**, ISBN-10: **172079278X**

<u>Law-7683</u>:

If both (**d**= D/A= Dividend Portion or Payout= <u>30.00%</u>), (**c**= C/X= Current Ratio= <u>1.20</u> times), (**P**= Procured Inventories= <u>6,497</u> millions USD), (**I**= L/E= Leverage or Gearing Ratio= <u>99.00%</u>), (**Q**= Quoted Longterm Liabilities= <u>35,351</u> millions USD), (**I**= Interest Expense= <u>513</u> millions USD), (**t**= T/B= Tax Rate= <u>30.00%</u>), (**F**= Fixed Cost= <u>29,740</u> millions USD), (**M**= Margin of Contribution= <u>41,548</u> millions USD), (**U**= Utilized or Starting Capital= <u>64,720</u> millions USD), and (**A**= After Tax Income= <u>7,899</u> millions USD), then it's (**q**= Quick or Acid Test Ratio planned), is:

$$q = c\text{-}P/(I\{U+[M\text{-}F\text{-}I][1\text{-}t]\text{-}Ad\}\text{-}Q)$$
$$= 1.2000\text{-}6,497/(0.9900\{64,720$$
$$+[41,548\text{-}29,750\text{-}513]$$
$$[1\text{-}0.3000]\text{-}7,899*0.3000\}$$
$$-35,351)$$
$$= \underline{1.01} \text{ times}$$

Corporate IFRS-GAAP (B/S-I/S), ISBN-13: **978-1720792789**, ISBN-10: **172079278X**

<u>Law-7684</u>:

If both (d= D/A=Dividend Portion or Payout= <u>30.00%</u>), (c= C/X= Current Ratio= <u>1.20</u> times), (V= Variable Cost= <u>9,746</u> millions USD), (p= 360 P/V= Procured Inventory Days= <u>240</u> Days), (l= L/E= Leverage or Gearing Ratio= <u>99.00%</u>), (q= [C-P]/X= Quick or Acid Test Ratio= <u>1.01</u> times), (l= Interest Expense= <u>513</u> millions USD), (t= T/B= Tax Rate= <u>30.00%</u>), (F= Fixed Cost= <u>29,740</u> millions USD), (M= Margin of Contribution= <u>41,548</u> millions USD), (U= Utilized or Starting Capital= <u>64,720</u> millions USD), and (A= After Tax Income= <u>7,899</u> millions USD), then it's (Q= Quoted Longterm Liabilities planned), is:

$$Q = l\{U+[M-F-I][1-t]-Ad\}$$
$$-Vp/\{360[c-q]\}$$
$$= 0.9900\{64,720+[41,548-29,750$$
$$-513][1-0.3000]-7,899$$
$$*0.3000\}-9,746*240$$
$$/\{360[1.2000-1.0100]\}$$
$$= 35,351 \text{ millions USD}$$

Corporate IFRS-GAAP (B/S-I/S), ISBN-13: **978-1720792789**, ISBN-10: **172079278X**

Law-7685:

If both (**d**= D/A= Dividend Portion or Payout= 30.00%), (**c**= C/X= Current Ratio= 1.20 times), (**V**= Variable Cost= 9,746 millions USD), (**p**= 360 P/V= Procured Inventory Days= 240 Days), (**Q**= Quoted Longterm Liabilities= 35,351 millions USD), (**q**= [C-P]/X= Quick or Acid Test Ratio= 1.01 times), (**I**= Interest Expense= 513 millions USD), (**t**= T/B= Tax Rate= 30.00%), (**F**= Fixed Cost= 29,740 millions USD), (**M**= Margin of Contribution= 41,548 millions USD), (**U**= Utilized or Starting Capital= 64,720 millions USD), and (**A**= After Tax Income= 7,899 millions USD), then it's (**I** = Leverage or Gearing Ratio planned), is:

$$I = (Q+Vp/\{360[c-q]\})/\{U+[M-F-I]$$
$$[1-t]-Ad\}$$
$$= (35,351+9,746*240/\{360[1.2000$$
$$-1.0100]\})/\{64,720+[41,548$$
$$-29,750-513][1-0.3000]$$
$$-7,899*0.3000\}$$
$$= 99.00\%$$

Law-7686:

If both (**d**= D/A= Dividend Portion or Payout= 30.00%), (**c**= C/X= Current Ratio= 1.20 times), (**V**= Variable Cost= 9,746 millions USD), (**p**= 360 P/V= Procured Inventory Days= 240 Days), (**Q**= Quoted Longterm Liabilities= 35,351 millions USD), (**q**= [C-P]/X= Quick or Acid Test Ratio= 1.01 times), (**I**= Interest Expense= 513 millions USD), (**t**= T/B= Tax Rate= 30.00%), (**F**= Fixed Cost= 29,740 millions USD), (**M**= Margin of Contribution= 41,548 millions USD), (**I**= L/E= Leverage or Gearing Ratio= 99.00%), and (**A**= After Tax Income= 7,899 millions USD), then it's (**U** = Utilized or Starting Capital planned), is:

$$U= Ad+(Q+Vp/\{360[c-q]\})/I-[M-F-I][1-t]$$
$$= 7,899*0.3000+(35,351+9,746$$
$$*240/\{360[1.2000-1.0100]\})$$
$$/0.9900-[41,548-29,750$$
$$-513][1-0.3000]$$
$$= 64,720 \text{ millions USD}$$

Corporate IFRS-GAAP (B/S-I/S), ISBN-13: **978-1720792789**, ISBN-10: **172079278X**

<u>Law-7687</u>:

If both (**d**= D/A= Dividend Portion or Payout= 30.00%), (**c**= C/X= Current Ratio= 1.20 times), (**V**= Variable Cost= 9,746 millions USD), (**p**= 360 P/V= Procured Inventory Days= 240 Days), (**Q**= Quoted Longterm Liabilities= 35,351 millions USD), (**q**= Quick or Acid Test Ratio= [C-P]/X= 1.01 times), (**I**= Interest Expense= 513 millions USD), (**t**= T/B= Tax Rate= 30.00%), (**F**= Fixed Cost= 29,740 millions USD), (**U**= Utilized or Starting Capital= 64,720 millions USD), (**I**= L/E= Leverage or Gearing Ratio= 99.00%), and (**A**= After Tax Income= 7,899 millions USD), then it's (**M** = Margin of Contribution planned), is:

$$M= F+I+[Ad-U+(Q+Vp/\{360[c-q]\})/I\,]$$
$$/[1-t]$$
$$= 29,750+513+[7,899*0.3000$$
$$-64,720+(35,351+9,746$$
$$*240/\{360[1.2000-1.0100]\})$$
$$/0.9900]/[1-0.3000]$$
$$= \underline{41,548} \text{ millions USD}$$

Corporate IFRS-GAAP (B/S-I/S), ISBN-13: **978-1720792789**, ISBN-10: **172079278X**

<u>Law-7688</u>:

If both (**d**= D/A= Dividend Portion or Payout= 30.00%), (**c**= C/X= Current Ratio= 1.20 times), (**V**= Variable Cost= 9,746 millions USD), (**p**= 360 P/V= Procured Inventory Days= 240 Days), (**Q**= Quoted Longterm Liabilities= 35,351 millions USD), (**q**= Quick or Acid Test Ratio= [C-P]/X= 1.01 times), (**I**= Interest Expense= 513 millions USD), (**t**= T/B= Tax Rate= 30.00%), (**M**= Margin of Contribution= 41,548 millions USD), (**U**= Utilized or Starting Capital= 64,720 millions USD), (**I**= L/E= Leverage or Gearing Ratio= 99.00%), and (**A**= After Tax Income= 7,899 millions USD), then it's (**F** = Fixed Cost planned), is:

$$F= M-I-[Ad-U+(Q+Vp/\{360[c-q]\})/I]/[1-t]$$
$$= 41{,}548-513-[7{,}899*0.3000-64{,}720$$
$$+ +(35{,}351+9{,}746*240/\{360$$
$$[1.2000-1.0100]\})/0.9900]$$
$$/[1-0.3000]$$
$$= 29{,}750 \text{ millions USD}$$

Corporate IFRS-GAAP (B/S-I/S), ISBN-13: **978-1720792789**, ISBN-10: **172079278X**

Law-7689:

If both (**d**= D/A= Dividend Portion or Payout= 30.00%), (**c**= C/X= Current Ratio= 1.20 times), (**V**= Variable Cost= 9,746 millions USD), (**p**= 360 P/V= Procured Inventory Days= 240 Days), (**Q**= Quoted Longterm Liabilities= 35,351 millions USD), (**q**= Quick or Acid Test Ratio= [C-P]/X= 1.01 times), (**M**= Margin of Contribution= 41,548 millions USD), (**F**= Fixed Cost= 29,740 millions USD), (**t**= T/B= Tax Rate= 30.00%), (**U**= Utilized or Starting Capital= 64,720 millions USD), (**I**= L/E= Leverage or Gearing Ratio= 99.00%), and (**A**= After Tax Income= 7,899 millions USD), then it's (**I** = Interest Expense planned), is:

$$\mathbf{I}= \mathbf{M} - \mathbf{F} - [\mathbf{Ad} - \mathbf{U} + (\mathbf{Q} + \mathbf{Vp}/\{360[\mathbf{c} - \mathbf{q}]\})/\mathbf{I}]/[1 - \mathbf{t}]$$
$$= 41,548 - 29,750 - [7,899*0.3000$$
$$-64,720 + (35,351 + 9,746*240$$
$$/\{360[1.2000 - 1.0100]\})$$
$$/0.9900]/[1 - 0.3000]$$
$$= \underline{513} \text{ millions USD}$$

Corporate IFRS-GAAP (B/S-I/S), ISBN-13: **978-1720792789**, ISBN-10: **172079278X**

Law-7690:

If both (**d**= D/A= Dividend Portion or Payout= 30.00%), (**c**= C/X= Current Ratio= 1.20 times), (**V**= Variable Cost= 9,746 millions USD), (**p**= 360 P/V= Procured Inventory Days= 240 Days), (**Q**= Quoted Longterm Liabilities= 35,351 millions USD), (**q**= Quick or Acid Test Ratio= [C-P]/X= 1.01 times), (**M**= Margin of Contribution= 41,548 millions USD), (**F**= Fixed Cost= 29,740 millions USD), (**I**= Interest Expense= 513 millions USD), (**U**= Utilized or Starting Capital= 64,720 millions USD), (**I**= L/E= Leverage or Gearing Ratio= 99.00%), and (**A**= After Tax Income= 7,899 millions USD), then it's (**t** = Tax Rate planned), is:

$$t= 1-[\mathbf{Ad}\text{-}\mathbf{U}+(\mathbf{Q}+\mathbf{Vp}/\{360[\mathbf{c}\text{-}\mathbf{q}]\})/\mathbf{I}]/[\mathbf{M}\text{-}\mathbf{F}\text{-}\mathbf{I}]$$
$$= 1-[7,899*0.3000-64,720+(35,351$$
$$+9,746*240/\{360[1.2000$$
$$-1.0100]\})/0.9900]/[41,548$$
$$-29,750-513]$$
$$= \underline{30.00\%}$$

Law-7691:

If both (**d**= D/A= Dividend Portion or Payout= 30.00%), (**c**= C/X= Current Ratio= 1.20 times), (**V**= Variable Cost= 9,746 millions USD), (**p**= 360 P/V= Procured Inventory Days= 240 Days), (**Q**= Quoted Longterm Liabilities= 35,351 millions USD), (**q**= Quick or Acid Test Ratio= [C-P]/X= 1.01 times), (**M**= Margin of Contribution= 41,548 millions USD), (**F**= Fixed Cost= 29,740 millions USD), (**I**= Interest Expense= 513 millions USD), (**U**= Utilized or Starting Capital= 64,720 millions USD), (**l**= L/E= Leverage or Gearing Ratio= 99.00%), and (**t**= T/B= Tax Rate= 30.00%), then it's (**A**= After Tax Income planned), is:

$$A= [U+[M\text{-}F\text{-}I][1\text{-}t]$$
$$-(Q+Vp/\{360[c\text{-}q]\})/l]/d$$
$$= [64,720+[41,548\text{-}29,750\text{-}513]$$
$$[1\text{-}0.3000]\text{-}(35,351+9,746$$
$$*240/\{360[1.2000$$
$$-1.0100]\})/0.9900]/0.3000$$
$$= 7,899 \text{ millions USD}$$

Corporate IFRS-GAAP (B/S-I/S), ISBN-13: **978-1720792789**, ISBN-10: **172079278X**

Law-7692:

If both (**A**= After Tax Income= 7,899 millions USD), (**c**= C/X= Current Ratio= 1.20 times), (**V**= Variable Cost= 9,746 millions USD), (**p**= 360 P/V= Procured Inventory Days= 240 Days), (**Q**= Quoted Longterm Liabilities= 35,351 millions USD), (**q**= [C-P]/X= Quick or Acid Test Ratio= 1.01 times), (**M**= Margin of Contribution= 41,548 millions USD), (**F**= Fixed Cost= 29,740 millions USD), (**I**= Interest Expense= 513 millions USD), (**U**= Utilized or Starting Capital= 64,720 millions USD), (**I**= L/E= Leverage or Gearing Ratio= 99.00%), and (**t**= T/B= Tax Rate= 30.00%), then it's (**d** = Dividend Portion or Payout planned), is:

$$d = [U+[M-F-I][1-t]-(Q+Vp/\{360[c-q]\})/I]/A$$
$$= [64,720+[41,548-29,750-513]$$
$$[1-0.3000]]-(35,351+9,746$$
$$*240/\{360[1.2000-1.0100]\})$$
$$/0.9900]/7,899$$
$$= 30.00\%$$

Corporate IFRS-GAAP (B/S-I/S), ISBN-13: **978-1720792789**, ISBN-10: **172079278X**

Law-7693:

If both (**A**= After Tax Income= 7,899 millions USD), (**c**= C/X= Current Ratio= 1.20 times), (**d**= D/A= Dividend Portion or Payout= 30.00%), (**p**= 360 P/V= Procured Inventory Days= 240 Days), (**Q**= Quoted Longterm Liabilities= 35,351 millions USD), (**q**= Quick or Acid Test Ratio= [C-P]/X= 1.01 times), (**M**= Margin of Contribution= 41,548 millions USD), (**F**= Fixed Cost= 29,740 millions USD), (**I**= Interest Expense= 513 millions USD), (**U**= Utilized or Starting Capital= 64,720 millions USD), (**l**= L/E= Leverage or Gearing Ratio= 99.00%), and (**t**= T/B= Tax Rate= 30.00%), then it's (**V** = Variable Cost planned), is:

$$\mathbf{V}= 360[\mathbf{c}\text{-}\mathbf{q}](\mathbf{l}\ \{\mathbf{U}+[\mathbf{M}\text{-}\mathbf{F}\text{-}\mathbf{I}][1\text{-}\mathbf{t}]\text{-}\mathbf{Ad}\}\text{-}\mathbf{Q})/\mathbf{p}$$

$$= 360[1.2000\text{-}1.0100](0.9900$$
$$\{64,720+[41,548\text{-}29,750$$
$$-513][1\text{-}0.3000]\text{-}7,899$$
$$*0.3000\}\text{-}35,351)/240$$

$$= 9,746 \text{ millions USD}$$

Corporate IFRS-GAAP (B/S-I/S), ISBN-13: **978-1720792789**, ISBN-10: **172079278X**

<u>Law-7694</u>:

If both (**A**= After Tax Income= <u>7,899</u> millions USD), (**c**= C/X= Current Ratio= <u>1.20</u> times), (**d**= D/A= Dividend Portion or Payout= <u>30.00%</u>), (**V**= Variable Cost= <u>9,746</u> millions USD), (**Q**= Quoted Longterm Liabilities= <u>35,351</u> millions USD), (**q**= [C-P]/X= Quick or Acid Test Ratio= <u>1.01</u> times), (**M**= Margin of Contribution= <u>41,548</u> millions USD), (**F**= Fixed Cost= <u>29,740</u> millions USD), (**I**= Interest Expense= <u>513</u> millions USD), (**U**= Utilized or Starting Capital= <u>64,720</u> millions USD), (**l**= L/E= Leverage or Gearing Ratio= <u>99.00%</u>), and (**t**= T/B= Tax Rate= <u>30.00%</u>), then it's (**p** = Procured Inventory Days planned), is:

$$p= 360[c\text{-}q](l \{U+[M\text{-}F\text{-}I][1\text{-}t]\text{-}Ad\}\text{-}Q)/V$$
$$= 360[1.2000\text{-}1.0100](0.9900$$
$$\{64,720+[41,548\text{-}29,750$$
$$-513][1\text{-}0.3000]\text{-}7,899$$
$$*0.3000\}\text{-}35,351)/9,746$$

$$= \underline{240} \text{ days}$$

<u>Law-7695</u>:

If both (**A**= After Tax Income= <u>7,899</u> millions USD),
(**p**= 360 P/V= Procured Inventory Days= <u>240</u> Days),
(**d**= D/A=Dividend Portion or Payout= <u>30.00%</u>), (**V**=
Variable Cost= <u>9,746</u> millions USD), (**Q**= Quoted
Longterm Liabilities= <u>35,351</u> millions USD), (**q**=
Quick or Acid Test Ratio= [C-P]/X= <u>1.01</u> times),
(**M**= Margin of Contribution= <u>41,548</u> millions USD),
(**F**= Fixed Cost= <u>29,740</u> millions USD), (**I**= Interest
Expense= <u>513</u> millions USD), (**U**= Utilized or
Starting Capital= <u>64,720</u> millions USD), (**I**= L/E=
Leverage or Gearing Ratio= <u>99.00%</u>), and (**t**= T/B=
Tax Rate= <u>30.00%</u>), then it's (**c** = Current Ratio
planned), is:

$$c= q+Vp/[360(I\{U+[M-F-I][1-t]-Ad\}-Q)]$$
$$= 1.0100+9,746*240/[360(0.9900$$
$$\{64,720+[41,548-29,750$$
$$-513][1-0.3000]-7,899$$
$$*0.3000\}-35,351)]$$
$$= \underline{1.20}\text{ times}$$

Corporate IFRS-GAAP (B/S-I/S), ISBN-13: **978-1720792789**, ISBN-10: **172079278X**

Law-7696:

If both (**A**= After Tax Income= 7,899 millions USD),
(**p**= 360P/V= Procured Inventory Days= 240 Days),
(**d**= D/A= Dividend Portion or Payout= 30.00%), (**V**=
Variable Cost= 9,746 millions USD), (**Q**= Quoted
Longterm Liabilities= 35,351 millions USD), (**c**=
C/X= Current Ratio= 1.20 times), (**M**= Margin of
Contribution= 41,548 millions USD), (**F**= Fixed
Cost= 29,740 millions USD), (**I**= Interest Expense=
513 millions USD), (**U**= Utilized or Starting Capital=
64,720 millions USD), (**I**= L/E= Leverage or Gearing
Ratio= 99.00%), and (**t**= T/B= Tax Rate= 30.00%),
then it's (**q**= Quick or Acid Test Ratio planned), is:

$$q= c\text{-}Vp/[360(I \{U+[M\text{-}F\text{-}I][1\text{-}t]\text{-}Ad\}\text{-}Q)]$$

$$= 1.2000\text{-}9,746*240/[360(0.9900$$
$$\{64,720+[41,548\text{-}29,750$$
$$-513][1\text{-}0.3000]\text{-}7,899$$
$$*0.3000\}\text{-}35,351)]$$

$$= \underline{1.01} \text{ times}$$

Corporate IFRS-GAAP (B/S-I/S), ISBN-13: **978-1720792789**, ISBN-10: **172079278X**

Law-7697:

If both (**A**= After Tax Income= 7,899 millions USD),
(**p**= 360 P/V= Procured Inventory Days= 240 Days),
(**d**= D/A= Dividend Portion or Payout= 30.00%), (**v**=
V/S= Variable Portion= 19.00%), (**q**= [C-P]/X=
Quick or Acid Test Ratio= 1.01 times), (**c**= C/X=
Current Ratio= 1.20 times), (**$**= Sales or Revenues=
51,294 millions USD), (**F**= Fixed Cost= 29,740
millions USD), (**I**= Interest Expense= 513 millions
USD), (**M**= Margin of Contribution= 41,548 millions
USD), (**U**= Utilized or Starting Capital= 64,720
millions USD), (**f**= L/E= Leverage or Gearing Ratio=
99.00%), and (**t**= T/B= Tax Rate= 30.00%), then it's
(**Q**= Quoted Longterm Liabilities planned), is:

$$Q= f\{U+[M-F-I][1-t]-Ad\}-\$vp/\{360[c-q]\}$$
$$= 0.9900\{64,720+[41,548-29,750$$
$$-513][1-0.3000]-7,899$$
$$*0.3000\}-51,294*0.1900$$
$$*240/\{360[1.2000-1.0100]\}$$
$$= 35,351 \text{ millions USD}$$

Corporate IFRS-GAAP (B/S-I/S), ISBN-13: **978-1720792789**, ISBN-10: **172079278X**

<u>Law-7698</u>:

If both (**A**= After Tax Income= <u>7,899</u> millions USD), (**p**= 360 P/V= Procured Inventory Days= <u>240</u> Days), (**d**= D/A= Dividend Portion or Payout= <u>30.00%</u>), (**v**= V/S= Variable Portion= <u>19.00%</u>), (**\$**= Sales or Revenues= <u>51,294</u> millions USD), (**q**= [C-P]/X= Quick or Acid Test Ratio= <u>1.01</u> times), (**c**= C/X= Current Ratio= <u>1.20</u> times), (**M**= Margin of Contribution= <u>41,548</u> millions USD), (**F**= Fixed Cost= <u>29,740</u> millions USD), (**I**= Interest Expense= <u>513</u> millions USD), (**M**= Margin of Contribution= <u>41,548</u> millions USD), (**U**= Utilized or Starting Capital= <u>64,720</u> millions USD), (**Q**= Quoted Longterm Liabilities= <u>35,351</u> millions USD), and (**t**= T/B= Tax Rate= <u>30.00%</u>), then it's (**I** = Leverage or Gearing Ratio planned), is:

$$I = (Q+\$vp/\{360[c-q]\})$$
$$/\{U+[M-F-I][1-t]-Ad\}$$
$$= (35,351+51,294*0.1900*240$$
$$/\{360[1.2000-1.0100]\})$$
$$/\{64,720+[41,548-29,750$$
$$-513][1-0.3000]-7,899$$
$$*0.3000\}$$
$$= 99.00\%$$

Corporate IFRS-GAAP (B/S-I/S), ISBN-13: **978-1720792789**, ISBN-10: **172079278X**

Law-7699:

If both (**A**= After Tax Income= 7,899 millions USD), (**p**= 360 P/V= Procured Inventory Days= 240 Days), (**d**= D/A= Dividend Portion or Payout= 30.00%), (**v**= V/S= Variable Portion= 19.00%), (**S**= Sales or Revenues= 51,294 millions USD), (**q**= [C-P]/X= Quick or Acid Test Ratio= 1.01 times), (**c**= C/X= Current Ratio= 1.20 times), (**M**= Margin of Contribution= 41,548 millions USD), (**F**= Fixed Cost= 29,740 millions USD), (**I**= Interest Expense= 513 millions USD), (**M**= Margin of Contribution= 41,548 millions USD), (**I**= L/E= Leverage or Gearing Ratio= 99.00%), (**Q**= Quoted Longterm Liabilities= 35,351 millions USD), and (**t**= T/B= Tax Rate= 30.00%), then it's (**U** = Utilized or Starting Capital planned), is:

$$U= (Q+Svp/\{360[c-q]\})/I$$
$$-\{M-F-I\}[1-t]-Ad\}$$
$$= (35,351+51,294*0.1900*240/\{360$$
$$[1.2000-1.0100]\})/0.9900$$
$$-\{[41,548-29,750-513]$$
$$[1-0.3000]-7,899*0.3000\}$$
$$= 64,720 \text{ millions USD}$$

Corporate IFRS-GAAP (B/S-I/S), ISBN-13: **978-1720792789**, ISBN-10: **172079278X**

<u>Law-7700</u>:

If both (**A**= After Tax Income= <u>7,899</u> millions USD), (**p**= 360 P/V= Procured Inventory Days= <u>240</u> Days), (**d**= D/A= Dividend Portion or Payout= <u>30.00%</u>), (**v**= V/S= Variable Portion= <u>19.00%</u>), (**S**= Sales or Revenues= <u>51,294</u> millions USD), (**q**= [C-P]/X= Quick or Acid Test Ratio= <u>1.01</u> times), (**c**= C/X= Current Ratio= <u>1.20</u> times), (**U**= Utilized or Starting Capital= <u>64,720</u> millions USD), (**F**= Fixed Cost= <u>29,740</u> millions USD), (**I**= Interest Expense= <u>513</u> millions USD), (**M**= Margin of Contribution= <u>41,548</u> millions USD), (**F**= L/E= Leverage or Gearing Ratio= <u>99.00%</u>), (**Q**= Quoted Longterm Liabilities= <u>35,351</u> millions USD), and (**t**= T/B= Tax Rate= <u>30.00%</u>), then it's (**M** = Margin of Contribution planned), is:

$$M= F+I+[Ad-U+(Q+Svp/\{360[c-q]\})/F]/[1-t]$$
$$= 29,750+513+[7,899*0.3000-64,720$$
$$+(35,351+51,294*0.1900$$
$$*240/\{360[1.2000-1.0100]\})$$
$$/0.9900]/[1-0.3000]$$
$$= \underline{41,548} \text{ millions USD}$$

Corporate IFRS-GAAP (B/S-I/S), ISBN-13: **978-1720792789**, ISBN-10: **172079278X**

<u>Law-7701</u>:

If both (**A**= After Tax Income= <u>7,899</u> millions USD), (**p**= 360 P/V= Procured Inventory Days= <u>240</u> Days), (**d**= D/A= Dividend Portion or Payout= <u>30.00%</u>), (**v**= V/S= Variable Portion= <u>19.00%</u>), (**S**= Sales or Revenues= <u>51,294</u> millions USD), (**q**= [C-P]/X= Quick or Acid Test Ratio= <u>1.01</u> times), (**c**= C/X= Current Ratio= <u>1.20</u> times), (**U**= Utilized or Starting Capital= <u>64,720</u> millions USD), (**M**= Margin of Contribution= <u>41,548</u> millions USD), (**I**= Interest Expense= <u>513</u> millions USD), (**M**= Margin of Contribution= <u>41,548</u> millions USD), (**l**= L/E= Leverage or Gearing Ratio= <u>99.00%</u>), (**Q**= Quoted Longterm Liabilities= <u>35,351</u> millions USD), and (**t**= T/B= Tax Rate= <u>30.00%</u>), then it's (**F** = Fixed Cost planned), is:

$$F = M\text{-}I\text{-}[Ad\text{-}U+(Q+Svp/\{360[c\text{-}q]\})/l]/[1\text{-}t]$$

$$= 41,548\text{-}513\text{-}[7,899*0.3000\text{-}64,720 +(35,351+51,294*0.1900 *240/\{360[1.2000\text{-}1.0100]\}) /0.9900]/[1\text{-}0.3000]$$

$$= \underline{29,750} \text{ millions USD}$$

Corporate IFRS-GAAP (B/S-I/S), ISBN-13: **978-1720792789**, ISBN-10: **172079278X**

Law-7702:

If both (**A**= After Tax Income= 7,899 millions USD),
(**p**= 360 P/V= Procured Inventory Days= 240 Days),
(**d**= D/A= Dividend Portion or Payout= 30.00%), (**v**=
V/S= Variable Portion= 19.00%), (**S**= Sales or
Revenues= 51,294 millions USD), (**q**= [C-P]/X=
Quick or Acid Test Ratio= 1.01 times), (**c**= C/X=
Current Ratio= 1.20 times), (**U**= Utilized or Starting
Capital= 64,720 millions USD), (**M**= Margin of
Contribution= 41,548 millions USD), (**F**= Fixed
Cost= 29,740 millions USD), (**M**= Margin of
Contribution= 41,548 millions USD), (**I**= L/E=
Leverage or Gearing Ratio= 99.00%), (**Q**= Quoted
Longterm Liabilities= 35,351 millions USD), and (**t**=
T/B= Tax Rate= 30.00%), then it's (**I** = Interest
Expense planned), is:

$$\mathbf{I} = \mathbf{M} - \mathbf{F} - [\mathbf{Ad} - \mathbf{U} + (\mathbf{Q} + \mathbf{Svp}/\{360[\mathbf{c} - \mathbf{q}]\})/\mathbf{I}]$$
$$/[1 - \mathbf{t}]$$

$$= 41,548 - 29,750 - [7,899*0.3000$$
$$-64,720 + (35,351 + 51,294$$
$$*0.1900*240/\{360[1.2000$$
$$-1.0100]\})/0.9900]$$
$$/[1 - 0.3000]$$

$$= 513 \text{ millions USD}$$

Corporate IFRS-GAAP (B/S-I/S), ISBN-13: **978-1720792789**, ISBN-10: **172079278X**

Law-7703:

If both (**A**= After Tax Income= 7,899 millions USD), (**p**= 360 P/V= Procured Inventory Days= 240 Days), (**d**= D/A= Dividend Portion or Payout= 30.00%), (**v**= V/S= Variable Portion= 19.00%), (**S**= Sales or Revenues= 51,294 millions USD), (**q**= [C-P]/X= Quick or Acid Test Ratio= 1.01 times), (**c**= C/X= Current Ratio= 1.20 times), (**U**= Utilized or Starting Capital= 64,720 millions USD), (**M**= Margin of Contribution= 41,548 millions USD), (**F**= Fixed Cost= 29,740 millions USD), (**M**= Margin of Contribution= 41,548 millions USD), (**I**= L/E= Leverage or Gearing Ratio= 99.00%), (**Q**= Quoted Longterm Liabilities= 35,351 millions USD), and (**I**= Interest Expense= 513 millions USD), then it's (**t** = Tax Rate planned), is:

$$t= 1-[\mathbf{Ad\text{-}U}+(\mathbf{Q}+\mathbf{Svp}/\{360[\mathbf{c\text{-}q}]\})/\mathbf{I}]/[\mathbf{M\text{-}F\text{-}I}]$$
$$= 1-[7,899*0.3000-64,720+(35,351$$
$$+51,294*0.1900*240/\{360$$
$$[1.2000-1.0100]\})/0.9900]$$
$$/[41,548-29,750-513]$$
$$= 30.00\%$$

Corporate IFRS-GAAP (B/S-I/S), ISBN-13: **978-1720792789**, ISBN-10: **172079278X**

<u>Law-7704</u>:

If both (**t**= T/B= Tax Rate= <u>30.00%</u>), (**p**= 360 P/V= Procured Inventory Days= <u>240</u> Days), (**d**= D/A= Dividend Portion or Payout= <u>30.00%</u>), (**v**= V/S= Variable Portion= <u>19.00%</u>), (**$**= Sales or Revenues= <u>51,294</u> millions USD), (**q**= [C-P]/X= Quick or Acid Test Ratio= <u>1.01</u> times), (**c**= C/X= Current Ratio= <u>1.20</u> times), (**U**= Utilized or Starting Capital= <u>64,720</u> millions USD), (**M**= Margin of Contribution= <u>41,548</u> millions USD), (**F**= Fixed Cost= <u>29,740</u> millions USD), (**M**= Margin of Contribution= <u>41,548</u> millions USD), (**l**= L/E= Leverage or Gearing Ratio= <u>99.00%</u>), (**Q**= Quoted Longterm Liabilities= <u>35,351</u> millions USD), and (**I**= Interest Expense= <u>513</u> millions USD), then it's (**A**= After Tax Income planned), is:

$$A= [U-(Q+\$vp/\{360[c-q]\})/I$$
$$+[M-F-I][1-t]]/d$$
$$= [64,720-(35,351+51,294*0.1900$$
$$*240/\{360[1.2000-1.0100]\})$$
$$/0.9900+[41,548-29,750-513]$$
$$[1-0.3000]]/0.3000$$
$$= \underline{7,899} \text{ millions USD}$$

Corporate IFRS-GAAP (B/S-I/S), ISBN-13: **978-1720792789**, ISBN-10: **172079278X**

<u>Law-7705</u>:

If both (**t**= T/B= Tax Rate= <u>30.00%</u>), (**p**= 360 P/V= Procured Inventory Days= <u>240</u> Days), (**A**= After Tax Income= <u>7,899</u> millions USD), (**v**= V/S= Variable Portion= <u>19.00%</u>), (**S**= Sales or Revenues= <u>51,294</u> millions USD), (**q**= [C-P]/X= Quick or Acid Test Ratio= <u>1.01</u> times), (**c**= C/X= Current Ratio= <u>1.20</u> times), (**U**= Utilized or Starting Capital= <u>64,720</u> millions USD), (**M**= Margin of Contribution= <u>41,548</u> millions USD), (**F**= Fixed Cost= <u>29,740</u> millions USD), (**M**= Margin of Contribution= <u>41,548</u> millions USD), (**I**= L/E= Leverage or Gearing Ratio= <u>99.00%</u>), (**Q**= Quoted Longterm Liabilities= <u>35,351</u> millions USD), and (**I**= Interest Expense= <u>513</u> millions USD), then it's (**d**= Dividend Portion or Payout planned), is:

$$
\begin{aligned}
\mathbf{d} &= \mathbf{[U\text{-}(Q\text{+}Svp}/\{360[\mathbf{c\text{-}q}]\})/\mathbf{I} \\
&\quad +[\mathbf{M\text{-}F\text{-}I}][1\text{-}\mathbf{t}]]/\mathbf{A} \\
&= [64{,}720\text{-}(35{,}351\text{+}51{,}294*0.1900 \\
&\qquad *240/\{360[1.2000\text{-}1.0100]\}) \\
&\qquad /0.9900\text{+}[41{,}548\text{-}29{,}750\text{-}513] \\
&\qquad [1\text{-}0.3000]]/7{,}899 \\
&= \underline{30.00\%}
\end{aligned}
$$

Corporate IFRS-GAAP (B/S-I/S), ISBN-13: **978-1720792789**, ISBN-10: **172079278X**

<u>Law-7706</u>:

If both (**t**= T/B= Tax Rate= <u>30.00%</u>), (**p**= 360 P/V= Procured Inventory Days= <u>240</u> Days), (**A**= After Tax Income= <u>7,899</u> millions USD), (**v**= V/S= Variable Portion= <u>19.00%</u>), (**q**= [C-P]/X= Quick or Acid Test Ratio= <u>1.01</u> times), (**c**= C/X= Current Ratio= <u>1.20</u> times), (**U**= Utilized or Starting Capital= <u>64,720</u> millions USD), (**M**= Margin of Contribution= <u>41,548</u> millions USD), (**F**= Fixed Cost= <u>29,740</u> millions USD), (**d**= D/A= Dividend Portion or Payout= <u>30.00%</u>), (**l**= L/E= Leverage or Gearing Ratio= <u>99.00%</u>), (**Q**= Quoted Longterm Liabilities= <u>35,351</u> millions USD), and (**I**= Interest Expense= <u>513</u> millions USD), then it's (**$**= Sales or Revenues planned), is:

$$\$= 360[\mathbf{c\text{-}q}](\mathbf{l}\,\{U+[M\text{-}F\text{-}I][1\text{-}t]\text{-}Ad\}\text{-}Q)$$
$$/[\mathbf{vp}]$$

$$= 360[1.2000\text{-}1.0100](0.9900$$
$$\{64,720+[41,548\text{-}29,750$$
$$-513][1\text{-}0.3000]\text{-}7,899$$
$$*0.3000\}\text{-}35,351)$$
$$/[0.1900*240]$$

$$= \underline{51,294}\ \text{millions USD}$$

Corporate IFRS-GAAP (B/S-I/S), ISBN-13: **978-1720792789**, ISBN-10: **172079278X**

Law-7707:

If both (**t**= T/B= Tax Rate= 30.00%), (**p**= 360 P/V= Procured Inventory Days= 240 Days), (**A**= After Tax Income= 7,899 millions USD), (**\$**= Sales or Revenues= 51,294 millions USD), (**q**= [C-P]/X= Quick or Acid Test Ratio= 1.01 times), (**c**= C/X= Current Ratio= 1.20 times), (**U**= Utilized or Starting Capital= 64,720 millions USD), (**M**= Margin of Contribution= 41,548 millions USD), (**F**= Fixed Cost= 29,740 millions USD), (**d**= D/A= Dividend Portion or Payout= 30.00%), (**l**= L/E= Leverage or Gearing Ratio= 99.00%), (**Q**= Quoted Longterm Liabilities= 35,351 millions USD), and (**I**= Interest Expense= 513 millions USD), then it's (**v**= Variable Portion planned), is:

$$v= 360[\textbf{c-q}](\textbf{l} \{\textbf{U}+[\textbf{M-F-I}][1\text{-}\textbf{t}]\text{-}\textbf{Ad}\}\text{-}\textbf{Q})$$
$$/[\textbf{\$p}]$$
$$= 360[1.2000\text{-}1.0100](0.9900$$
$$\{64,720+[41,548\text{-}29,750$$
$$-513][1\text{-}0.3000]\text{-}7,899$$
$$*0.3000\}\text{-}35,351)$$
$$/[51,294*240]$$
$$= 19.00\% \text{ millions USD}$$

Corporate IFRS-GAAP (B/S-I/S), ISBN-13: **978-1720792789**, ISBN-10: **172079278X**

Law-7708:

If both (**t**= T/B= Tax Rate= 30.00%), (**v**= V/S= Variable Portion= 19.00%), (**A**= After Tax Income= 7,899 millions USD), (**S**= Sales or Revenues= 51,294 millions USD), (**q**= [C-P]/X= Quick or Acid Test Ratio= 1.01 times), (**c**= C/X= Current Ratio= 1.20 times), (**U**= Utilized or Starting Capital= 64,720 millions USD), (**M**= Margin of Contribution= 41,548 millions USD), (**F**= Fixed Cost= 29,740 millions USD), (**d**= D/A= Dividend Portion or Payout= 30.00%), (**l**= L/E= Leverage or Gearing Ratio= 99.00%), (**Q**= Quoted Longterm Liabilities= 35,351 millions USD), and (**I**= Interest Expense= 513 millions USD), then it's (**p**= Procured Inventory Days planned), is:

$$p= 360[\textbf{c-q}](\textbf{\textit{l}} \{\textbf{U}+[\textbf{M-F-I}][1\text{-}\textbf{t}]\text{-}\textbf{Ad}\}\text{-}\textbf{Q})$$
$$/[\textbf{Sv}]$$
$$= 360[1.2000\text{-}1.0100](0.9900$$
$$\{64,720+[41,548\text{-}29,750$$
$$\text{-}513][1\text{-}0.3000]\text{-}7,899$$
$$*0.3000\}\text{-}35,351)$$
$$/[51,294*0.1900]$$
$$= \underline{240} \text{ days}$$

Corporate IFRS-GAAP (B/S-I/S), ISBN-13: **978-1720792789**, ISBN-10: **172079278X**

<u>Law-7709</u>:

If both (**t**= T/B= Tax Rate= <u>30.00%</u>), (**v**= V/S= Variable Portion= <u>19.00%</u>), (**A**= After Tax Income= <u>7,899</u> millions USD), (**S**= Sales or Revenues= <u>51,294</u> millions USD), (**q**= [C-P]/X= Quick or Acid Test Ratio= <u>1.01</u> times), (**p**= 360 P/V= Procured Inventory Days= <u>240</u> Days), (**U**= Utilized or Starting Capital= <u>64,720</u> millions USD), (**M**= Margin of Contribution= <u>41,548</u> millions USD), (**F**= Fixed Cost= <u>29,740</u> millions USD), (**d**= D/A= Dividend Portion or Payout= <u>30.00%</u>), (**l**= L/E= Leverage or Gearing Ratio= <u>99.00%</u>), (**Q**= Quoted Longterm Liabilities= <u>35,351</u> millions USD), and (**I**= Interest Expense= <u>513</u> millions USD), then it's (**c**= Current Ratio planned), is:

$$c= q+Svp/[360(l \{U+[M-F-I][1-t]-Ad\}-Q)]$$
$$= 1.0100+51,294*0.1900*240/[360$$
$$(0.9900\{64,720+[41,548$$
$$-29,750-513][1-0.3000]$$
$$-7,899*0.3000\}-35,351)]$$
$$= \underline{1.20} \text{ times}$$

Corporate IFRS-GAAP (B/S-I/S), ISBN-13: **978-1720792789**, ISBN-10: **172079278X**

<u>Law-7710</u>:

If both (**t**= T/B= Tax Rate= <u>30.00%</u>), (**v**= V/S= Variable Portion= <u>19.00%</u>), (**A**= After Tax Income= <u>7,899</u> millions USD), (**S**= Sales or Revenues= <u>51,294</u> millions USD), (**c**= C/X= Current Ratio= <u>1.20</u> times), (**p**= 360 P/V= Procured Inventory Days= <u>240</u> Days), (**U**= Utilized or Starting Capital= <u>64,720</u> millions USD), (**M**= Margin of Contribution= <u>41,548</u> millions USD), (**F**= Fixed Cost= <u>29,740</u> millions USD), (**d**= D/A= Dividend Portion or Payout= <u>30.00%</u>), (**l**= L/E= Leverage or Gearing Ratio= <u>99.00%</u>), (**Q**= Quoted Longterm Liabilities= <u>35,351</u> millions USD), and (**I**= Interest Expense= <u>513</u> millions USD), then it's (**q**= Quick or Acid Test Ratio planned), is:

$$q= c\text{-}Svp/[360(\mathbf{l}\{U+[M\text{-}F\text{-}I][1\text{-}t]\text{-}Ad\}\text{-}Q)]$$
$$= 1.2000\text{-}51,294*0.1900*240/[360$$
$$(0.9900\{64,720+[41,548$$
$$-29,750\text{-}513][1\text{-}0.3000]$$
$$-7,899*0.3000\}\text{-}35,351)]$$
$$= \underline{1.01}\ \text{times}$$

Corporate IFRS-GAAP (B/S-I/S), ISBN-13: **978-1720792789**, ISBN-10: **172079278X**

Law-7711:

If both (**t**= T/B= Tax Rate= 30.00%), (**v**= V/S= Variable Portion= 19.00%), (**A**= After Tax Income= 7,899 millions USD), (**S'**= Sales of Past Year= 48,851 millions USD), (**s**= [S/S']-1= Sales Growth= 5.00%), (**c**= C/X= Current Ratio= 1.20 times), (**p**= 360 P/V= Procured Inventory Days= 240 Days), (**U**= Utilized or Starting Capital= 64,720 millions USD), (**M**= Margin of Contribution= 41,548 millions USD), (**F**= Fixed Cost= 29,740 millions USD), (**d**= D/A= Dividend Portion or Payout= 30.00%), (**l**= L/E= Leverage or Gearing Ratio= 99.00%), (**q**= [C-P]/X= Quick or Acid Test Ratio= 1.01 times), and (**I**= Interest Expense= 513 millions USD), then it's (**Q**= Quoted Longterm Liabilities planned), is:

$$Q = l\{U+[M-F-I][1-t]-Ad\}-S'vp[1+s]$$
$$/\{360[c-q]\}$$
$$= 0.9900\{64,720+[41,548-29,750$$
$$-513][1-0.3000]-7,899$$
$$*0.3000\}-48,851*0.1900$$
$$*240[1+0.0500]/\{360$$
$$[1.2000-1.0100]\}$$
$$= \underline{35,351} \text{ millions USD}$$

Corporate IFRS-GAAP (B/S-I/S), ISBN-13: **978-1720792789**, ISBN-10: **172079278X**

<u>Law-7712</u>:

If both (**t**= T/B= Tax Rate= <u>30.00%</u>), (**v**= V/S= Variable Portion= <u>19.00%</u>), (**A**= After Tax Income= <u>7,899</u> millions USD), (**S'**= Sales of Past Year= <u>48,851</u> millions USD), (**s**= [S/S']-1= Sales Growth= <u>5.00%</u>), (**c**= C/X= Current Ratio= <u>1.20</u> times), (**p**= 360 P/V= Procured Inventory Days= <u>240</u> Days), (**U**= Utilized or Starting Capital= <u>64,720</u> millions USD), (**M**= Margin of Contribution= <u>41,548</u> millions USD), (**F**= Fixed Cost= <u>29,740</u> millions USD), (**d**= D/A= Dividend Portion or Payout= <u>30.00%</u>), (**Q**= Quoted Longterm Liabilities= <u>35,351</u> millions USD), (**q**= [C-P]/X= Quick or Acid Test Ratio= <u>1.01</u> times), and (**I**= Interest Expense= <u>513</u> millions USD), then it's (**l** = Leverage or Gearing Ratio planned), is:

$$l = (Q+S'vp[1+s]/\{360[c-q]\})$$
$$/\{U+[M-F-I][1-t]-Ad\}$$
$$= (35,351+48,851*0.1900*240$$
$$[1+0.0500]/\{360[1.2000$$
$$-1.0100]\})/\{64,720+[41,548$$
$$-29,750-513][1-0.3000]$$
$$-7,899*0.3000\}$$
$$= \underline{99.00\%}$$

Corporate IFRS-GAAP (B/S-I/S), ISBN-13: **978-1720792789**, ISBN-10: **172079278X**

Law-7713:

If both (**t**= T/B= Tax Rate= 30.00%), (**v**= V/S= Variable Portion= 19.00%), (**A**= After Tax Income= 7,899 millions USD), (**S'**= Sales of Past Year= 48,851 millions USD), (**s**= [S/S']-1= Sales Growth= 5.00%), (**c**= C/X= Current Ratio= 1.20 times), (**p**= 360 P/V= Procured Inventory Days= 240 Days), (**l**= L/E= Leverage or Gearing Ratio= 99.00%), (**M**= Margin of Contribution= 41,548 millions USD), (**F**= Fixed Cost= 29,740 millions USD), (**d**= D/A= Dividend Portion or Payout= 30.00%), (**Q**= Quoted Longterm Liabilities= 35,351 millions USD), (**q**= [C-P]/X= Quick or Acid Test Ratio= 1.01 times), and (**I**= Interest Expense= 513 millions USD), then it's (**U**= Utilized or Starting Capital planned), is:

$$U= (Q+S'vp[1+s]/\{360[c-q]\})/I$$
$$-\{M-F-I\}[1-t]-Ad\}$$
$$= (35,351+48,851*0.1900*240$$
$$[1+0.0500]/\{360[1.2000$$
$$-1.0100]\})/0.9900-\{[41,548$$
$$-29,750-513][1-0.3000]$$
$$-7,899*0.3000\}$$
$$= 64,720 \text{ millions USD}$$

Corporate IFRS-GAAP (B/S-I/S), ISBN-13: **978-1720792789**, ISBN-10: **172079278X**

Law-7714:

If both (**t**= T/B= Tax Rate= 30.00%), (**v**= V/S=
Variable Portion= 19.00%), (**A**= After Tax Income=
7,899 millions USD), (**S'**= Sales of Past Year= 48,851
millions USD), (**s**= [S/S']-1= Sales Growth= 5.00%),
(**c**= C/X= Current Ratio= 1.20 times), (**p**= 360 P/V=
Procured Inventory Days= 240 Days), (**l**= L/E=
Leverage or Gearing Ratio= 99.00%), (**U**= Utilized or
Starting Capital= 64,720 millions USD), (**F**= Fixed
Cost= 29,740 millions USD), (**d**= D/A= Dividend
Portion or Payout= 30.00%), (**Q**= Quoted Longterm
Liabilities= 35,351 millions USD), (**q**= [C-P]/X=
Quick or Acid Test Ratio= 1.01 times), and (**I**=
Interest Expense= 513 millions USD), then it's (**M**=
Margin of Contribution planned), is:

$$M= F+I+[(Ad-U+Q+S'vp[1+s]$$
$$/\{360[c-q]\})/l]/[1-t]$$
$$= 29,750+513+[7,899*0.3000-64,720$$
$$+ (35,351+48,851*0.1900$$
$$*240[1+0.0500]/\{360[1.2000$$
$$-1.0100]\})/0.9900]/[1-0.3000]$$
$$= 41,548 \text{ millions USD}$$

Corporate IFRS-GAAP (B/S-I/S), ISBN-13: **978-1720792789**, ISBN-10: **172079278X**

Law-7715:

If both (**t**= T/B= Tax Rate= 30.00%), (**v**= V/S= Variable Portion= 19.00%), (**A**= After Tax Income= 7,899 millions USD), (**S'**= Sales of Past Year= 48,851 millions USD), (**s**= [S/S']-1= Sales Growth= 5.00%), (**c**= C/X= Current Ratio= 1.20 times), (**p**= 360 P/V= Procured Inventory Days= 240 Days), (**l**= L/E= Leverage or Gearing Ratio= 99.00%), (**U**= Utilized or Starting Capital= 64,720 millions USD), (**M**= Margin of Contribution= 41,548 millions USD), (**d**= D/A= Dividend Portion or Payout= 30.00%), (**Q**= Quoted Longterm Liabilities= 35,351 millions USD), (**q**= Quick or Acid Test Ratio= [C-P]/X= 1.01 times), and (**I**= Interest Expense= 513 millions USD), then it's (**F**= Fixed Cost planned), is:

$$F= M\text{-}I\text{-}[(Ad\text{-}U+Q+S'vp[1+s]$$
$$/\{360[c\text{-}q]\})/I\,]/[1\text{-}t]$$
$$= 41{,}548\text{-}513\text{-}[7{,}899*0.3000\text{-}64{,}720$$
$$+ (35{,}351+48{,}851*0.1900*240$$
$$[1+0.0500]/\{360[1.2000$$
$$-1.0100]\})/0.9900]/[1\text{-}0.3000]$$
$$= \underline{29{,}750} \text{ millions USD}$$

Corporate IFRS-GAAP (B/S-I/S), ISBN-13: **978-1720792789**, ISBN-10: **172079278X**

Law-7716:

If both (**t**= T/B= Tax Rate= 30.00%), (**v**= V/S=
Variable Portion= 19.00%), (**A**= After Tax Income=
7,899 millions USD), (**S'**= Sales of Past Year= 48,851
millions USD), (**s**= [S/S']-1= Sales Growth= 5.00%),
(**c**= C/X= Current Ratio= 1.20 times), (**p**= 360 P/V=
Procured Inventory Days= 240 Days), (**l**= L/E=
Leverage or Gearing Ratio= 99.00%), (**U**= Utilized or
Starting Capital= 64,720 millions USD), (**M**= Margin
of Contribution= 41,548 millions USD), (**d**= D/A=
Dividend Portion or Payout= 30.00%), (**Q**= Quoted
Longterm Liabilities= 35,351 millions USD), (**q**=
Quick or Acid Test Ratio= [C-P]/X= 1.01 times), and
(**F**= Fixed Cost= 29,740 millions USD), then it's (**F**=
Fixed Cost planned), is:

$$\mathbf{l= M\text{-}F\text{-}[(Ad\text{-}U+Q+S'vp}[1+s]$$
$$/\{360[\mathbf{c\text{-}q}]\})/\mathbf{l}\,]/[1\text{-}\mathbf{t}]$$
$$= 41,548\text{-}29,750\text{-}[7,899*0.3000$$
$$-64,720+ (35,351+48,851$$
$$*0.1900*240[1+0.0500]$$
$$/\{360[1.2000\text{-}1.0100]\})$$
$$/0.9900]/[1\text{-}0.3000]$$
$$= \underline{513} \text{ millions USD}$$

Corporate IFRS-GAAP (B/S-I/S), ISBN-13: **978-1720792789**, ISBN-10: **172079278X**

Law-7717:

If both (**I**= Interest Expense= 513 millions USD), (**v**= V/S= Variable Portion= 19.00%), (**A**= After Tax Income= 7,899 millions USD), (**S'**= Sales of Past Year= 48,851 millions USD), (**s**= [S/S']-1= Sales Growth= 5.00%), (**c**= C/X= Current Ratio= 1.20 times), (**p**= 360 P/V= Procured Inventory Days= 240 Days), (**l**= L/E= Leverage or Gearing Ratio= 99.00%), (**U**= Utilized or Starting Capital= 64,720 millions USD), (**M**= Margin of Contribution= 41,548 millions USD), (**d**= D/A= Dividend Portion or Payout= 30.00%), (**Q**= Quoted Longterm Liabilities= 35,351 millions USD), (**q**= [C-P]/X= Quick or Acid Test Ratio= 1.01 times), and (**F**= Fixed Cost= 29,740 millions USD), then it's (**t**= Tax Rate planned), is:

$$t= 1-[(\mathbf{Ad}-\mathbf{U}+\mathbf{Q}+\mathbf{S'vp}[1+s]/\{360[\mathbf{c}-\mathbf{q}]\})/\mathbf{I}]$$
$$/[\mathbf{M}-\mathbf{F}-\mathbf{I}]$$

$$= 1-[7,899*0.3000-64,720+(35,351$$
$$+48,851*0.1900*240$$
$$[1+0.0500]/\{360[1.2000$$
$$-1.0100]\})/0.9900]/[41,548$$
$$-29,750-513]$$

$$= 30.00\%$$

Corporate IFRS-GAAP (B/S-I/S), ISBN-13: **978-1720792789**, ISBN-10: **172079278X**

<u>Law-7718</u>:

If both (**I**= Interest Expense= <u>513</u> millions USD), (**v**= V/S= Variable Portion= <u>19.00%</u>), (**t**= T/B= Tax Rate= <u>30.00%</u>), (**S'**= Sales of Past Year= <u>48,851</u> millions USD), (**s**= [S/S']-1= Sales Growth= <u>5.00%</u>), (**c**= C/X= Current Ratio= <u>1.20</u> times), (**p**= 360 P/V= Procured Inventory Days= <u>240</u> Days), (**I**= L/E= Leverage or Gearing Ratio= <u>99.00%</u>), (**U**= Utilized or Starting Capital= <u>64,720</u> millions USD), (**M**= Margin of Contribution= <u>41,548</u> millions USD), (**d**= D/A= Dividend Portion or Payout= <u>30.00%</u>), (**Q**= Quoted Longterm Liabilities= <u>35,351</u> millions USD), (**q**= Quick or Acid Test Ratio= [C-P]/X= <u>1.01</u> times), and (**F**= Fixed Cost= <u>29,740</u> millions USD), then it's (**A**= After Tax Income planned), is:

$$A= [[M\text{-}F\text{-}I][1\text{-}t]+U\text{-}(Q+S'vp[1+s]$$
$$/\{360[c\text{-}q]\})/I\,]/d$$
$$= [[41,548\text{-}29,750\text{-}513][1\text{-}0.3000]$$
$$+64,720\text{-}(35,351+48,851$$
$$*0.1900*240[1+0.0500]$$
$$/\{360[1.2000\text{-}1.0100]\})$$
$$/0.9900]/0.3000$$
$$= \underline{7,899} \text{ millions USD}$$

Corporate IFRS-GAAP (B/S-I/S), ISBN-13: **978-1720792789**, ISBN-10: **172079278X**

Law-7719:

If both (**I**= Interest Expense= 513 millions USD), (**v**= V/S= Variable Portion= 19.00%), (**t**= T/B= Tax Rate= 30.00%), (**S'**= Sales of Past Year= 48,851 millions USD), (**s**= [S/S']-1= Sales Growth= 5.00%), (**c**= C/X= Current Ratio= 1.20 times), (**p**= 360 P/V= Procured Inventory Days= 240 Days), (**l**= L/E= Leverage or Gearing Ratio= 99.00%), (**U**= Utilized or Starting Capital= 64,720 millions USD), (**M**= Margin of Contribution= 41,548 millions USD), (**A**= After Tax Income= 7,899 millions USD), (**Q**= Quoted Longterm Liabilities= 35,351 millions USD), (**q**= Quick or Acid Test Ratio= [C-P]/X= 1.01 times), and (**F**= Fixed Cost= 29,740 millions USD), then it's (**d**= Dividend Portion or Payout planned), is:

$$d= [[M-F-I][1-t]+U-(Q+S'vp[1+s]$$
$$/\{360[c-q]\})/l]/A$$
$$= [[41,548-29,750-513][1-0.3000]$$
$$+64,720-(35,351+48,851$$
$$*0.1900*240[1+0.0500]$$
$$/\{360[1.2000-1.0100]\})$$
$$/0.9900]/7,899$$
$$= 30.00\%$$

Corporate IFRS-GAAP (B/S-I/S), ISBN-13: **978-1720792789**, ISBN-10: **172079278X**

<u>Law-7720</u>:

If both (**I**= Interest Expense= <u>513</u> millions USD), (**v**= V/S= Variable Portion= <u>19.00%</u>), (**t**= T/B= Tax Rate= <u>30.00%</u>), (**d**= D/A= Dividend Portion or Payout= <u>30.00%</u>), (**s**= [S/S']-1= Sales Growth= <u>5.00%</u>), (**c**= C/X= Current Ratio= <u>1.20</u> times), (**p**= 360 P/V= Procured Inventory Days= <u>240</u> Days), (**l**= L/E= Leverage or Gearing Ratio= <u>99.00%</u>), (**U**= Utilized or Starting Capital= <u>64,720</u> millions USD), (**M**= Margin of Contribution= <u>41,548</u> millions USD), (**A**= After Tax Income= <u>7,899</u> millions USD), (**Q**= Quoted Longterm Liabilities= <u>35,351</u> millions USD), (**q**= [C-P]/X= Quick or Acid Test Ratio= <u>1.01</u> times), and (**F**= Fixed Cost= <u>29,740</u> millions USD), then it's (**S'**= Sales Past), must be:

$$S'= [360[c\text{-}q](l\,\{U+[M\text{-}F\text{-}I][1\text{-}t]\text{-}Ad\}\text{-}Q]$$
$$/\{vp[1+s]\}$$
$$= [360[1.2000\text{-}1.0100](0.9900\{64,720$$
$$+[41,548\text{-}29,750\text{-}513]$$
$$[1\text{-}0.3000]\text{-}7,899*0.3000\}$$
$$-35,351]/\{0.1900*240$$
$$[1+0.0500]\}$$
$$= 48,851 \text{ millions USD}$$

Corporate IFRS-GAAP (B/S-I/S), ISBN-13: **978-1720792789**, ISBN-10: **172079278X**

Law-7721:

If both (**I**= Interest Expense= 513 millions USD), (**S'**= Sales of Past Year= 48,851 millions USD), (**t**= T/B= Tax Rate= 30.00%), (**d**= D/A= Dividend Portion or Payout= 30.00%), (**s**= [S/S']-1= Sales Growth= 5.00%), (**c**= C/X= Current Ratio= 1.20 times), (**p**= 360 P/V= Procured Inventory Days= 240 Days), (**l**= L/E= Leverage or Gearing Ratio= 99.00%), (**U**= Utilized or Starting Capital= 64,720 millions USD), (**M**= Margin of Contribution= 41,548 millions USD), (**A**= After Tax Income= 7,899 millions USD), (**Q**= Quoted Longterm Liabilities= 35,351 millions USD), (**q**= [C-P]/X= Quick or Acid Test Ratio= 1.01 times), and (**F**= Fixed Cost= 29,740 millions USD), then it's (**v**= Variable Portion planned), is:

$$v= [360[\textbf{c-q}](\textbf{l} \{\textbf{U+[M-F-I]}[1-\textbf{t}]-\textbf{Ad}\}-\textbf{Q}] / \{\textbf{S'p}[1+\textbf{s}]\}$$

$$= [360[1.2000-1.0100](0.9900\{64,720 +[41,548-29,750-513] [1-0.3000]- 7,899*0.3000\} -35,351]/\{48,851*240 [1+0.0500]\}$$

$$= 19.00\%$$

Corporate IFRS-GAAP (B/S-I/S), ISBN-13: **978-1720792789**, ISBN-10: **172079278X**

Law-7722:

If both (**I**= Interest Expense= 513 millions USD), (**S'**= Sales of Past Year= 48,851 millions USD), (**t**= T/B= Tax Rate= 30.00%), (**d**= D/A= Dividend Portion or Payout= 30.00%), (**s**= [S/S']-1= Sales Growth= 5.00%), (**c**= C/X= Current Ratio= 1.20 times), (**v**= V/S= Variable Portion= 19.00%), (**I**= L/E= Leverage or Gearing Ratio= 99.00%), (**U**= Utilized or Starting Capital= 64,720 millions USD), (**M**= Margin of Contribution= 41,548 millions USD), (**A**= After Tax Income= 7,899 millions USD), (**Q**= Quoted Longterm Liabilities= 35,351 millions USD), (**q**= Quick or Acid Test Ratio= [C-P]/X= 1.01 times), and (**F**= Fixed Cost= 29,740 millions USD), then it's (**p**= Procured Inventory Days planned), is:

$$p= [360[c-q](I\{U+[M-F-I][1-t]-Ad\}-Q] /\{S'v[1+s]\}$$

$$= [360[1.2000-1.0100](0.9900\{64,720 +[41,548-29,750-513]$$
$$[1-0.3000]- 7,899*0.3000\}$$
$$-35,351]/\{48,851*0.1900$$
$$[1+0.0500]\}$$

$$= \underline{240} \text{ days}$$

Corporate IFRS-GAAP (B/S-I/S), ISBN-13: **978-1720792789**, ISBN-10: **172079278X**

Law-7723:

If both (**I**= Interest Expense= 513 millions USD), (**S'**= Sales of Past Year= 48,851 millions USD), (**t**= T/B= Tax Rate= 30.00%), (**d**= D/A= Dividend Portion or Payout= 30.00%), (**p**= 360 P/V= Procured Inventory Days= 240 Days), (**c**= C/X= Current Ratio= 1.20 times), (**v**= V/S= Variable Portion= 19.00%), (**l**= L/E= Leverage or Gearing Ratio= 99.00%), (**U**= Utilized or Starting Capital= 64,720 millions USD), (**M**= Margin of Contribution= 41,548 millions USD), (**A**= After Tax Income= 7,899 millions USD), (**Q**= Quoted Longterm Liabilities= 35,351 millions USD), (**q**= [C-P]/X= Quick or Acid Test Ratio= 1.01 times), and (**F**= Fixed Cost= 29,740 millions USD), then it's (**s**= Sales Growth planned), is:

$$s= [360[c-q](l \{U+[M-F-I][1-t]-Ad\}-Q]$$
$$/[S'vp]-1$$
$$= [360[1.2000-1.0100](0.9900\{64,720$$
$$+[41,548-29,750-513]$$
$$[1-0.3000]- 7,899*0.3000\}$$
$$-35,351]/[48,851*0.1900$$
$$*240]-1$$

$$= 5.00\%$$

Corporate IFRS-GAAP (B/S-I/S), ISBN-13: **978-1720792789**, ISBN-10: **172079278X**

Law-7724:

If both (**I**= Interest Expense= 513 millions USD), (**S'**= Sales of Past Year= 48,851 millions USD), (**t**= T/B= Tax Rate= 30.00%), (**d**= D/A= Dividend Portion or Payout= 30.00%), (**p**= 360 P/V= Procured Inventory Days= 240 Days), (**s**= [S/S']-1= Sales Growth= 5.00%), (**v**= V/S= Variable Portion= 19.00%), (**I**= L/E= Leverage or Gearing Ratio= 99.00%), (**U**= Utilized or Starting Capital= 64,720 millions USD), (**M**= Margin of Contribution= 41,548 millions USD), (**A**= After Tax Income= 7,899 millions USD), (**Q**= Quoted Longterm Liabilities= 35,351 millions USD), (**q**= [C-P]/X= Quick or Acid Test Ratio= 1.01 times), and (**F**= Fixed Cost= 29,740 millions USD), then it's (**c**= Current Ratio planned), is:

$$c= q+S'vp[1+s]/[360(I\{U+[M-F-I]$$
$$[1-t]-Ad\}-Q)]$$
$$= 1.0100+48,851*0.1900*240$$
$$[1+0.0500]/[360(0.9900$$
$$\{64,720+[41,548-29,750$$
$$-513][1-0.3000]-7,899$$
$$*0.3000\}-35,351)]$$
$$= 1.20 \text{ times}$$

Corporate IFRS-GAAP (B/S-I/S), ISBN-13: **978-1720792789**, ISBN-10: **172079278X**

Law-7725:

If both (**I**= Interest Expense= 513 millions USD), (**S'**= Sales of Past Year= 48,851 millions USD), (**t**= T/B= Tax Rate= 30.00%), (**d**= D/A= Dividend Portion or Payout= 30.00%), (**p**= 360 P/V= Procured Inventory Days= 240 Days), (**s**= [S/S']-1= Sales Growth= 5.00%), (**v**= V/S= Variable Portion= 19.00%), (**l**= L/E= Leverage or Gearing Ratio= 99.00%), (**U**= Utilized or Starting Capital= 64,720 millions USD), (**M**= Margin of Contribution= 41,548 millions USD), (**A**= After Tax Income= 7,899 millions USD), (**Q**= Quoted Longterm Liabilities= 35,351 millions USD), (**c**= C/X= Current Ratio= 1.20 times), and (**F**= Fixed Cost= 29,740 millions USD), then it's (**q**= Quick or Acid Test Ratio planned), is:

$$q= c\text{-}S'vp[1+s]/[360(\boldsymbol{l}\{U+[M\text{-}F\text{-}I]$$
$$[1\text{-}t]\text{-}Ad\}\text{-}Q)]$$
$$= 1.2000\text{-}48,851*0.1900*240$$
$$[1+0.0500]/[360(0.9900$$
$$\{64,720+[41,548\text{-}29,750$$
$$\text{-}513][1\text{-}0.3000]\text{-}7,899$$
$$*0.3000\}\text{-}35,351)]$$
$$= 1.01 \text{ times}$$

Corporate IFRS-GAAP (B/S-I/S), ISBN-13: **978-1720792789**, ISBN-10: **172079278X**

Law-7726:

If both (**I**= Interest Expense= 513 millions USD), (**t**= T/B= Tax Rate= 30.00%), (**d**= D/A= Dividend Portion or Payout= 30.00 (**I**= L/E= Leverage or Gearing Ratio= 99.00%), (**U**= Utilized or Starting Capital= 64,720 millions USD), (**M**= Margin of Contribution= 41,548 millions USD), (**X**= Xpress or Current Debt= 34,196 millions USD), and (**F**= Fixed Cost= 29,740 millions USD), then it's (**Q**= Quoted Longterm Liabilities planned), is:

$$Q= I\{U+[M-F-I][1-t][1-d]\}-X$$
$$= 0.9900\{64,720+[41,548-29,750$$
$$-513][1-0.3000][1-0.3000]\}$$
$$-34,196$$
$$= \underline{35,351} \text{ millions USD}$$

Corporate IFRS-GAAP (B/S-I/S), ISBN-13: **978-1720792789**, ISBN-10: **172079278X**

<u>Law-7727</u>:

If both (**I**= Interest Expense= <u>513</u> millions USD), (**t**= T/B= Tax Rate= <u>30.00%</u>), (**Q**= Quoted Longterm Liabilities= <u>35,351</u> millions USD), (**d**= D/A= Dividend Portion or Payout= <u>30.00%</u>), (**U**= Utilized or Starting Capital= <u>64,720</u> millions USD), (**M**= Margin of Contribution= <u>41,548</u> millions USD), (**X**= Xpress or Current Debt= <u>34,196</u> millions USD), and (**F**= Fixed Cost= <u>29,740</u> millions USD), then it's (**I** = Leverage or Gearing Ratio planned), is:

$$I = [Q+X]/\{U+[M-F-I][1-t][1-d]\}$$
$$= [35,351+34,196]/\{64,720+[41,548$$
$$-29,750-513][1-0.3000]$$
$$[1-0.3000]\}$$
$$= \underline{99.00\%}$$

Corporate IFRS-GAAP (B/S-I/S), ISBN-13: **978-1720792789**, ISBN-10: **172079278X**

Law-7728:

If both (**I**= Interest Expense= 513 millions USD), (**t**= T/B= Tax Rate= 30.00%), (**Q**= Quoted Longterm Liabilities= 35,351 millions USD), (**d**= D/A= Dividend Portion or Payout= 30.00%), (**I**= L/E= Leverage or Gearing Ratio= 99.00%), (**M**= Margin of Contribution= 41,548 millions USD), (**X**= Xpress or Current Debt= 34,196 millions USD), and (**F**= Fixed Cost= 29,740 millions USD), then it's (**U** = Utilized or Starting Capital planned), is:

$$\mathbf{U}= [\mathbf{Q}+\mathbf{X}]/\mathbf{I} -[\mathbf{M}-\mathbf{F}-\mathbf{I}][1-\mathbf{t}][1-\mathbf{d}]$$
$$= [35,351+34,196]/0.9900-[41,548$$
$$-29,750-513][1-0.3000]$$
$$[1-0.3000]$$
$$= 64,720 \text{ millions USD}$$

Corporate IFRS-GAAP (B/S-I/S), ISBN-13: **978-1720792789**, ISBN-10: **172079278X**

Law-7729:

If both (**I**= Interest Expense= 513 millions USD), (**t**= T/B= Tax Rate= 30.00%), (**Q**= Quoted Longterm Liabilities= 35,351 millions USD), (**d**= D/A= Dividend Portion or Payout= 30.00%), (**I**= L/E= Leverage or Gearing Ratio= 99.00%), (**U**= Utilized or Starting Capital= 64,720 millions USD), (**X**= Xpress or Current Debt= 34,196 millions USD), and (**F**= Fixed Cost= 29,740 millions USD), then it's (**M** = Margin of Contribution planned), is:

$$M= F+I+\{[Q+X]/I -U\}\{[1-t][1-d]\}$$
$$= 29,750+513+\{[35,351+34,196]$$
$$/0.9900-64,720\}/\{[1-0.3000]$$
$$[1-0.3000]\}$$
$$= 41,548 \text{ millions USD}$$

Corporate IFRS-GAAP (B/S-I/S), ISBN-13: **978-1720792789**, ISBN-10: **172079278X**

Law-7730:

If both (**I**= Interest Expense= 513 millions USD), (**t**= T/B= Tax Rate= 30.00%), (**Q**= Quoted Longterm Liabilities= 35,351 millions USD), (**d**= D/A= Dividend Portion or Payout= 30.00%), (**I**= L/E= Leverage or Gearing Ratio= 99.00%), (**U**= Utilized or Starting Capital= 64,720 millions USD), (**X**= Xpress or Current Debt= 34,196 millions USD), and (**M**= Margin of Contribution= 41,548 millions USD), then it's (**F**= Fixed Cost planned), is:

$$F= M-I-\{[Q+X]/I-U\}\{[1-t][1-d]\}$$
$$= 41,548-513-\{[35,351+34,196]$$
$$/0.9900-64,720\}/\{[1-0.3000]$$
$$[1-0.3000]\}$$
$$= 29,750 \text{ millions USD}$$

Corporate IFRS-GAAP (B/S-I/S), ISBN-13: **978-1720792789**, ISBN-10: **172079278X**

Law-7731:

If both (**F**= Fixed Cost= 29,740 millions USD), (**t**= T/B= Tax Rate= 30.00%), (**Q**= Quoted Longterm Liabilities= 35,351 millions USD), (**d**= D/A= Dividend Portion or Payout= 30.00%), (**l**= L/E= Leverage or Gearing Ratio= 99.00%), (**U**= Utilized or Starting Capital= 64,720 millions USD), (**X**= Xpress or Current Debt= 34,196 millions USD), and (**M**= Margin of Contribution= 41,548 millions USD), then it's (**I**= Interest Expense planned), is:

$$I= M\text{-}F\text{-}\{[Q+X]/l\text{-}U\}\{[1\text{-}t][1\text{-}d]\}$$
$$= 41,548\text{-}29,750\text{-}\{[35,351+34,196]$$
$$/0.9900\text{-}64,720\}/\{[1\text{-}0.3000]$$
$$[1\text{-}0.3000]\}$$
$$= 513 \text{ millions USD}$$

Corporate IFRS-GAAP (B/S-I/S), ISBN-13: **978-1720792789**, ISBN-10: **172079278X**

Law-7732:

If both (**F**= Fixed Cost= 29,740 millions USD), (**I**= Interest Expense= 513 millions USD), (**Q**= Quoted Longterm Liabilities= 35,351 millions USD), (**d**= D/A= Dividend Portion or Payout= 30.00%), (**I**= L/E= Leverage or Gearing Ratio= 99.00%), (**U**= Utilized or Starting Capital= 64,720 millions USD), (**X**= Xpress or Current Debt= 34,196 millions USD), and (**M**= Margin of Contribution= 41,548 millions USD), then it's (**t**= Tax Rate planned), is:

$$t= 1-\{[Q+X]/I-U\}/\{[1-d][M-F-I]\}$$
$$= 1-\{35,351+34,196]/0.9900$$
$$-64,720\}/\{[1-0.3000]$$
$$[41,548-29,750-513]\}$$
$$= 30.00\%$$

Corporate IFRS-GAAP (B/S-I/S), ISBN-13: **978-1720792789**, ISBN-10: **172079278X**

Law-7733:

If both (**F**= Fixed Cost= 29,740 millions USD), (**I**= Interest Expense= 513 millions USD), (**Q**= Quoted Longterm Liabilities= 35,351 millions USD), (**t**= T/B= Tax Rate= 30.00%), (**I**= L/E= Leverage or Gearing Ratio= 99.00%), (**U**= Utilized or Starting Capital= 64,720 millions USD), (**X**= Xpress or Current Debt= 34,196 millions USD), and (**M**= Margin of Contribution= 41,548 millions USD), then it's (**d**= Dividend Portion or Payout planned), is:

$$d= 1-\{[\mathbf{Q}+\mathbf{X}]/\mathbf{I}-\mathbf{U}\}/\{[1-\mathbf{t}][\mathbf{M}-\mathbf{F}-\mathbf{I}]\}$$
$$= 1-\{35,351+34,196]/0.9900$$
$$-64,720\}/\{[1-0.3000]$$
$$[41,548-29,750-513]\}$$
$$= 30.00\%$$

Corporate IFRS-GAAP (B/S-I/S), ISBN-13: **978-1720792789**, ISBN-10: **172079278X**

<u>Law-7734</u>:

If both (**F**= Fixed Cost= <u>29,740</u> millions USD), (**I**= Interest Expense= <u>513</u> millions USD), (**Q**= Quoted Longterm Liabilities= <u>35,351</u> millions USD), (**t**= T/B= Tax Rate= <u>30.00%</u>), (**l**= L/E= Leverage or Gearing Ratio= <u>99.00%</u>), (**U**= Utilized or Starting Capital= <u>64,720</u> millions USD), (**d**= D/A= Dividend Portion or Payout= <u>30.00%</u>), and (**M**= Margin of Contribution= <u>41,548</u> millions USD), then it's (**X**= Xpress or Current Debt planned), is:

$$X = l\{U+[M-F-I][1-t][1-d]\}-Q$$
$$= 0.9900\{64,720+[41,548$$
$$-29,750-513][1-0.3000$$
$$[1-0.3000]\}-35,351$$
$$= \underline{34,196} \text{ millions USD}$$

Corporate IFRS-GAAP (B/S-I/S), ISBN-13: **978-1720792789**, ISBN-10: **172079278X**

Law-7735:

If both (**F**= Fixed Cost= 29,740 millions USD), (**I**= Interest Expense= 513 millions USD), (**c**= C/X= Current Ratio= 1.20 times), (**P**= Procured Inventories= 6,497 millions USD), (**q**= [C-P]/X= Quick or Acid Test Ratio= 1.01 times), (**t**= T/B= Tax Rate= 30.00%), (**l**= L/E= Leverage or Gearing Ratio= 99.00%), (**U**= Utilized or Starting Capital= 64,720 millions USD), (**d**= D/A= Dividend Portion or Payout= 30.00%), and (**M**= Margin of Contribution= 41,548 millions USD), then it's (**Q**= Quoted Longterm Liabilities planned), is:

$$Q= I\{U+[M\text{-}F\text{-}I][1\text{-}t][1\text{-}d]\}\text{-}P/[c\text{-}q]$$
$$= 0.9900\{64,720+[41,548\text{-}29,750$$
$$-513][1\text{-}0.3000[1\text{-}0.3000]\}$$
$$-6,497/[1.2000\text{-}1.0100]$$
$$= 35,351 \text{ millions USD}$$

Corporate IFRS-GAAP (B/S-I/S), ISBN-13: **978-1720792789**, ISBN-10: **172079278X**

<u>Law-7736</u>:

If both (**F**= Fixed Cost= <u>29,740</u> millions USD), (**I**= Interest Expense= <u>513</u> millions USD), (**c**= C/X= Current Ratio= <u>1.20</u> times), (**P**= Procured Inventories= <u>6,497</u> millions USD), (**q**= [C-P]/X= Quick or Acid Test Ratio= <u>1.01</u> times), (**t**= T/B= Tax Rate= <u>30.00%</u>), (**Q**= Quoted Longterm Liabilities= <u>35,351</u> millions USD), (**U**= Utilized or Starting Capital= <u>64,720</u> millions USD), (**d**= D/A= Dividend Portion or Payout= <u>30.00%</u>), and (**M**= Margin of Contribution= <u>41,548</u> millions USD), then it's (**I** = Leverage or Gearing Ratio planned), is:

$$I = \{Q+P/[c\text{-}q]\}/\{U+[M\text{-}F\text{-}I][1\text{-}t][1\text{-}d]\}$$
$$= \{35,351+6,497/[1.2000\text{-}1.0100]\}$$
$$/\{64,720+[41,548\text{-}29,750$$
$$-513][1\text{-}0.3000[1\text{-}0.3000]\}$$
$$= \underline{99.00\%}$$

Corporate IFRS-GAAP (B/S-I/S), ISBN-13: **978-1720792789**, ISBN-10: **172079278X**

Law-7737:

If both (**F**= Fixed Cost= 29,740 millions USD), (**I**= Interest Expense= 513 millions USD), (**c**= C/X= Current Ratio= 1.20 times), (**P**= Procured Inventories= 6,497 millions USD), (**q**= [C-P]/X= Quick or Acid Test Ratio= 1.01 times), (**t**= T/B= Tax Rate= 30.00%), (**Q**= Quoted Longterm Liabilities= 35,351 millions USD), (**I**= L/E= Leverage or Gearing Ratio= 99.00%), (**d**= D/A= Dividend Portion or Payout= 30.00%), and (**M**= Margin of Contribution= 41,548 millions USD), then it's (**U**= Utilized or Starting Capital planned), is:

$$U= \{Q+P/[c\text{-}q]\}/I - [M\text{-}F\text{-}I][1\text{-}t][1\text{-}d]$$
$$= \{35,351+6,497/[1.2000\text{-}1.0100]\}$$
$$/0.9900 - \{[41,548\text{-}29,750$$
$$-513][1\text{-}0.3000[1\text{-}0.3000]\}$$
$$= \underline{64,720} \text{ millions USD}$$

Corporate IFRS-GAAP (B/S-I/S), ISBN-13: **978-1720792789**, ISBN-10: **172079278X**

Law-7738:

If both (**F**= Fixed Cost= 29,740 millions USD), (**I**= Interest Expense= 513 millions USD), (**c**= C/X= Current Ratio= 1.20 times), (**P**= Procured Inventories= 6,497 millions USD), (**q**= [C-P]/X= Quick or Acid Test Ratio= 1.01 times), (**t**= T/B= Tax Rate= 30.00%), (**Q**= Quoted Longterm Liabilities= 35,351 millions USD), (**l**= L/E= Leverage or Gearing Ratio= 99.00%), (**d**= D/A= Dividend Portion or Payout= 30.00%), and (**U**= Utilized or Starting Capital= 64,720 millions USD), then it's (**M**= Margin of Contribution planned), is:

$$M = F+I+(\{Q+P/[c-q]\}/l - U)/\{[1-d][1-t]\}$$
$$= 29,750+513+(\{35,351+6,497$$
$$/[1.2000-1.0100]\}/0.9900$$
$$-64,720)/\{[1-0.3000$$
$$[1-0.3000]\}$$
$$= 41,548 \text{ millions USD}$$

Corporate IFRS-GAAP (B/S-I/S), ISBN-13: **978-1720792789**, ISBN-10: **172079278X**

Law-7739:

If both (**M**= Margin of Contribution= 41,548 millions USD), (**I**= Interest Expense= 513 millions USD), (**c**= C/X= Current Ratio= 1.20 times), (**P**= Procured Inventories= 6,497 millions USD), (**q**= [C-P]/X= Quick or Acid Test Ratio= 1.01 times), (**t**= T/B= Tax Rate= 30.00%), (**Q**= Quoted Longterm Liabilities= 35,351 millions USD), (**I**= L/E= Leverage or Gearing Ratio= 99.00%), (**d**= D/A= Dividend Portion or Payout= 30.00%), and (**U**= Utilized or Starting Capital= 64,720 millions USD), then it's (**F**= Fixed Cost planned), is:

$$F = M-I-(\{Q+P/[c-q]\}/I-U)/\{[1-d][1-t]\}$$
$$= 41,548-513-(\{35,351+6,497/[1.2000$$
$$-1.0100]\}/0.9900-64,720)$$
$$/\{[1-0.3000[1-0.3000]]\}$$
$$= 29,750 \text{ millions USD}$$

Corporate IFRS-GAAP (B/S-I/S), ISBN-13: **978-1720792789**, ISBN-10: **172079278X**

Law-7740:

If both (**M**= Margin of Contribution= 41,548 millions
USD), (**F**= Fixed Cost= 29,740 millions USD), (**c**=
C/X= Current Ratio= 1.20 times), (**P**= Procured
Inventories= 6,497 millions USD), (**q**= [C-P]/X=
Quick or Acid Test Ratio= 1.01 times), (**t**= T/B= Tax
Rate= 30.00%), (**Q**= Quoted Longterm Liabilities=
35,351 millions USD), (**I**= L/E= Leverage or Gearing
Ratio= 99.00%), (**d**= D/A= Dividend Portion or
Payout= 30.00%), and (**U**= Utilized or Starting
Capital= 64,720 millions USD), then it's (**I**= Interest
Expense planned), is:

$$I= M-F-(\{Q+P/[c-q]\}/I-U)/\{[1-d][1-t]\}$$
$$= 41,548-29,750-(\{35,351+6,497$$
$$/[1.2000-1.0100]\}/0.9900$$
$$-64,720)/\{[1-0.3000$$
$$[1-0.3000]\}$$
$$= 513 \text{ millions USD}$$

Law-7741:

If both (**M**= Margin of Contribution= 41,548 millions USD), (**F**= Fixed Cost= 29,740 millions USD), (**c**= C/X= Current Ratio= 1.20 times), (**P**= Procured Inventories= 6,497 millions USD), (**q**= [C-P]/X= Quick or Acid Test Ratio= 1.01 times), (**I**= Interest Expense= 513 millions USD), (**Q**= Quoted Longterm Liabilities= 35,351 millions USD), (**l**= L/E= Leverage or Gearing Ratio= 99.00%), (**d**= D/A= Dividend Portion or Payout= 30.00%), and (**U**= Utilized or Starting Capital= 64,720 millions USD), then it's (**t**= Tax Rate planned), is:

$$t= 1-(\{Q+P/[c-q]\}/l-U)/\{[1-d][M-F-I]\}$$
$$= 1-(\{35,351+6,497/[1.2000$$
$$-1.0100]\}/0.9900-64,720)$$
$$/\{[1-0.3000][41,548$$
$$-29,750-513]\}$$
$$= 30.00\%$$

Corporate IFRS-GAAP (B/S-I/S), ISBN-13: **978-1720792789**, ISBN-10: **172079278X**

Law-7742:

If both (**M**= Margin of Contribution= 41,548 millions USD), (**F**= Fixed Cost= 29,740 millions USD), (**c**= C/X= Current Ratio= 1.20 times), (**P**= Procured Inventories= 6,497 millions USD), (**q**= [C-P]/X= Quick or Acid Test Ratio= 1.01 times), (**I**= Interest Expense= 513 millions USD), (**Q**= Quoted Longterm Liabilities= 35,351 millions USD), (**I**= L/E= Leverage or Gearing Ratio= 99.00%), (**t**= T/B= Tax Rate= 30.00%), and (**U**= Utilized or Starting Capital= 64,720 millions USD), then it's (**d**= Dividend Portion or Payout planned), is:

$$d= 1-(\{Q+P/[c-q]\}/I-U)/\{[1-t][M-F-I]\}$$
$$= 1-(\{35,351+6,497/[1.2000-1.0100]\}$$
$$/0.9900-64,720)/\{[1-0.3000]$$
$$[41,548-29,750-513]\}$$
$$= 30.00\%$$

Corporate IFRS-GAAP (B/S-I/S), ISBN-13: **978-1720792789**, ISBN-10: **172079278X**

Law-7743:

If both (**M**= Margin of Contribution= <u>41,548</u> millions USD), (**F**= Fixed Cost= <u>29,740</u> millions USD), (**c**= C/X= Current Ratio= <u>1.20</u> times), (**d**= D/A= Dividend Portion or Payout= <u>30.00%</u>), (**q**= [C-P]/X= Quick or Acid Test Ratio= <u>1.01</u> times), (**I**= Interest Expense= <u>513</u> millions USD), (**Q**= Quoted Longterm Liabilities= <u>35,351</u> millions USD), (**I**= L/E= Leverage or Gearing Ratio= <u>99.00%</u>), (**t**= T/B= Tax Rate= <u>30.00%</u>), and (**U**= Utilized or Starting Capital= <u>64,720</u> millions USD), then it's (**P**= Procured Inventories planned), is:

$$P= [c-q](I\{U+[M-F-I][1-d][1-t]\}-Q)$$
$$= [1.2000-1.0100](0.9900\{64,720$$
$$+[41,548-29,750-513]$$
$$[1-0.3000][1-0.3000]\}$$
$$-35,351)$$
$$= \underline{6,497} \text{ millions USD}$$

Corporate IFRS-GAAP (B/S-I/S), ISBN-13: **978-1720792789**, ISBN-10: **172079278X**

Law-7744:

If both (**M**= Margin of Contribution= 41,548 millions
USD), (**F**= Fixed Cost= 29,740 millions USD), (**P**=
Procured Inventories= 6,497 millions USD), (**d**=
D/A= Dividend Portion or Payout= 30.00%), (**q**=
Quick or Acid Test Ratio= [C-P]/X= 1.01 times), (**I**=
Interest Expense= 513 millions USD), (**Q**= Quoted
Longterm Liabilities= 35,351 millions USD), (**I**=
L/E= Leverage or Gearing Ratio= 99.00%), (**t**= T/B=
Tax Rate= 30.00%), and (**U**= Utilized or Starting
Capital= 64,720 millions USD), then it's (**c**= Current
Ratio planned), is:

$$c= q+P/(I \{U+[M-F-I][1-d][1-t]\}-Q)$$
$$= 1.0100+6,497/(0.9900\{64.720$$
$$+[41,548-29,750-513]$$
$$[1-0.3000][1-0.3000]\}$$
$$-35,351)$$
$$= 1.20 \text{ times}$$

Corporate IFRS-GAAP (B/S-I/S), ISBN-13: **978-1720792789**, ISBN-10: **172079278X**

Law-7745:

If both (**M**= Margin of Contribution= 41,548 millions USD), (**F**= Fixed Cost= 29,740 millions USD), (**P**= Procured Inventories= 6,497 millions USD), (**d**= D/A= Dividend Portion or Payout= 30.00%), (**c**= C/X= Current Ratio= 1.20 times), (**I**= Interest Expense= 513 millions USD), (**Q**= Quoted Longterm Liabilities= 35,351 millions USD), (**I**= L/E= Leverage or Gearing Ratio= 99.00%), (**t**= T/B= Tax Rate= 30.00%), and (**U**= Utilized or Starting Capital= 64,720 millions USD), then it's (**q**= Quick or Acid Test Ratio planned), is:

$$q= c\text{-}P/(I\{U+[M\text{-}F\text{-}I][1\text{-}d][1\text{-}t]\}\text{-}Q)$$
$$= 1.2000\text{-}6,497/(0.9900\{64,720$$
$$+[41,548\text{-}29,750\text{-}513]$$
$$[1\text{-}0.3000][1\text{-}0.3000]\}$$
$$-35,351)$$

$$= 1.01 \text{ times}$$

Corporate IFRS-GAAP (B/S-I/S), ISBN-13: **978-1720792789**, ISBN-10: **172079278X**

Law-7746:

If both (**M**= Margin of Contribution= <u>41,548</u> millions
USD), (**F**= Fixed Cost= <u>29,740</u> millions USD), (**V**=
Variable Cost= <u>9,746</u> millions USD), (**p**= 360 P/V=
Procured Inventory Days= <u>240</u> Days), (**d**= D/A=
Dividend Portion or Payout= <u>30.00%</u>), (**c**= C/X=
Current Ratio= <u>1.20</u> times), (**I**= Interest Expense=
<u>513</u> millions USD), (**q**= [C-P]/X= Quick or Acid Test
Ratio= <u>1.01</u> times), (**I**= L/E= Leverage or Gearing
Ratio= <u>99.00%</u>), (**t**= T/B= Tax Rate= <u>30.00%</u>), and
(**U**= Utilized or Starting Capital= <u>64,720</u> millions
USD), then it's (**Q**= Quoted Longterm Liabilities
planned), is:

$$Q = I\{U+[M-F-I][1-d][1-t]\}$$
$$-Vp/\{360[c-q]\}$$
$$= 0.9900\{64,720+[41,548-29,750$$
$$-513][1-0.3000][1-0.3000]\}$$
$$-9,746*240/\{360[1.2000$$
$$-1.0100]\}$$
$$= \underline{35,351} \text{ millions USD}$$

Corporate IFRS-GAAP (B/S-I/S), ISBN-13: **978-1720792789**, ISBN-10: **172079278X**

<u>Law-7747</u>:

If both (**M**= Margin of Contribution= <u>41,548</u> millions USD), (**F**= Fixed Cost= <u>29,740</u> millions USD), (**V**= Variable Cost= <u>9,746</u> millions USD), (**p**= 360 P/V= Procured Inventory Days= <u>240</u> Days), (**d**= D/A= Dividend Portion or Payout= <u>30.00%</u>), (**c**= C/X= Current Ratio= <u>1.20</u> times), (**I**= Interest Expense= <u>513</u> millions USD), (**q**= [C-P]/X= Quick or Acid Test Ratio= <u>1.01</u> times), (**Q**= Quoted Longterm Liabilities= <u>35,351</u> millions USD), (**t**= T/B= Tax Rate= <u>30.00%</u>), and (**U**= Utilized or Starting Capital= <u>64,720</u> millions USD), then it's (**I** = Leverage or Gearing Ratio planned), is:

$$I = (\mathbf{Q} + \mathbf{Vp}/\{360[\mathbf{c}\text{-}\mathbf{q}]\})$$
$$/\{\mathbf{U} + [\mathbf{M}\text{-}\mathbf{F}\text{-}\mathbf{I}][1\text{-}\mathbf{d}][1\text{-}\mathbf{t}]\}$$
$$= (35,351 + 9,746*240/\{360[1.2000$$
$$-1.0100]\})/\{64,720 + [41,548$$
$$-29,750 - 513][1 - 0.3000]$$
$$[1 - 0.3000]\}$$
$$= \underline{99.00\%}$$

Corporate IFRS-GAAP (B/S-I/S), ISBN-13: **978-1720792789**, ISBN-10: **172079278X**

<u>Law-7748</u>:

If both (**M**= Margin of Contribution= <u>41,548</u> millions USD), (**F**= Fixed Cost= <u>29,740</u> millions USD), (**V**= Variable Cost= <u>9,746</u> millions USD), (**p**= 360 P/V= Procured Inventory Days= <u>240</u> Days), (**d**= D/A= Dividend Portion or Payout= <u>30.00%</u>), (**c**= C/X= Current Ratio= <u>1.20</u> times), (**I**= Interest Expense= <u>513</u> millions USD), (**q**= [C-P]/X= Quick or Acid Test Ratio= <u>1.01</u> times), (**Q**= Quoted Longterm Liabilities= <u>35,351</u> millions USD), (**t**= T/B= Tax Rate= <u>30.00%</u>), and (**I**= L/E= Leverage or Gearing Ratio= <u>99.00%</u>), then it's (**U**= Utilized or Starting Capital planned), is:

$$U= (Q+Vp/\{360[c\text{-}q]\})/I$$
$$-[M\text{-}F\text{-}I][1\text{-}d][1\text{-}t]$$
$$= (35,351+9,746*240/\{360[1.2000$$
$$-1.0100]\})/0.9900-[41,548$$
$$-29,750-513][1-0.3000]$$
$$[1-0.3000]$$
$$= \underline{64,720} \text{ millions USD}$$

Corporate IFRS-GAAP (B/S-I/S), ISBN-13: **978-1720792789**, ISBN-10: **172079278X**

Law-7749:

If both (**U**= Utilized or Starting Capital= 64,720 millions USD), (**F**= Fixed Cost= 29,740 millions USD), (**V**= Variable Cost= 9,746 millions USD), (**p**= 360 P/V= Procured Inventory Days= 240 Days), (**d**= D/A= Dividend Portion or Payout= 30.00%), (**c**= C/X= Current Ratio= 1.20 times), (**I**= Interest Expense= 513 millions USD), (**q**= [C-P]/X= Quick or Acid Test Ratio= 1.01 times), (**Q**= Quoted Longterm Liabilities= 35,351 millions USD), (**t**= T/B= Tax Rate= 30.00%), and (**l**= L/E= Leverage or Gearing Ratio= 99.00%), then it's (**M**= Margin of Contribution planned), is:

$$M= F+I+[(Q+Vp/\{360[c-q]\})/l-U]$$
$$/\{[1-d][1-t]\}$$
$$= 29,750+513+[(35,351+9,746$$
$$*240/\{360[1.2000-1.0100]\})$$
$$/0.9900-64,720]/\{[1-0.3000]$$
$$[1-0.3000]\}$$
$$= 41,548 \text{ millions USD}$$

Corporate IFRS-GAAP (B/S-I/S), ISBN-13: **978-1720792789**, ISBN-10: **172079278X**

Law-7750:

If both (**U**= Utilized or Starting Capital= 64,720 millions USD), (**M**= Margin of Contribution= 41,548 millions USD), (**V**= Variable Cost= 9,746 millions USD), (**p**= 360 P/V= Procured Inventory Days= 240 Days), (**d**= D/A= Dividend Portion or Payout= 30.00%), (**c**= C/X= Current Ratio= 1.20 times), (**I**= Interest Expense= 513 millions USD), (**q**= [C-P]/X= Quick or Acid Test Ratio= 1.01 times), (**Q**= Quoted Longterm Liabilities= 35,351 millions USD), (**t**= T/B= Tax Rate= 30.00%), and (**I**= L/E= Leverage or Gearing Ratio= 99.00%), then it's (**F**= Fixed Cost planned), is:

$$F= M-I-[(Q+Vp/\{360[c-q]\})/I-U]$$
$$/\{[1-d][1-t]\}$$
$$= 41,548-513-[(35,351+9,746*240$$
$$/\{360[1.2000-1.0100]\})$$
$$/0.9900-64,720]/\{[1-0.3000]$$
$$[1-0.3000]\}$$
$$= 29,750 \text{ millions USD}$$

Corporate IFRS-GAAP (B/S-I/S), ISBN-13: **978-1720792789**, ISBN-10: **172079278X**

Law-7751:

If both (**U**= Utilized or Starting Capital= 64,720 millions USD), (**M**= Margin of Contribution= 41,548 millions USD), (**F**= Fixed Cost= 29,740 millions USD), (**V**= Variable Cost= 9,746 millions USD), (**p**= 360 P/V= Procured Inventory Days= 240 Days), (**d**= D/A= Dividend Portion or Payout= 30.00%), (**c**= C/X= Current Ratio= 1.20 times), (**q**= [C-P]/X= Quick or Acid Test Ratio= 1.01 times), (**Q**= Quoted Longterm Liabilities= 35,351 millions USD), (**t**= T/B= Tax Rate= 30.00%), and (**I**= L/E= Leverage or Gearing Ratio= 99.00%), then it's (**I**= Fixed Cost planned), is:

$$I= M-F-[(Q+Vp/\{360[c-q]\})/I-U]$$
$$/\{[1-d][1-t]\}$$
$$= 41,548-29,750-[(35,351+9,746$$
$$*240/\{360[1.2000-1.0100]\})$$
$$/0.9900-64,720]/\{[1-0.3000]$$
$$[1-0.3000]\}$$
$$= 513 \text{ millions USD}$$

Corporate IFRS-GAAP (B/S-I/S), ISBN-13: **978-1720792789**, ISBN-10: **172079278X**

Law-7752:

If both (**U**= Utilized or Starting Capital= 64,720 millions USD), (**M**= Margin of Contribution= 41,548 millions USD), (**F**= Fixed Cost= 29,740 millions USD), (**V**= Variable Cost= 9,746 millions USD), (**p**= 360 P/V= Procured Inventory Days= 240 Days), (**d**= D/A= Dividend Portion or Payout= 30.00%), (**c**= C/X= Current Ratio= 1.20 times), (**q**= [C-P]/X= Quick or Acid Test Ratio= 1.01 times), (**Q**= Quoted Longterm Liabilities= 35,351 millions USD), (**I**= Interest Expense= 513 millions USD), and (**I**= L/E= Leverage or Gearing Ratio= 99.00%), then it's (**t**= Tax Rate planned), is:

$$t = 1 - [(Q + Vp / \{360[c-q]\}) / I - U]$$
$$/ \{[1-d][M-F-I]$$
$$= 1 - [(35,351 + 9,746 * 240 / \{360$$
$$[1.2000 - 1.0100]\}) / 0.9900$$
$$-64,720] / \{[1 - 0.3000]$$
$$[41,548 - 29,750 - 513]\}$$
$$= 30.00\%$$

Corporate IFRS-GAAP (B/S-I/S), ISBN-13: **978-1720792789**, ISBN-10: **172079278X**

Law-7753:

If both (**U**= Utilized or Starting Capital= 64,720 millions USD), (**M**= Margin of Contribution= 41,548 millions USD), (**F**= Fixed Cost= 29,740 millions USD), (**V**= Variable Cost= 9,746 millions USD), (**p**= 360 P/V= Procured Inventory Days= 240 Days), (**t**= T/B= Tax Rate= 30.00%), (**c**= C/X= Current Ratio= 1.20 times), (**q**= [C-P]/X= Quick or Acid Test Ratio= 1.01 times), (**Q**= Quoted Longterm Liabilities= 35,351 millions USD), (**I**= Interest Expense= 513 millions USD), and (**I**= L/E= Leverage or Gearing Ratio= 99.00%), then it's (**d**= Dividend Portion or Payout planned), is:

$$d= 1-[(\mathbf{Q+Vp}/\{360[\mathbf{c-q}]\})/\mathbf{I}-\mathbf{U}]$$
$$/\{[1-\mathbf{t}][\mathbf{M-F-I}]$$
$$= 1-[(35,351+9,746*240/\{360$$
$$[1.2000-1.0100]\})/0.9900$$
$$-64,720]/\{[1-0.3000]$$
$$[41,548-29,750-513]\}$$
$$= 30.00\%$$

Corporate IFRS-GAAP (B/S-I/S), ISBN-13: **978-1720792789**, ISBN-10: **172079278X**

<u>Law-7754</u>:

If both (**U**= Utilized or Starting Capital= <u>64,720</u> millions USD), (**M**= Margin of Contribution= <u>41,548</u> millions USD), (**F**= Fixed Cost= <u>29,740</u> millions USD), (**d**= D/A= Dividend Portion or Payout= <u>30.00%</u>), (**p**= 360 P/V= Procured Inventory Days= <u>240</u> Days), (**t**= T/B= Tax Rate= <u>30.00%</u>), (**c**= C/X= Current Ratio= <u>1.20</u> times), (**q**= [C-P]/X= Quick or Acid Test Ratio= <u>1.01</u> times), (**Q**= Quoted Longterm Liabilities= <u>35,351</u> millions USD), (**I**= Interest Expense= <u>513</u> millions USD), and (**F**= L/E= Leverage or Gearing Ratio= <u>99.00%</u>), then it's (**V**= Variable Cost planned), is:

$$V= [c-q]360(I\{U+[M-F-I][1-t][1-d]\}-Q)/p$$
$$= 360[1.2000-1.0100](0.9900\{64,720$$
$$+[41,548-29,750-513]$$
$$[1-0.3000][1-0.3000]\}$$
$$-35,351)/240$$
$$= \underline{9,746} \text{ millions USD}$$

Corporate IFRS-GAAP (B/S-I/S), ISBN-13: **978-1720792789**, ISBN-10: **172079278X**

Law-7755:

If both (**U**= Utilized or Starting Capital= 64,720
millions USD), (**M**= Margin of Contribution= 41,548
millions USD), (**F**= Fixed Cost= 29,740 millions
USD), (**d**= D/A= Dividend Portion or Payout=
30.00%), (**V**= Variable Cost= 9,746 millions USD),
(**t**= T/B= Tax Rate= 30.00%), (**c**= C/X= Current
Ratio= 1.20 times), (**q**= [C-P]/X= Quick or Acid Test
Ratio= 1.01 times), (**Q**= Quoted Longterm
Liabilities= 35,351 millions USD), (**I**= Interest
Expense= 513 millions USD), and (**l**= L/E= Leverage
or Gearing Ratio= 99.00%), then it's (**p**= Procured
Inventory Days planned), is:

$$p= [c\text{-}q]360(l\{U+[M\text{-}F\text{-}I][1\text{-}t][1\text{-}d]\}\text{-}Q)/V$$
$$= 360[1.2000\text{-}1.0100](0.9900\{64,720$$
$$+[41,548\text{-}29,750\text{-}513]$$
$$[1\text{-}0.3000][1\text{-}0.3000]\}$$
$$-35,351)/9,746$$

$$= \underline{240} \text{ days}$$

Corporate IFRS-GAAP (B/S-I/S), ISBN-13: **978-1720792789**, ISBN-10: **172079278X**

Law-7756:

If both (**U**= Utilized or Starting Capital= 64,720 millions USD), (**M**= Margin of Contribution= 41,548 millions USD), (**F**= Fixed Cost= 29,740 millions USD), (**d**= D/A= Dividend Portion or Payout= 30.00%), (**V**= Variable Cost= 9,746 millions USD), (**t**= T/B= Tax Rate= 30.00%), (**p**= 360 P/V= Procured Inventory Days= 240 Days), (**q**= [C-P]/X= Quick or Acid Test Ratio= 1.01 times), (**Q**= Quoted Longterm Liabilities= 35,351 millions USD), (**I**= Interest Expense= 513 millions USD), and (**I**= L/E= Leverage or Gearing Ratio= 99.00%), then it's (**c**= Current Ratio planned), is:

$$c = q + Vp/[360(I\{U+[M-F-I]$$
$$[1-t][1-d]\}-Q)]$$
$$= 1.0100+9,746*240/[360(0.9900$$
$$\{64,720+[41,548-29,750$$
$$-513][1-0.3000][1-0.3000]\}$$
$$-35,351)]$$
$$= 1.20 \text{ times}$$

Corporate IFRS-GAAP (B/S-I/S), ISBN-13: **978-1720792789**, ISBN-10: **172079278X**

<u>Law-7757</u>:

If both (**U**= Utilized or Starting Capital= <u>64,720</u> millions USD), (**M**= Margin of Contribution= <u>41,548</u> millions USD), (**F**= Fixed Cost= <u>29,740</u> millions USD), (**d**= D/A= Dividend Portion or Payout= <u>30.00%</u>), (**V**= Variable Cost= <u>9,746</u> millions USD), (**t**= T/B= Tax Rate= <u>30.00%</u>), (**p**= 360 P/V= Procured Inventory Days= <u>240</u> Days), (**c**= C/X= Current Ratio= <u>1.20</u> times), (**Q**= Quoted Longterm Liabilities= <u>35,351</u> millions USD), (**I**= Interest Expense= <u>513</u> millions USD), and (**F**= L/E= Leverage or Gearing Ratio= <u>99.00%</u>), then it's (**q**= Quick or Acid Test Ratio planned), is:

$$q= c\text{-}Vp/[360(I\,\{U+[M\text{-}F\text{-}I]$$
$$[1\text{-}t][1\text{-}d]\}\text{-}Q)]$$
$$= 1.2000\text{-}9.746*240/[360(0.9900$$
$$\{64,720+[41,548\text{-}29,750$$
$$-513][1\text{-}0.3000]$$
$$[1\text{-}0.3000]\}\text{-}35,351)]$$
$$= \underline{1.01} \text{ times}$$

Corporate IFRS-GAAP (B/S-I/S), ISBN-13: **978-1720792789**, ISBN-10: **172079278X**

<u>Law-7758</u>:

If both (**U**= Utilized or Starting Capital= <u>64,720</u> millions USD), (**M**= Margin of Contribution= <u>41,548</u> millions USD), (**F**= Fixed Cost= <u>29,740</u> millions USD), (**d**= D/A= Dividend Portion or Payout= <u>30.00%</u>), (**$**= Sales or Revenues= <u>51,294</u> millions USD), (**v**= V/S= Variable Portion= <u>19.00%</u>), (**t**= T/B= Tax Rate= <u>30.00%</u>), (**p**= 360 P/V= Procured Inventory Days= <u>240</u> Days), (**c**= C/X= Current Ratio= <u>1.20</u> times), (**q**= [C-P]/X= Quick or Acid Test Ratio= <u>1.01</u> times), (**I**= Interest Expense= <u>513</u> millions USD), and (**I**= L/E= Leverage or Gearing Ratio= <u>99.00%</u>), then it's (**Q**= Quoted Longterm Liabilities planned), is:

$$Q= I\{U+[M-F-I][1-t][1-d]\}$$
$$-\$vp/\{360[c-q]\}$$
$$= 0.9900\{64,720+[41,548-29,750$$
$$-513][1-0.3000][1-0.3000]\}$$
$$-51,294*0.1900*240$$
$$/\{360[1.2000-1.0100]\}$$
$$= \underline{35,351} \text{ millions USD}$$

Corporate IFRS-GAAP (B/S-I/S), ISBN-13: **978-1720792789**, ISBN-10: **172079278X**

<u>Law-7759</u>:

If both (**U**= Utilized or Starting Capital= <u>64,720</u> millions USD), (**M**= Margin of Contribution= <u>41,548</u> millions USD), (**F**= Fixed Cost= <u>29,740</u> millions USD), (**d**= D/A= Dividend Portion or Payout= <u>30.00%</u>), (**t**= T/B= Tax Rate= <u>30.00%</u>), (**v**= V/S= Variable Portion= <u>19.00%</u>), (**$**= Sales or Revenues= <u>51,294</u> millions USD), (**p**= 360 P/V= Procured Inventory Days= <u>240</u> Days), (**c**= C/X= Current Ratio= <u>1.20</u> times), (**q**= [C-P]/X= Quick or Acid Test Ratio= <u>1.01</u> times), (**l**= Interest Expense= <u>513</u> millions USD), and (**Q**= Quoted Longterm Liabilities= <u>35,351</u> millions USD), then it's (**/** = Leverage or Gearing Ratio planned), is:

$$/ = (\mathbf{Q} + \mathbf{\$vp} / \{360[\mathbf{c}\text{-}\mathbf{q}]\})$$
$$/ \{\mathbf{U} + [\mathbf{M}\text{-}\mathbf{F}\text{-}\mathbf{I}][1\text{-}\mathbf{t}][1\text{-}\mathbf{d}]\}$$
$$= (35,351 + 51,294 * 0.1900 * 240$$
$$/ \{360[1.2000 - 1.0100]\})$$
$$/ \{64,720 + [41,548 - 29,750$$
$$-513][1 - 0.3000][1 - 0.3000]\}$$
$$= \underline{99.00\%}$$

Corporate IFRS-GAAP (B/S-I/S), ISBN-13: **978-1720792789**, ISBN-10: **172079278X**

<u>Law-7760</u>:

If both (I= L/E= Leverage or Gearing Ratio= <u>99.00%</u>), (M= Margin of Contribution= <u>41,548</u> millions USD), (F= Fixed Cost= <u>29,740</u> millions USD), (d= D/A= Dividend Portion or Payout= <u>30.00%</u>), (t= T/B= Tax Rate= <u>30.00%</u>), (v= V/S= Variable Portion= <u>19.00%</u>), (S= Sales or Revenues= <u>51,294</u> millions USD), (p= 360 P/V= Procured Inventory Days= <u>240</u> Days), (c= C/X= Current Ratio= <u>1.20</u> times), (q= [C-P]/X= Quick or Acid Test Ratio= <u>1.01</u> times), (I= Interest Expense= <u>513</u> millions USD), and (Q= Quoted Longterm Liabilities= <u>35,351</u> millions USD), then it's (U= Utilized or Starting Capital planned), is:

$$U = (Q + Svp / \{360[c-q]\}) / I$$
$$- \{[M-F-I][1-t][1-d]\}$$
$$= (35,351 + 51,294*0.1900*240$$
$$/ \{360[1.2000-1.0100]\})$$
$$/0.9900 - [41,548-29,750$$
$$-513][1-0.3000]$$
$$[1-0.3000]$$
$$= \underline{64,720} \text{ millions USD}$$

Corporate IFRS-GAAP (B/S-I/S), ISBN-13: **978-1720792789**, ISBN-10: **172079278X**

Law-7761:

If both (**I**= L/E= Leverage or Gearing Ratio= 99.00%), (**F**= Fixed Cost= 29,740 millions USD), (**d**= D/A= Dividend Portion or Payout= 30.00%), (**t**= T/B= Tax Rate= 30.00%), (**v**= V/S= Variable Portion= 19.00%), (**U**= Utilized or Starting Capital= 64,720 millions USD), (**S**= Sales or Revenues= 51,294 millions USD), (**p**= 360 P/V= Procured Inventory Days= 240 Days), (**c**= C/X= Current Ratio= 1.20 times), (**q**= [C-P]/X= Quick or Acid Test Ratio= 1.01 times), (**I**= Interest Expense= 513 millions USD), and (**Q**= Quoted Longterm Liabilities= 35,351 millions USD), then it's (**M**= Margin of Contribution planned), is:

$$M= F+I+[(Q+Svp/\{360[c-q]\})/I -U]$$
$$/\{[1-t][1-d]\}$$
$$= 29,750+513 +[(35,351+51,294$$
$$*0.1900*240/\{360[1.2000$$
$$-1.0100]\})/0.9900-64,720]$$
$$/\{[1-0.3000][1-0.3000]\}$$
$$= 41,548 \text{ millions USD}$$

Corporate IFRS-GAAP (B/S-I/S), ISBN-13: **978-1720792789**, ISBN-10: **172079278X**

Law-7762:

If both (I= L/E= Leverage or Gearing Ratio= 99.00%), (M= Margin of Contribution= 41,548 millions USD), (d= D/A= Dividend Portion or Payout= 30.00%), (t= T/B= Tax Rate= 30.00%), (v= V/S= Variable Portion= 19.00%), (U= Utilized or Starting Capital= 64,720 millions USD), (S= Sales or Revenues= 51,294 millions USD), (p= 360 P/V= Procured Inventory Days= 240 Days), (c= C/X= Current Ratio= 1.20 times), (q= [C-P]/X= Quick or Acid Test Ratio= 1.01 times), (I= Interest Expense= 513 millions USD), and (Q= Quoted Longterm Liabilities= 35,351 millions USD), then it's (F= Fixed Cost planned), is:

$$F= M-I-[(Q+Svp/\{360[c-q]\})/I-U]$$
$$/\{[1-t][1-d]\}$$
$$= 41,548-513-[(35,351+51,294$$
$$*0.1900*240/\{360[1.2000$$
$$-1.0100]\})/0.9900-64,720]$$
$$/\{[1-0.3000][1-0.3000]\}$$
$$= 29,750 \text{ millions USD}$$

Corporate IFRS-GAAP (B/S-I/S), ISBN-13: **978-1720792789**, ISBN-10: **172079278X**

Law-7763:

If both (**I**= L/E= Leverage or Gearing Ratio= 99.00%), (**M**= Margin of Contribution= 41,548 millions USD), (**d**= D/A= Dividend Portion or Payout= 30.00%), (**t**= T/B= Tax Rate= 30.00%), (**v**= V/S= Variable Portion= 19.00%), (**U**= Utilized or Starting Capital= 64,720 millions USD), (**S**= Sales or Revenues= 51,294 millions USD), (**p**= 360 P/V= Procured Inventory Days= 240 Days), (**c**= C/X= Current Ratio= 1.20 times), (**q**= [C-P]/X= Quick or Acid Test Ratio= 1.01 times), (**F**= Fixed Cost= 29,740 millions USD), and (**Q**= Quoted Longterm Liabilities= 35,351 millions USD), then it's (**I**= Interest Expense planned), is:

$$I= M-F-[(Q+Svp/\{360[c-q]\})/I-U]$$
$$/\{[1-t][1-d]\}$$
$$= 41,548-29,750-[(35,351+51,294$$
$$*0.1900*240/\{360[1.2000$$
$$-1.0100]\})/0.9900-64,720]$$
$$/\{[1-0.3000][1-0.3000]\}$$
$$= 513 \text{ millions USD}$$

Corporate IFRS-GAAP (B/S-I/S), ISBN-13: **978-1720792789**, ISBN-10: **172079278X**

Law-7764:

If both (**I**= L/E= Leverage or Gearing Ratio=
99.00%), (**M**= Margin of Contribution= 41,548
millions USD), (**d**= D/A= Dividend Portion or
Payout= 30.00%), (**I**= Interest Expense= 513 millions
USD), (**v**= V/S= Variable Portion= 19.00%), (**U**=
Utilized or Starting Capital= 64,720 millions USD),
(**$**= Sales or Revenues= 51,294 millions USD), (**p**=
360 P/V= Procured Inventory Days= 240 Days), (**c**=
C/X= Current Ratio= 1.20 times), (**q**= [C-P]/X=
Quick or Acid Test Ratio= 1.01 times), (**F**= Fixed
Cost= 29,740 millions USD), and (**Q**= Quoted
Longterm Liabilities= 35,351 millions USD), then
it's (**t**= Tax Rate planned), is:

$$t= 1-[(Q+\$vp/\{360[c-q]\})/I-U]$$
$$/\{[1-d][M-F-I]\}$$
$$= 1-[(35,351+51,294*0.1900*240$$
$$/\{360[1.2000-1.0100]\})$$
$$/0.9900-64,720]$$
$$/\{[1-0.3000][41,548$$
$$-29,750-513]\}$$
$$= 30.00\%$$

Corporate IFRS-GAAP (B/S-I/S), ISBN-13: **978-1720792789**, ISBN-10: **172079278X**

Law-7765:

If both (**I**= L/E= Leverage or Gearing Ratio= 99.00%), (**M**= Margin of Contribution= 41,548 millions USD), (**t**= T/B= Tax Rate= 30.00%), (**I**= Interest Expense= 513 millions USD), (**v**= V/S= Variable Portion= 19.00%), (**U**= Utilized or Starting Capital= 64,720 millions USD), (**S**= Sales or Revenues= 51,294 millions USD), (**p**= 360 P/V= Procured Inventory Days= 240 Days), (**c**= C/X= Current Ratio= 1.20 times), (**q**= [C-P]/X= Quick or Acid Test Ratio= 1.01 times), (**F**= Fixed Cost= 29,740 millions USD), and (**Q**= Quoted Longterm Liabilities= 35,351 millions USD), then it's (**d**= Dividend Portion planned), is:

$$d = 1-[(Q+Svp/\{360[c-q]\})/I-U]$$
$$/\{[1-t][M-F-I]\}$$
$$= 1-[(35,351+51,294*0.1900*240$$
$$/\{360[1.2000-1.0100]\})$$
$$/0.9900-64,720]$$
$$/\{[1-0.3000][41,548$$
$$-29,750-513]\}$$
$$= 30.00\%$$

Corporate IFRS-GAAP (B/S-I/S), ISBN-13: **978-1720792789**, ISBN-10: **172079278X**

Law-7766:

If both (**I**= L/E= Leverage or Gearing Ratio= 99.00%), (**M**= Margin of Contribution= 41,548 millions USD), (**t**= T/B= Tax Rate= 30.00%), (**I**= Interest Expense= 513 millions USD), (**v**= V/S= Variable Portion= 19.00%), (**U**= Utilized or Starting Capital= 64,720 millions USD), (**d**= D/A= Dividend Portion or Payout= 30.00%), (**p**= 360P/V= Procured Inventory Days= 240 Days), (**c**= C/X= Current Ratio= 1.20 times), (**q**= [C-P]/X= Quick or Acid Test Ratio= 1.01 times), (**F**= Fixed Cost= 29,740 millions USD), and (**Q**= Quoted Longterm Liabilities= 35,351 millions USD), then it's (**S**= Sales or Revenues planned), is:

$$\mathbf{S}= 360[\mathbf{c\text{-}q}](\mathbf{I}\{\mathbf{U}+[\mathbf{M\text{-}F\text{-}I}][1\text{-}\mathbf{t}][1\text{-}\mathbf{d}]\}\text{-}\mathbf{Q})/[\mathbf{vp}]$$

$$= 360[1.2000\text{-}1.0100](0.9900$$
$$\{64,720+[41,548\text{-}29,750$$
$$-513][1\text{-}0.3000][1\text{-}0.3000]\}$$
$$-35,351)/[0.1900*240]$$

$$= \underline{51,294} \text{ millions USD}$$

Corporate IFRS-GAAP (B/S-I/S), ISBN-13: **978-1720792789**, ISBN-10: **172079278X**

Law-7767:

If both (I= L/E= Leverage or Gearing Ratio=
99.00%), (M= Margin of Contribution= 41,548
millions USD), (t= T/B= Tax Rate= 30.00%), (I=
Interest Expense= 513 millions USD), (S= Sales or
Revenues= 51,294 millions USD), (U= Utilized or
Starting Capital= 64,720 millions USD), (d= D/A=
Dividend Portion or Payout= 30.00%), (p= 360 P/V=
Procured Inventory Days= 240 Days), (c= C/X=
Current Ratio= 1.20 times), (q= [C-P]/X= Quick or
Acid Test Ratio= 1.01 times), (F= Fixed Cost=
29,740 millions USD), and (Q= Quoted Longterm
Liabilities= 35,351 millions USD), then it's (v=
Variable Portion planned), is:

$$v= 360[c\text{-}q](I\,\{U+[M\text{-}F\text{-}I][1\text{-}t][1\text{-}d]\}\text{-}Q)\,/[Sp]$$

$$= 360[1.2000\text{-}1.0100](0.9900$$
$$\{64,720+[41,548\text{-}29,750$$
$$\text{-}513][1\text{-}0.3000][1\text{-}0.3000]\}$$
$$\text{-}35,351)/[51,294*240]$$

$$= 19.00\%$$

Corporate IFRS-GAAP (B/S-I/S), ISBN-13: **978-1720792789**, ISBN-10: **172079278X**

Law-7768:

If both (**I**= L/E= Leverage or Gearing Ratio=
99.00%), (**M**= Margin of Contribution= 41,548
millions USD), (**t**= T/B= Tax Rate= 30.00%), (**I**=
Interest Expense= 513 millions USD), (**S**= Sales or
Revenues= 51,294 millions USD), (**U**= Utilized or
Starting Capital= 64,720 millions USD), (**d**= D/A=
Dividend Portion or Payout= 30.00%), (**v**= V/S=
Variable Portion= 19.00%), (**c**= C/X= Current Ratio=
1.20 times), (**q**= [C-P]/X= Quick or Acid Test Ratio=
1.01 times), (**F**= Fixed Cost= 29,740 millions USD),
and (**Q**= Quoted Longterm Liabilities= 35,351
millions USD), then it's (**p**= Procured Inventory
Days planned), is:

$$p= 360[\textbf{c-q}](\textbf{I} \{\textbf{U}+[\textbf{M-F-I}][1-\textbf{t}][1-\textbf{d}]\}-\textbf{Q})$$
$$/[\textbf{Sv}]$$
$$= 360[1.2000-1.0100](0.9900$$
$$\{64,720+[41,548-29,750$$
$$-513][1-0.3000][1-0.3000]\}$$
$$-35,351)/[51,294*0.1900]$$
$$= \underline{240} \text{ days}$$

Corporate IFRS-GAAP (B/S-I/S), ISBN-13: **978-1720792789**, ISBN-10: **172079278X**

Law-7769:

If both (I= L/E= Leverage or Gearing Ratio= 99.00%), (M= Margin of Contribution= 41,548 millions USD), (t= T/B= Tax Rate= 30.00%), (I= Interest Expense= 513 millions USD), (S= Sales or Revenues= 51,294 millions USD), (U= Utilized or Starting Capital= 64,720 millions USD), (d= D/A= Dividend Portion or Payout= 30.00%), (v= V/S= Variable Portion= 19.00%), (p= 360 P/V= Procured Inventory Days= 240 Days), (q= [C-P]/X= Quick or Acid Test Ratio= 1.01 times), (F= Fixed Cost= 29,740 millions USD), and (Q= Quoted Longterm Liabilities= 35,351 millions USD), then it's (c= Current Ratio planned), is:

$$c= q+Svp/[360(I \{U+[M-F-I]$$
$$[1-t][1-d]\}-Q)]$$
$$= 1.0100+51,294*0.1900*240$$
$$/[360(0.9900\{64,720$$
$$+[41,548-29,750-513]$$
$$[1-0.3000][1-0.3000]\}$$
$$-35,351)]$$
$$= 1.20 \text{ times}$$

Corporate IFRS-GAAP (B/S-I/S), ISBN-13: **978-1720792789**, ISBN-10: **172079278X**

Law-7770:

If both (**I**= L/E= Leverage or Gearing Ratio= 99.00%), (**M**= Margin of Contribution= 41,548 millions USD), (**t**= T/B= Tax Rate= 30.00%), (**I**= Interest Expense= 513 millions USD), (**S**= Sales or Revenues= 51,294 millions USD), (**U**= Utilized or Starting Capital= 64,720 millions USD), (**d**= D/A= Dividend Portion or Payout= 30.00%), (**v**= V/S= Variable Portion= 19.00%), (**p**= 360 P/V= Procured Inventory Days= 240 Days), (**c**= C/X= Current Ratio= 1.20 times), (**F**= Fixed Cost= 29,740 millions USD), and (**Q**= Quoted Longterm Liabilities= 35,351 millions USD), then it's (**q**= Quick or Acid Test Ratio planned), is:

$$q= c\text{-}Svp/[360(I\{U+[M\text{-}F\text{-}I][1\text{-}t][1\text{-}d]\}\text{-}Q)]$$
$$= 1.2000\text{-}51,294*0.1900*240/[360$$
$$(0.9900\{64,720+[41,548$$
$$-29,750\text{-}513][1\text{-}0.3000]$$
$$[1\text{-}0.3000]\}\text{-}35,351)]$$
$$= \underline{1.01} \text{ times}$$

Law-7771:

If both ($\textbf{\textit{F}}$= L/E= Leverage or Gearing Ratio= 99.00%), ($\textbf{M}$= Margin of Contribution= 41,548 millions USD), ($\textbf{t}$= T/B= Tax Rate= 30.00%), ($\textbf{I}$= Interest Expense= 513 millions USD), ($\textbf{\textit{S}'}$= Sales of Past Year= 48,851 millions USD), ($\textbf{s}$= [S/S']-1= Sales Growth= 5.00%), ($\textbf{U}$= Utilized or Starting Capital= 64,720 millions USD), ($\textbf{d}$= D/A= Dividend Portion or Payout= 30.00%), ($\textbf{v}$= V/S= Variable Portion= 19.00%), ($\textbf{p}$= 360 P/V= Procured Inventory Days= 240 Days), ($\textbf{c}$= C/X= Current Ratio= 1.20 times), ($\textbf{F}$= Fixed Cost= 29,740 millions USD), and ($\textbf{q}$= [C-P]/X= Quick or Acid Test Ratio= 1.01 times), then it's ($\textbf{Q}$= Quoted Longterm Liabilities planned), is:

$$\textbf{Q}= \textbf{\textit{F}}\,\{\textbf{U}+[\textbf{M}\textbf{-}\textbf{F}\textbf{-}\textbf{I}][1\textbf{-}\textbf{t}][1\textbf{-}\textbf{d}]\}$$
$$-\textbf{\textit{S}'}\textbf{v}\textbf{p}[1+\textbf{s}]/\{360[\textbf{c}\textbf{-}\textbf{q}]\}$$
$$= 0.9900\{64,720+[41,548-29,750$$
$$-513][1-0.3000][1-0.3000]\}$$
$$-48,851*0.1900*240$$
$$[1+0.0500]/\{360[1.2000$$
$$-1.0100]\}$$
$$= 35,351 \text{ millions USD}$$

Corporate IFRS-GAAP (B/S-I/S), ISBN-13: **978-1720792789**, ISBN-10: **172079278X**

Law-7772:

If both (**Q**= Quoted Longterm Liabilities= 35,351 millions USD), (**M**= Margin of Contribution= 41,548 millions USD), (**t**= T/B= Tax Rate= 30.00%), (**I**= Interest Expense= 513 millions USD), (**S'**= Sales of Past Year= 48,851 millions USD), (**s**= [S/S']-1= Sales Growth= 5.00%), (**U**= Utilized or Starting Capital= 64,720 millions USD), (**d**= D/A= Dividend Portion or Payout= 30.00%), (**v**= V/S= Variable Portion= 19.00%), (**p**= 360 P/V= Procured Inventory Days= 240 Days), (**c**= C/X= Current Ratio= 1.20 times), (**F**= Fixed Cost= 29,740 millions USD), and (**q**= [C-P]/X= Quick or Acid Test Ratio= 1.01 times), then it's (**I** = Leverage or gearing Ratio planned), is:

$$\mathbf{I} = (\mathbf{Q} + \mathbf{S'vp}[1+\mathbf{s}]/\{360[\mathbf{c}-\mathbf{q}]\})$$
$$/\{\mathbf{U}+[\mathbf{M}-\mathbf{F}-\mathbf{I}][1-\mathbf{t}][1-\mathbf{d}]\}$$
$$= (35,351+48,851*0.1900*240$$
$$[1+0.0500]/\{360[1.2000$$
$$-1.0100]\})/\{64,720+[41,548$$
$$-29,750-513][1-0.3000]$$
$$[1-0.3000]\}$$
$$= 99.00\%$$

Corporate IFRS-GAAP (B/S-I/S), ISBN-13: **978-1720792789**, ISBN-10: **172079278X**

Law-7773:

If both (**Q**= Quoted Longterm Liabilities= 35,351 millions USD), (**M**= Margin of Contribution= 41,548 millions USD), (**t**= T/B= Tax Rate= 30.00%), (**I**= Interest Expense= 513 millions USD), (**S'**= Sales of Past Year= 48,851 millions USD), (**s**= [S/S']-1= Sales Growth= 5.00%), (**l**= L/E= Leverage or Gearing Ratio= 99.00%), (**d**= D/A= Dividend Portion or Payout= 30.00%), (**v**= V/S= Variable Portion= 19.00%), (**p**= 360 P/V= Procured Inventory Days= 240 Days), (**c**= C/X= Current Ratio= 1.20 times), (**F**= Fixed Cost= 29,740 millions USD), and (**q**= Quick or Acid Test Ratio= [C-P]/X= 1.01 times), then it's (**U**= Utilized or Starting Capital planned), is:

$$\mathbf{U}= (\mathbf{Q}+\mathbf{S'vp}[1+\mathbf{s}]/\{360[\mathbf{c}-\mathbf{q}]\})/\mathbf{l}$$
$$-\{[\mathbf{M}-\mathbf{F}-\mathbf{I}][1-\mathbf{t}][1-\mathbf{d}]\}$$
$$= (35,351+48,851*0.1900*240$$
$$[1+0.0500]/\{360[1.2000$$
$$-1.0100]\})/0.9900-\{[41,548$$
$$-29,750-513][1-0.3000]$$
$$[1-0.3000]\}$$
$$= \underline{64,720} \text{ millions USD}$$

Corporate IFRS-GAAP (B/S-I/S), ISBN-13: **978-1720792789**, ISBN-10: **172079278X**

Law-7774:

If both (Q= Quoted Longterm Liabilities= 35,351
millions USD), (U= Utilized or Starting Capital=
64,720 millions USD), (t= T/B= Tax Rate= 30.00%),
(I= Interest Expense= 513 millions USD), (S'= Sales
of Past Year= 48,851 millions USD), (s= [S/S']-1=
Sales Growth= 5.00%), (F= L/E= Leverage or
Gearing Ratio= 99.00%), (d= D/A= Dividend Portion
or Payout= 30.00%), (v= V/S= Variable Portion=
19.00%), (p= 360 P/V= Procured Inventory Days=
240 Days), (c= C/X= Current Ratio= 1.20 times), (F=
Fixed Cost= [C-P]/X= 29,740 millions USD), and
(q= Quick or Acid Test Ratio= 1.01 times), then it's
(M= Margin of Contribution planned), is:

$$M= F+I+[(Q+S'vp[1+s]/\{360[c-q]\})/I-U]$$
$$/\{[1-t][1-d]\}$$
$$= 29,750+513 +[(35,351+48,851$$
$$*0.1900*240[1+0.0500]$$
$$/\{360[1.2000-1.0100]\})$$
$$/0.9900-64,720]$$
$$/\{[1-0.3000][1-0.3000]\}$$
$$= 41,548 \text{ millions USD}$$

Corporate IFRS-GAAP (B/S-I/S), ISBN-13: **978-1720792789**, ISBN-10: **172079278X**

<u>Law-7775</u>:

If both (**Q**= Quoted Longterm Liabilities= <u>35,351</u> millions USD), (**U**= Utilized or Starting Capital= <u>64,720</u> millions USD), (**t**= T/B= Tax Rate= <u>30.00%</u>), (**I**= Interest Expense= <u>513</u> millions USD), (**S'**= Sales of Past Year= <u>48,851</u> millions USD), (**s**= [S/S']-1= Sales Growth= <u>5.00%</u>), (**l**= L/E= Leverage or Gearing Ratio= <u>99.00%</u>), (**d**= D/A= Dividend Portion or Payout= <u>30.00%</u>), (**v**= V/S= Variable Portion= <u>19.00%</u>), (**p**= 360 P/V= Procured Inventory Days= <u>240</u> Days), (**c**= C/X= Current Ratio= <u>1.20</u> times), (**M**= Margin of Contribution= <u>41,548</u> millions USD), and (**q**= [C-P]/X= Quick or Acid Test Ratio= <u>1.01</u> times), then it's (**F**= Fixed Cost planned), is:

$$F= M-I-[(Q+S'vp[1+s]/\{360[c-q]\})/l-U]$$
$$/\{[1-t][1-d]\}$$
$$= 41,548-513-[(35,351+48,851$$
$$*0.1900*240[1+0.0500]$$
$$/\{360[1.2000-1.0100]\})$$
$$/0.9900-64,720]$$
$$/\{[1-0.3000][1-0.3000]\}$$
$$= \underline{29,750} \text{ millions USD}$$

Corporate IFRS-GAAP (B/S-I/S), ISBN-13: **978-1720792789**, ISBN-10: **172079278X**

Law-7776:

If both (Q= Quoted Longterm Liabilities= 35,351 millions USD), (U= Utilized or Starting Capital= 64,720 millions USD), (t= T/B= Tax Rate= 30.00%), (F= Fixed Cost= 29,740 millions USD), (S'= Sales of Past Year= 48,851 millions USD), (s= [S/S']-1= Sales Growth= 5.00%), (I= L/E= Leverage or Gearing Ratio= 99.00%), (d= D/A= Dividend Portion or Payout= 30.00%), (v= V/S= Variable Portion= 19.00%), (p= 360 P/V= Procured Inventory Days= 240 Days), (c= C/X= Current Ratio= 1.20 times), (M= Margin of Contribution= 41,548 millions USD), and (q= [C-P]/X= Quick or Acid Test Ratio= 1.01 times), then it's (I = Interest Expense planned), is:

$$I= M\text{-}F\text{-}[(Q+S'vp[1+s]/\{360[c\text{-}q]\})/I \text{-}U]$$
$$/\{[1\text{-}t][1\text{-}d]\}$$
$$= 41,548\text{-}29,750\text{-}[(35,351+48,851$$
$$*0.1900*240[1+0.0500]$$
$$/\{360[1.2000\text{-}1.0100]\})$$
$$/0.9900\text{-}64,720]$$
$$/\{[1\text{-}0.3000][1\text{-}0.3000]\}$$
$$= 513 \text{ millions USD}$$

Corporate IFRS-GAAP (B/S-I/S), ISBN-13: **978-1720792789**, ISBN-10: **172079278X**

Law-7777:

If both (**Q**= Quoted Longterm Liabilities= 35,351 millions USD), (**U**= Utilized or Starting Capital= 64,720 millions USD), (**I**= Interest Expense= 513 millions USD), (**F**= Fixed Cost= 29,740 millions USD), (**$'**= Sales of Past Year= 48,851 millions USD), (**s**= [S/S']-1= Sales Growth= 5.00%), (**I**= L/E= Leverage or Gearing Ratio= 99.00%), (**d**= D/A= Dividend Portion or Payout= 30.00%), (**v**= V/S= Variable Portion= 19.00%), (**p**= 360 P/V= Procured Inventory Days= 240 Days), (**c**= C/X= Current Ratio= 1.20 times), (**M**= Margin of Contribution= 41,548 millions USD), and (**q**= [C-P]/X= Quick or Acid Test Ratio= 1.01 times), then it's (**t**= Tax Rate planned), is:

$$t= 1-[(\mathbf{Q}+\mathbf{\$'vp}[1+\mathbf{s}]/\{360[\mathbf{c}-\mathbf{q}]\})/\mathbf{I} -\mathbf{U}]$$
$$/\{[1-\mathbf{d}][\mathbf{M}-\mathbf{F}-\mathbf{I}]\}$$
$$= 1-[(35,351+48,851*0.1900*240$$
$$[1+0.0500]/\{360[1.2000$$
$$-1.0100]\})/0.9900-64,720]$$
$$/\{[1-0.3000][41,548$$
$$-29,750-513]\}$$
$$= 30.00\%$$

Corporate IFRS-GAAP (B/S-I/S), ISBN-13: **978-1720792789**, ISBN-10: **172079278X**

Law-7778:

If both (**Q**= Quoted Longterm Liabilities= 35,351 millions USD), (**U**= Utilized or Starting Capital= 64,720 millions USD), (**I**= Interest Expense= 513 millions USD), (**F**= Fixed Cost= 29,740 millions USD), (**S'**= Sales of Past Year= 48,851 millions USD), (**s**= [S/S']-1= Sales Growth= 5.00%), (**l**= L/E= Leverage or Gearing Ratio= 99.00%), (**t**= T/B= Tax Rate= 30.00%), (**v**= V/S= Variable Portion= 19.00%), (**p**= 360 P/V= Procured Inventory Days= 240 Days), (**c**= C/X= Current Ratio= 1.20 times), (**M**= Margin of Contribution= 41,548 millions USD), and (**q**= [C-P]/X= Quick or Acid Test Ratio= 1.01 times), then it's (**d** = Dividend Portion or Payout planned), is:

$$d= 1-[(Q+S'vp[1+s]/\{360[c-q]\})/I-U]$$
$$/\{[1-t][M-F-I]\}$$

$$= 1-[(35,351+48,851*0.1900*240$$
$$[1+0.0500]/\{360[1.2000$$
$$-1.0100]\})/0.9900-64,720]$$
$$/\{[1-0.3000][41,548-29,750$$
$$-513]\}$$

$$= \underline{30.00\%}$$

Corporate IFRS-GAAP (B/S-I/S), ISBN-13: **978-1720792789**, ISBN-10: **172079278X**

Law-7779:

If both (**Q**= Quoted Longterm Liabilities= 35,351 millions USD), (**U**= Utilized or Starting Capital= 64,720 millions USD), (**I**= Interest Expense= 513 millions USD), (**F**= Fixed Cost= 29,740 millions USD), (**d**= D/A= Dividend Portion or Payout= 30.00%), (**s**= [S/S']-1= Sales Growth= 5.00%), (**l**= L/E= Leverage or Gearing Ratio= 99.00%), (**t**= T/B= Tax Rate= 30.00%), (**v**= V/S= Variable Portion= 19.00%), (**p**= 360 P/V= Procured Inventory Days= 240 Days), (**c**= C/X= Current Ratio= 1.20 times), (**M**= Margin of Contribution= 41,548 millions USD), and (**q**= [C-P]/X= Quick or Acid Test Ratio= 1.01 times), then it's (**S'** = Sales Past), must be:

$$S' = 360[\mathbf{c}\text{-}\mathbf{q}](\mathbf{l} \{U+[\mathbf{M}\text{-}\mathbf{F}\text{-}\mathbf{I}][1\text{-}\mathbf{d}][1\text{-}\mathbf{t}]\}\text{-}Q\}$$
$$/\{\mathbf{vp}[1+\mathbf{s}]\}$$
$$= 360[1.2000\text{-}1.0100](0.9900$$
$$\{64,720+[41,548\text{-}29,750$$
$$-513][1\text{-}0.3000][1\text{-}0.3000]\}$$
$$- 35,351)/\{0.1900*240$$
$$[1+0.0500]\}$$
$$= \underline{48,851} \text{ millions USD}$$

Corporate IFRS-GAAP (B/S-I/S), ISBN-13: **978-1720792789**, ISBN-10: **172079278X**

Law-7780:

If both (**Q**= Quoted Longterm Liabilities= 35,351 millions USD), (**U**= Utilized or Starting Capital= 64,720 millions USD), (**I**= Interest Expense= 513 millions USD), (**F**= Fixed Cost= 29,740 millions USD), (**d**= D/A= Dividend Portion or Payout= 30.00%), (**s**= [S/S']-1= Sales Growth= 5.00%), (**l**= L/E= Leverage or Gearing Ratio= 99.00%), (**t**= T/B= Tax Rate= 30.00%), (**S'**= Sales of Past Year= 48,851 millions USD), (**p**= 360 P/V= Procured Inventory Days= 240 Days), (**c**= C/X= Current Ratio= 1.20 times), (**M**= Margin of Contribution= 41,548 millions USD), and (**q**= [C-P]/X= Quick or Acid Test Ratio= 1.01 times), then it's (**v** = Variable Portion planned), is:

$$\mathbf{v}= 360[\mathbf{c}\text{-}\mathbf{q}](\mathbf{l}\,\{\mathbf{U}+[\mathbf{M}\text{-}\mathbf{F}\text{-}\mathbf{I}][1\text{-}\mathbf{d}][1\text{-}\mathbf{t}]\}\text{-}\mathbf{Q}\}$$
$$/\{\mathbf{S'p}[1+\mathbf{s}]\}$$
$$= 360[1.2000\text{-}1.0100](0.9900$$
$$\{64,720+[41,548\text{-}29,750$$
$$-513][1\text{-}0.3000][1\text{-}0.3000]\}$$
$$- 35,351)/\{48,851*240$$
$$[1+0.0500]\}$$
$$= \underline{19.00\%}$$

Corporate IFRS-GAAP (B/S-I/S), ISBN-13: **978-1720792789**, ISBN-10: **172079278X**

Law-7781:

If both (**Q**= Quoted Longterm Liabilities= 35,351 millions USD), (**U**= Utilized or Starting Capital= 64,720 millions USD), (**I**= Interest Expense= 513 millions USD), (**F**= Fixed Cost= 29,740 millions USD), (**d**= D/A= Dividend Portion or Payout= 30.00%), (**s**= [S/S']-1= Sales Growth= 5.00%), (**l**= L/E= Leverage or Gearing Ratio= 99.00%), (**t**= T/B= Tax Rate= 30.00%), (**S'**= Sales of Past Year= 48,851 millions USD), (**v**= V/S= Variable Portion= 19.00%), (**c**= C/X= Current Ratio= 1.20 times), (**M**= Margin of Contribution= 41,548 millions USD), and (**q**= Quick or Acid Test Ratio= [C-P]/X=1.01 times), then it's (**p**= Procured Inventory Days planned), is:

$$p= 360[\mathbf{c}\text{-}\mathbf{q}](\mathbf{l}\ \{\mathbf{U}+[\mathbf{M}\text{-}\mathbf{F}\text{-}\mathbf{I}][1\text{-}\mathbf{d}][1\text{-}\mathbf{t}]\}\text{-}\mathbf{Q}\}$$
$$/\{\mathbf{S'v}[1+\mathbf{s}]\}$$
$$= 360[1.2000\text{-}1.0100](0.9900$$
$$\{64,720+[41,548\text{-}29,750$$
$$\text{-}513][1\text{-}0.3000][1\text{-}0.3000]\}$$
$$\text{-}35,351)/\{48,851*0.1900$$
$$[1+0.0500]\}$$
$$= \underline{240}\ \text{days}$$

Corporate IFRS-GAAP (B/S-I/S), ISBN-13: **978-1720792789**, ISBN-10: **172079278X**

Law-7782:

If both (**Q**= Quoted Longterm Liabilities= 35,351 millions USD), (**U**= Utilized or Starting Capital= 64,720 millions USD), (**I**= Interest Expense= 513 millions USD), (**F**= Fixed Cost= 29,740 millions USD), (**d**= D/A= Dividend Portion or Payout= 30.00%), (**p**= 360 P/V= Procured Inventory Days= 240 Days), (**l**= L/E= Leverage or Gearing Ratio= 99.00%), (**t**= T/B= Tax Rate= 30.00%), (**S'**= Sales of Past Year= 48,851 millions USD), (**v**= V/S= Variable Portion= 19.00%), (**c**= C/X= Current Ratio= 1.20 times), (**M**= Margin of Contribution= 41,548 millions USD), and (**q**= [C-P]/X= Quick or Acid Test Ratio= 1.01 times), then it's (**s** = Sales Growth planned), is:

$$s= 360[\textbf{c-q}](\textbf{l} \{\textbf{U}+[\textbf{M-F-I}][1-\textbf{d}][1-\textbf{t}]\}-\textbf{Q}) / [\textbf{S'vp}]-1$$

$$= 360[1.2000-1.0100](0.9900 \{64,720+[41,548-29,750 -513][1-0.3000][1-0.3000]\} - 35,351)/[48,851*0.1900 *240]-1$$

$$= \underline{5.00\%}$$

Corporate IFRS-GAAP (B/S-I/S), ISBN-13: **978-1720792789**, ISBN-10: **172079278X**

Law-7783:

If both (**Q**= Quoted Longterm Liabilities= 35,351 millions USD), (**U**= Utilized or Starting Capital= 64,720 millions USD), (**I**= Interest Expense= 513 millions USD), (**F**= Fixed Cost= 29,740 millions USD), (**d**= D/A= Dividend Portion or Payout= 30.00%), (**p**= 360 P/V= Procured Inventory Days= 240 Days), (**l**= L/E= Leverage or Gearing Ratio= 99.00%), (**t**= T/B= Tax Rate= 30.00%), (**S'**= Sales of Past Year= 48,851 millions USD), (**v**= V/S= Variable Portion= 19.00%), (**s**= [S/S']-1= Sales Growth= 5.00%), (**M**= Margin of Contribution= 41,548 millions USD), and (**q**= [C-P]/X= Quick or Acid Test Ratio= 1.01 times), then it's (**c** = Current Ratio planned), is:

$$c= q+S'vp[1+s]/[360(I \{U+[M-F-I]$$
$$[1-d][1-t]\}-Q\}]$$
$$= 1.0100+48,851*0.1900*240$$
$$[1+0.0500]/[360(0.9900$$
$$\{64,720+[41,548-29,750$$
$$-513][1-0.3000][1-0.3000]\}$$
$$-35,351)]]$$
$$= 1.20 \text{ times}$$

Corporate IFRS-GAAP (B/S-I/S), ISBN-13: **978-1720792789**, ISBN-10: **172079278X**

<u>Law-7784</u>:

If both (**Q**= Quoted Longterm Liabilities= <u>35,351</u> millions USD), (**U**= Utilized or Starting Capital= <u>64,720</u> millions USD), (**I**= Interest Expense= <u>513</u> millions USD), (**F**= Fixed Cost= <u>29,740</u> millions USD), (**d**= D/A= Dividend Portion or Payout= <u>30.00%</u>), (**p**= 360P/V= Procured Inventory Days= <u>240</u> Days), (**l**= L/E= Leverage or Gearing Ratio= <u>99.00%</u>), (**t**= T/B= Tax Rate= <u>30.00%</u>), (**S'**= Sales of Past Year= <u>48,851</u> millions USD), (**v**= V/S= Variable Portion= <u>19.00%</u>), (**s**= [S/S']-1= Sales Growth= <u>5.00%</u>), (**M**= Margin of Contribution= <u>41,548</u> millions USD), and (**c**= C/X= Current Ratio= <u>1.20</u> times), then it's (**q** = Quick or Acid Test Ratio planned), is :

$$q= c\text{-}S'vp[1+s]/[360(l\ \{U+[M\text{-}F\text{-}I] [1\text{-}d][1\text{-}t]\}\text{-}Q\}]$$

$$= 1.2000\text{-}48,851*0.1900*240 [1+0.0500]/[360(0.9900 \{64,720+[41,548\text{-}29,750 \text{-}513][1\text{-}0.3000] [1\text{-}0.3000]\}\text{-}35,351)]]$$

$$= \underline{1.01}\ \text{times}$$

Corporate IFRS-GAAP (B/S-I/S), ISBN-13: **978-1720792789**, ISBN-10: **172079278X**

Law-7785:

If both (**X**= Xpress or Current Debt= 34,196 millions
USD), (**U**= Utilized or Starting Capital= 64,720
millions USD), (**i**= I/S= Interest Portion= 1.00%),
(**F**= Fixed Cost= 29,740 millions USD), (**D**=
Dividend Paid= 2,370 millions USD), (**I**= L/E=
Leverage or Gearing Ratio= 99.00%), (**S**= Sales or
Revenues= 51,294 millions USD), (**T**= Tax Paid=
3,385 millions USD), and (**M**= Margin of
Contribution= 41,548 millions USD), then it's (**Q**=
Quoted Longterm Liabilities planned), is:

$$Q= I\,[U+M-F-Si-T-D]-X$$
$$= 0.9900[64,720+41,548-29,750$$
$$-51,294*0.0100-3,385$$
$$-2,370]-34,196$$
$$= 35,351 \text{ millions USD}$$

Corporate IFRS-GAAP (B/S-I/S), ISBN-13: **978-1720792789**, ISBN-10: **172079278X**

Law-7786:

If both (**X**= Xpress or Current Debt= 34,196 millions USD), (**U**= Utilized or Starting Capital= 64,720 millions USD), (**i**= I/S= Interest Portion= 1.00%), (**F**= Fixed Cost= 29,740 millions USD), (**D**= Dividend Paid= 2,370 millions USD), (**Q**= Quoted Longterm Liabilities= 35,351 millions USD), (**S**= Sales or Revenues= 51,294 millions USD), (**T**= Tax Paid= 3,385 millions USD), and (**M**= Margin of Contribution= 41,548 millions USD), then it's (**I** = Leverage or Gearing Ratio planned), is:

$$I = [Q+X]/[U+M-F-Si-T-D]$$
$$= [35,351+34,196]/[64,720+41,548$$
$$-29,750-51,294*0.0100$$
$$-3,385-2,370]$$
$$= \underline{99.00\%}$$

Corporate IFRS-GAAP (B/S-I/S), ISBN-13: **978-1720792789**, ISBN-10: **172079278X**

Law-7787:

If both (**X**= Xpress or Current Debt= 34,196 millions
USD), (**I**= L/E= Leverage or Gearing Ratio=
99.00%), (**i**= I/S= Interest Portion= 1.00%), (**F**=
Fixed Cost= 29,740 millions USD), (**D**= Dividend
Paid= 2,370 millions USD), (**Q**= Quoted Longterm
Liabilities= 35,351 millions USD), (**$**= Sales or
Revenues= 51,294 millions USD), (**T**= Tax Paid=
3,385 millions USD), and (**M**= Margin of
Contribution= 41,548 millions USD), then it's (**U**=
Utilized or Starting Capital planned), is:

$$U= [Q+X]/I -[M-F-Si-T-D]$$
$$= [35,351+34,196]/0.9900-[41,548$$
$$-29,750-51,294*0.0100$$
$$-3,385-2,370]$$
$$= 64,720 \text{ millions USD}$$

Corporate IFRS-GAAP (B/S-I/S), ISBN-13: **978-1720792789**, ISBN-10: **172079278X**

Law-7788:

If both (**X**= Xpress or Current Debt= <u>34,196</u> millions USD), (**I**= L/E= Leverage or Gearing Ratio= <u>99.00%</u>), (**i**= I/S= Interest Portion= <u>1.00%</u>), (**F**= Fixed Cost= <u>29,740</u> millions USD), (**D**= Dividend Paid= <u>2,370</u> millions USD), (**Q**= Quoted Longterm Liabilities= <u>35,351</u> millions USD), (**S**= Sales or Revenues= <u>51,294</u> millions USD), (**T**= Tax Paid= <u>3,385</u> millions USD), and (**U**= Utilized or Starting Capital= <u>64,720</u> millions USD), then it's (**M** = Margin of Contribution planned), is:

$$M = F + Si + T + D - U + [Q + X]/I$$
$$= 29,750 + 51,294*0.0100 + 3,385$$
$$+ 2,370 - 64,720 + [35,351$$
$$+ 34,196]/0.9900$$
$$= \underline{41,548} \text{ millions USD}$$

Law-7789:

If both (**X**= Xpress or Current Debt= 34,196 millions USD), (**I**= L/E= Leverage or Gearing Ratio= 99.00%), (**i**= I/S= Interest Portion= 1.00%), (**M**= Margin of Contribution= 41,548 millions USD), (**D**= Dividend Paid= 2,370 millions USD), (**Q**= Quoted Longterm Liabilities= 35,351 millions USD), (**S**= Sales or Revenues= 51,294 millions USD), (**T**= Tax Paid= 3,385 millions USD), and (**U**= Utilized or Starting Capital= 64,720 millions USD), then it's (**F** = Fixed Cost planned), is:

$$F= U+M\text{-}Si\text{-}T\text{-}D\text{-}[Q+X]/I$$

$$= 64{,}720+41{,}548\text{-}51{,}294*0.0100$$
$$-3{,}385\text{-}2{,}370\text{-}[35{,}351$$
$$+34{,}196]/0.9900$$
$$= 29{,}750 \text{ millions USD}$$

Corporate IFRS-GAAP (B/S-I/S), ISBN-13: **978-1720792789**, ISBN-10: **172079278X**

Law-7790:

If both (**X**= Xpress or Current Debt= 34,196 millions USD), (**I**= L/E= Leverage or Gearing Ratio= 99.00%), (**i**= I/S= Interest Portion= 1.00%), (**M**= Margin of Contribution= 41,548 millions USD), (**D**= Dividend Paid= 2,370 millions USD), (**Q**= Quoted Longterm Liabilities= 35,351 millions USD), (**F**= Fixed Cost= 29,740 millions USD), (**T**= Tax Paid= 3,385 millions USD), and (**U**= Utilized or Starting Capital= 64,720 millions USD), then it's (**$** = Sales or Revenues planned), is:

$$\$= \{U+M\text{-}F\text{-}T\text{-}D\text{-}[Q+X]/I\}/i$$

$$= \{64,720+41,548\text{-}29,750\text{-}3,385$$
$$-2,370\text{-}[35,351+34,196]$$
$$/0.9900\}/0.0100$$

$$= \underline{51,294} \text{ millions USD}$$

Corporate IFRS-GAAP (B/S-I/S), ISBN-13: **978-1720792789**, ISBN-10: **172079278X**

Law-7791:

If both (**X**= Xpress or Current Debt= 34,196 millions
USD), (**I**= L/E= Leverage or Gearing Ratio=
99.00%), (**S**= Sales or Revenues= 51,294 millions
USD), (**M**= Margin of Contribution= 41,548 millions
USD), (**D**= Dividend Paid= 2,370 millions USD),
(**Q**= Quoted Longterm Liabilities= 35,351 millions
USD), (**F**= Fixed Cost= 29,740 millions USD), (**T**=
Tax Paid= 3,385 millions USD), and (**U**= Utilized or
Starting Capital= 64,720 millions USD), then it's (**i** =
Interest Portion planned), is:

$$i= \{U+M-F-T-D-[Q+X]/I \}/S$$
$$= \{64,720+41,548-29,750-3,385$$
$$-2,370-[35,351+34,196]$$
$$/0.9900\}/51,294$$
$$= 1.00\%$$

Corporate IFRS-GAAP (B/S-I/S), ISBN-13: **978-1720792789**, ISBN-10: **172079278X**

<u>Law-7792</u>:

If both (**X**= Xpress or Current Debt= <u>34,196</u> millions USD), (**I**= L/E= Leverage or Gearing Ratio= <u>99.00%</u>), (**S**= Sales or Revenues= <u>51,294</u> millions USD), (**M**= Margin of Contribution= <u>41,548</u> millions USD), (**D**= Dividend Paid= <u>2,370</u> millions USD), (**Q**= Quoted Longterm Liabilities= <u>35,351</u> millions USD), (**F**= Fixed Cost= <u>29,740</u> millions USD), (**i**= I/S= Interest Portion= <u>1.00%</u>), and (**U**= Utilized or Starting Capital= <u>64,720</u> millions USD), then it's (**T** = Tax Paid planned), is:

$$T= U+M-F-Si-D-[Q+X]/I$$

$$= 64,720+41,548-29,750-51,294$$
$$*0.0100-2,370-[35,351$$
$$+34,196]/0.9900\}$$

$$= \underline{3,385} \text{ millions USD}$$

Corporate IFRS-GAAP (B/S-I/S), ISBN-13: **978-1720792789**, ISBN-10: **172079278X**

Law-7793:

If both (**X**= Xpress or Current Debt= 34,196 millions
USD), (**I**= L/E= Leverage or Gearing Ratio=
99.00%), (**S**= Sales or Revenues= 51,294 millions
USD), (**M**= Margin of Contribution= 41,548 millions
USD), (**T**= Tax Paid= 3,385 millions USD), (**Q**=
Quoted Longterm Liabilities= 35,351 millions USD),
(**F**= Fixed Cost= 29,740 millions USD), (**i**= I/S=
Interest Portion= 1.00%), and (**U**= Utilized or
Starting Capital= 64,720 millions USD), then it's (**D**
= Dividend Paid planned), is:

$$D= U+M-F-Si-T-[Q+X]/I$$
$$= 64,720+41,548-29,750-51,294$$
$$*0.0100-3,385-[35,351$$
$$+34,196]/0.9900\}$$
$$= 2,370 \text{ millions USD}$$

Corporate IFRS-GAAP (B/S-I/S), ISBN-13: **978-1720792789**, ISBN-10: **172079278X**

Law-7794:

If both (**D**= Dividend Paid= 2,370 millions USD), (**I**= L/E= Leverage or Gearing Ratio= 99.00%), (**S**= Sales or Revenues= 51,294 millions USD), (**M**= Margin of Contribution= 41,548 millions USD), (**T**= Tax Paid= 3,385 millions USD), (**Q**= Quoted Longterm Liabilities= 35,351 millions USD), (**F**= Fixed Cost= 29,740 millions USD), (**i**= I/S= Interest Portion= 1.00%), and (**U**= Utilized or Starting Capital= 64,720 millions USD), then it's (**X** = Xpress or Current Debt planned), is:

$$X= I[U+M-F-Si-T-D]-Q$$

$$= 64,720+41,548-29,750-51,294$$
$$*0.0100-3,385-[35,351$$
$$+34,196]/0.9900\}$$

$$= 34,196 \text{ millions USD}$$

Corporate IFRS-GAAP (B/S-I/S), ISBN-13: **978-1720792789**, ISBN-10: **172079278X**

Law-7795:

If both (**c**= C/X= Current Ratio= 1.20 times), (**P**=
Procured Inventories= 6,497 millions USD), (**q**=
Quick or Acid Test Ratio= [C-P]/X= 1.01 times), (**D**=
Dividend Paid= 2,370 millions USD), (**F**= L/E=
Leverage or Gearing Ratio= 99.00%), (**S**= Sales or
Revenues= 51,294 millions USD), (**M**= Margin of
Contribution= 41,548 millions USD), (**T**= Tax Paid=
3,385 millions USD), (**F**= Fixed Cost= 29,740
millions USD), (**i**= I/S= Interest Portion= 1.00%), and
(**U**= Utilized or Starting Capital= 64,720 millions
USD), then it's (**Q** = Quoted Longterm Liabilities
planned), is:

$$Q = I[U+M-F-Si-T-D]-P/[c-q]$$
$$= 0.9900[64,720+41,548-29,750$$
$$-51,294*0.0100-3,385$$
$$-2,370]-6,497$$
$$/[1.2000-1.0100]$$
$$= 35,351 \text{ millions USD}$$

Corporate IFRS-GAAP (B/S-I/S), ISBN-13: **978-1720792789**, ISBN-10: **172079278X**

<u>Law-7796</u>:

If both (**c**= C/X= Current Ratio= <u>1.20</u> times), (**P**= Procured Inventories= <u>6,497</u> millions USD), (**q**= Quick or Acid Test Ratio= [C-P]/X= <u>1.01</u> times), (**D**= Dividend Paid= <u>2,370</u> millions USD), (**Q**= Quoted Longterm Liabilities= <u>35,351</u> millions USD), (**S**= Sales or Revenues= <u>51,294</u> millions USD), (**M**= Margin of Contribution= <u>41,548</u> millions USD), (**T**= Tax Paid= <u>3,385</u> millions USD), (**F**= Fixed Cost= <u>29,740</u> millions USD), (**i**= I/S= Interest Portion= <u>1.00%</u>), and (**U**= Utilized or Starting Capital= <u>64,720</u> millions USD), then it's (**I** = Leverage or Gearing Ratio planned), is:

$$I = \{Q+P/[c-q]\}/[U+M-F-Si-T-D]$$
$$= \{35,351+6,497/[1.2000-1.0100]\}$$
$$/[64,720+41,548-29,750$$
$$-51,294*0.0100-3,385$$
$$-2,370]$$
$$= \underline{99.00\%}$$

Corporate IFRS-GAAP (B/S-I/S), ISBN-13: **978-1720792789**, ISBN-10: **172079278X**

Law-7797:

If both (**c**= C/X= Current Ratio= 1.20 times), (**P**= Procured Inventories= 6,497 millions USD), (**q**= Quick or Acid Test Ratio= [C-P]/X= 1.01 times), (**D**= Dividend Paid= 2,370 millions USD), (**Q**= Quoted Longterm Liabilities= 35,351 millions USD), (**S**= Sales or Revenues= 51,294 millions USD), (**M**= Margin of Contribution= 41,548 millions USD), (**T**= Tax Paid= 3,385 millions USD), (**F**= Fixed Cost= 29,740 millions USD), (**i**= I/S= Interest Portion= 1.00%), and (**I**= L/E= Leverage or Gearing Ratio= 99.00%), then it's (**U**= Utilized or Starting Capital planned), is:

$$U = F + Si + T + D - M + \{Q + P/[c-q]\}/I$$
$$= 29,750 + 51,294 * 0.0100 + 3,385$$
$$+ 2,370 - 41,548 + \{35,351$$
$$+ 6,497/[1.2000 - 1.0100]\}$$
$$/0.9900$$
$$= 64,720 \text{ millions USD}$$

Corporate IFRS-GAAP (B/S-I/S), ISBN-13: **978-1720792789**, ISBN-10: **172079278X**

Law-7798:

If both (**c**= C/X= Current Ratio= 1.20 times), (**P**= Procured Inventories= 6,497 millions USD), (**q**= Quick or Acid Test Ratio= [C-P]/X= 1.01 times), (**D**= Dividend Paid= 2,370 millions USD), (**Q**= Quoted Longterm Liabilities= 35,351 millions USD), (**S**= Sales or Revenues= 51,294 millions USD), (**U**= Utilized or Starting Capital= 64,720 millions USD), (**T**= Tax Paid= 3,385 millions USD), (**F**= Fixed Cost= 29,740 millions USD), (**i**= I/S= Interest Portion= 1.00%), and (**l**= L/E= Leverage or Gearing Ratio= 99.00%), then it's (**M**= Utilized or Starting Capital planned), is :

$$M= F+Si+T+D-U+\{Q+P/[c-q]\}/l$$
$$= 29,750+51,294*0.0100+3,385$$
$$+2,370-64,720+\{35,351$$
$$+6,497/[1.2000-1.0100]\}$$
$$/0.9900$$
$$= \underline{41,548} \text{ millions USD}$$

Corporate IFRS-GAAP (B/S-I/S), ISBN-13: **978-1720792789**, ISBN-10: **172079278X**

<u>Law-7799</u>:

If both (**c**= C/X= Current Ratio= <u>1.20</u> times), (**P**= Procured Inventories= <u>6,497</u> millions USD), (**q**= Quick or Acid Test Ratio= [C-P]/X= <u>1.01</u> times), (**D**= Dividend Paid= <u>2,370</u> millions USD), (**Q**= Quoted Longterm Liabilities= <u>35,351</u> millions USD), (**S**= Sales or Revenues= <u>51,294</u> millions USD), (**M**= Margin of Contribution= <u>41,548</u> millions USD), (**T**= Tax Paid= <u>3,385</u> millions USD), (**U**= Utilized or Starting Capital= <u>64,720</u> millions USD), (**i**= I/S= Interest Portion= <u>1.00%</u>), and (**I**= L/E= Leverage or Gearing Ratio= <u>99.00%</u>), then it's (**F**= Fixed Cost planned), is :

$$F= U+M-Si-T-D-\{Q+P/[c-q]\}/I$$
$$= 64,720+41,548-51,294*0.0100$$
$$-3,385-2,370-\{35,351$$
$$+6,497/[1.2000-1.0100]\}$$
$$/0.9900$$
$$= \underline{29,750} \text{ millions USD}$$

Corporate IFRS-GAAP (B/S-I/S), ISBN-13: **978-1720792789**, ISBN-10: **172079278X**

<u>Law-7800</u>:

If both (**c**= C/X= Current Ratio= <u>1.20</u> times), (**P**= Procured Inventories= <u>6,497</u> millions USD), (**q**= Quick or Acid Test Ratio= [C-P]/X= <u>1.01</u> times), (**D**= Dividend Paid= <u>2,370</u> millions USD), (**Q**= Quoted Longterm Liabilities= <u>35,351</u> millions USD), (**F**= Fixed Cost= <u>29,740</u> millions USD), (**M**= Margin of Contribution= <u>41,548</u> millions USD), (**T**= Tax Paid= <u>3,385</u> millions USD), (**U**= Utilized or Starting Capital= <u>64,720</u> millions USD), (**i**= I/S= Interest Portion= <u>1.00%</u>), and (**l**= L/E= Leverage or Gearing Ratio= <u>99.00%</u>), then it's (**S**= Sales or Revenues planned), is :

$$S= (U+M-F-T-D-\{Q+P/[c-q]\}/l\,)/i$$
$$= (64{,}720+41{,}548-29{,}750-3{,}385$$
$$-2{,}370-\{35{,}351+6{,}497$$
$$/[1.2000-1.0100]\}/0.9900)$$
$$/0.0100$$
$$= \underline{51{,}294} \text{ millions USD}$$

Corporate IFRS-GAAP (B/S-I/S), ISBN-13: **978-1720792789**, ISBN-10: **172079278X**

Law-7801:

If both (**c**= C/X= Current Ratio= 1.20 times), (**P**= Procured Inventories= 6,497 millions USD), (**q**= Quick or Acid Test Ratio= [C-P]/X= 1.01 times), (**D**= Dividend Paid= 2,370 millions USD), (**Q**= Quoted Longterm Liabilities= 35,351 millions USD), (**F**= Fixed Cost= 29,740 millions USD), (**M**= Margin of Contribution= 41,548 millions USD), (**T**= Tax Paid= 3,385 millions USD), (**U**= Utilized or Starting Capital= 64,720 millions USD), (**$**= Sales or Revenues= 51,294 millions USD), and (**I**= L/E= Leverage or Gearing Ratio= 99.00%), then it's (**i**= Interest Portion planned), is:

$$i= (U+M-F-T-D-\{Q+P/[c-q]\}/I\,)/\$$$
$$= (64,720+41,548-29,750-3,385$$
$$-2,370-\{35,351+6,497$$
$$/[1.2000-1.0100]\}/0.9900)$$
$$/\,51,294$$
$$= 1.00\%$$

Corporate IFRS-GAAP (B/S-I/S), ISBN-13: **978-1720792789**, ISBN-10: **172079278X**

<u>Law-7802</u>:

If both (**c**= C/X= Current Ratio= <u>1.20</u> times), (**P**= Procured Inventories= <u>6,497</u> millions USD), (**q**= Quick or Acid Test Ratio= [C-P]/X= <u>1.01</u> times), (**D**= Dividend Paid= <u>2,370</u> millions USD), (**Q**= Quoted Longterm Liabilities= <u>35,351</u> millions USD), (**F**= Fixed Cost= <u>29,740</u> millions USD), (**M**= Margin of Contribution= <u>41,548</u> millions USD), (**i**= I/S= Interest Portion= <u>1.00%</u>), (**U**= Utilized or Starting Capital= <u>64,720</u> millions USD), (**S**= Sales or Revenues= <u>51,294</u> millions USD), and (**l**= L/E= Leverage or Gearing Ratio= <u>99.00%</u>), then it's (**T**= Tax Paid planned), is :

$$T= U+M-F-Si-D-\{Q+P/[c-q]\}/l$$
$$= 64,720+41,548-29,750-51,294$$
$$*0.0100-2,370-\{35,351$$
$$+6,497/[1.200-1.0100]\}$$
$$/0.9900$$
$$= \underline{3,386} \text{ millions USD}$$

Corporate IFRS-GAAP (B/S-I/S), ISBN-13: **978-1720792789**, ISBN-10: **172079278X**

Law-7803:

If both (**c**= C/X= Current Ratio= 1.20 times), (**P**= Procured Inventories= 6,497 millions USD), (**q**= Quick or Acid Test Ratio= [C-P]/X= 1.01 times), (**T**= Tax Paid= 3,385 millions USD), (**Q**= Quoted Longterm Liabilities= 35,351 millions USD), (**F**= Fixed Cost= 29,740 millions USD), (**M**= Margin of Contribution= 41,548 millions USD), (**i**= I/S= Interest Portion= 1.00%), (**U**= Utilized or Starting Capital= 64,720 millions USD), (**$**= Sales or Revenues= 51,294 millions USD), and (**I**= L/E= Leverage or Gearing Ratio= 99.00%), then it's (**D**= Dividend Paid planned), is:

$$D= U+M-F-\$i-T-\{Q+P/[c-q]\}/I$$
$$= 64,720+41,548-29,750-51,294$$
$$*0.0100-3,385-\{35,351$$
$$+6,497/[1.200-1.0100]\}$$
$$/0.9900$$

= 2,370 millions USD

Corporate IFRS-GAAP (B/S-I/S), ISBN-13: **978-1720792789**, ISBN-10: **172079278X**

Law-7804:

If both (**c**= C/X= Current Ratio= <u>1.20</u> times), (**D**= Dividend Paid= <u>2,370</u> millions USD), (**q**= [C-P]/X= Quick or Acid Test Ratio= <u>1.01</u> times), (**T**= Tax Paid= <u>3,385</u> millions USD), (**Q**= Quoted Longterm Liabilities= <u>35,351</u> millions USD), (**F**= Fixed Cost= <u>29,740</u> millions USD), (**M**= Margin of Contribution= <u>41,548</u> millions USD), (**i**= I/S= Interest Portion= <u>1.00%</u>), (**U**= Utilized or Starting Capital= <u>64,720</u> millions USD), (**$**= Sales or Revenues= <u>51,294</u> millions USD), and (**I**= L/E= Leverage or Gearing Ratio= <u>99.00%</u>), then it's (**P**= Procured Inventories planned), is:

$$P= [c-q]\{I[U+M-F-Si-T-D]-Q\}$$
$$= [1.200-1.0100]\{0.9900\{64,720$$
$$+41,548-29,750-51,294$$
$$*0.0100-3,385-2,370$$
$$-35,351\}$$
$$= \underline{6,497} \text{ millions USD}$$

Corporate IFRS-GAAP (B/S-I/S), ISBN-13: **978-1720792789**, ISBN-10: **172079278X**

Law-7805:

If both (**P**= Procured Inventories= 6,497 millions USD), (**D**= Dividend Paid= 2,370 millions USD), (**q**= [C-P]/X= Quick or Acid Test Ratio= 1.01 times), (**T**= Tax Paid= 3,385 millions USD), (**Q**= Quoted Longterm Liabilities= 35,351 millions USD), (**F**= Fixed Cost= 29,740 millions USD), (**M**= Margin of Contribution= 41,548 millions USD), (**i**= I/S= Interest Portion= 1.00%), (**U**= Utilized or Starting Capital= 64,720 millions USD), (**S**= Sales or Revenues= 51,294 millions USD), and (**f**= L/E= Leverage or Gearing Ratio= 99.00%), then it's (**c**= Current Ratio planned), is:

$$c= q+P/\{f\,[U+M-F-Si-T-D]-Q\}$$
$$= 1.0100+6,497/\{0.9900\{64,720$$
$$+41,548-29,750-51,294$$
$$*0.0100-3,385-2,370$$
$$-35,351\}$$
$$= 1.20 \text{ times}$$

Corporate IFRS-GAAP (B/S-I/S), ISBN-13: **978-1720792789**, ISBN-10: **172079278X**

<u>Law-7806</u>:

If both (**P**= Procured Inventories= <u>6,497</u> millions USD), (**D**= Dividend Paid= <u>2,370</u> millions USD), (**c**= C/X= Current Ratio= <u>1.20</u> times), (**T**= Tax Paid= <u>3,385</u> millions USD), (**Q**= Quoted Longterm Liabilities= <u>35,351</u> millions USD), (**F**= Fixed Cost= <u>29,740</u> millions USD), (**M**= Margin of Contribution= <u>41,548</u> millions USD), (**i**= I/S= Interest Portion= <u>1.00%</u>), (**U**= Utilized or Starting Capital= <u>64,720</u> millions USD), (**S**= Sales or Revenues= <u>51,294</u> millions USD), and (**I**= L/E= Leverage or Gearing Ratio= <u>99.00%</u>), then it's (**q**= Quick or Acid Test Ratio planned), is :

$$q= c\text{-}P/\{I\,[U\text{+}M\text{-}F\text{-}Si\text{-}T\text{-}D]\text{-}Q\}$$
$$= 1.2000\text{-}6,497/\{0.9900\{64,720$$
$$+41,548\text{-}29,750\text{-}51,294$$
$$*0.0100\text{-}3,385\text{-}2,370$$
$$-35,351\}$$
$$= \underline{1.01}\ \text{times}$$

Corporate IFRS-GAAP (B/S-I/S), ISBN-13: **978-1720792789**, ISBN-10: **172079278X**

Law-7807:

If both (**p**= 360 P/V= Procured Inventory Days= 240 Days), (**V**= Variable Cost= 9,746 millions USD), (**D**= Dividend Paid= 2,370 millions USD), (**c**= C/X= Current Ratio= 1.20 times), (**T**= Tax Paid= 3,385 millions USD), (**q**= [C-P]/X= Quick or Acid Test Ratio= 1.01 times), (**F**= Fixed Cost= 29,740 millions USD), (**M**= Margin of Contribution= 41,548 millions USD), (**i**= I/S= Interest Portion= 1.00%), (**U**= Utilized or Starting Capital= 64,720 millions USD), (**$**= Sales or Revenues= 51,294 millions USD), and (**⌐**= L/E= Leverage or Gearing Ratio= 99.00%), then it's (**Q**= Quoted Longterm Liabilities planned), is :

$$Q= I\,[U+M\text{-}F\text{-}Si\text{-}T\text{-}D]\text{-}Vp/\{360[c\text{-}q]\}$$
$$= 0.9900\{64,720+41,548-29,750$$
$$-51,294*0.0100-3,385$$
$$-2,370-9,746*240/\{360$$
$$[1.2000-1.0100]\}$$
$$= \underline{35,351}\text{ millions USD}$$

Corporate IFRS-GAAP (B/S-I/S), ISBN-13: **978-1720792789**, ISBN-10: **172079278X**

Law-7808:

If both (**p**= 360 P/V= Procured Inventory Days= 240
Days), (**V**= Variable Cost= 9,746 millions USD), (**D**=
Dividend Paid= 2,370 millions USD), (**c**= C/X=
Current Ratio= 1.20 times), (**T**= Tax Paid= 3,385
millions USD), (**q**= [C-P]/X= Quick or Acid Test
Ratio= 1.01 times), (**F**= Fixed Cost= 29,740 millions
USD), (**M**= Margin of Contribution= 41,548 millions
USD), (**i**= I/S= Interest Portion= 1.00%), (**U**=
Utilized or Starting Capital= 64,720 millions USD),
(**S**= Sales or Revenues= 51,294 millions USD),and
(**Q**= Quoted Longterm Liabilities= 35,351 millions
USD), then it's (**I** = Leverage or Gearing Ratio
planned), is:

$$I = (Q+Vp/\{360[c-q]\})/[U+M-F-Si-T-D]$$
$$= (35,351+9,746*240/\{360[1.2000$$
$$-1.0100]\})/\{64,720+41,548$$
$$-29,750-51,294*0.0100$$
$$-3,385-2,370]$$
$$= 99.00\%$$

Corporate IFRS-GAAP (B/S-I/S), ISBN-13: **978-1720792789**, ISBN-10: **172079278X**

Law-7809:

If both (**p**= 360 P/V= Procured Inventory Days= 240 Days), (**V**= Variable Cost= 9,746 millions USD), (**D**= Dividend Paid= 2,370 millions USD), (**c**= C/X= Current Ratio= 1.20 times), (**T**= Tax Paid= 3,385 millions USD), (**q**= [C-P]/X= Quick or Acid Test Ratio= 1.01 times), (**F**= Fixed Cost= 29,740 millions USD), (**M**= Margin of Contribution= 41,548 millions USD), (**i**= I/S= Interest Portion= 1.00%), (**I**= L/E= Leverage or Gearing Ratio= 99.00%), (**S**= Sales or Revenues= 51,294 millions USD), and (**Q**= Quoted Longterm Liabilities= 35,351 millions USD), then it's (**U**= Utilized or Starting Capital planned), is :

$$U= F+Si+T+D-M-(Q+Vp/\{360[c-q]\})/I$$
$$= 29,750+51,294*0.0100+3,385$$
$$+2,370-41,548+(35,351$$
$$+9,746*240/\{360[1.2000$$
$$-1.0100]\})/0.9900$$
$$= 64,720 \text{ millions USD}$$

Corporate IFRS-GAAP (B/S-I/S), ISBN-13: **978-1720792789**, ISBN-10: **172079278X**

Law-7810:

If both (**p**= 360 P/V= Procured Inventory Days= 240
Days), (**V**= Variable Cost= 9,746 millions USD), (**D**=
Dividend Paid= 2,370 millions USD), (**c**= C/X=
Current Ratio= 1.20 times), (**T**= Tax Paid= 3,385
millions USD), (**q**= [C-P]/X= Quick or Acid Test
Ratio= 1.01 times), (**F**= Fixed Cost= 29,740 millions
USD), (**U**= Utilized or Starting Capital= 64,720
millions USD), (**i**= I/S= Interest Portion= 1.00%), (**f**=
L/E= Leverage or Gearing Ratio= 99.00%), (**$**= Sales
or Revenues= 51,294 millions USD), and (**Q**=
Quoted Longterm Liabilities= 35,351 millions USD),
then it's (**M**= Margin of Contribution planned), is:

$$M= F+\$i+T+D-U-(Q+Vp/\{360[c-q]\})/f$$
$$= 29,750+51,294*0.0100+3,385$$
$$+2,370-64,720(35,351$$
$$+9,746*240/\{360[1.2000$$
$$-1.0100]\})/0.9900$$
$$= 41,548 \text{ millions USD}$$

Corporate IFRS-GAAP (B/S-I/S), ISBN-13: **978-1720792789**, ISBN-10: **172079278X**

Law-7811:

If both (**p**= 360 P/V= Procured Inventory Days= 240 Days), (**V**= Variable Cost= 9,746 millions USD), (**D**= Dividend Paid= 2,370 millions USD), (**c**= C/X= Current Ratio= 1.20 times), (**T**= Tax Paid= 3,385 millions USD), (**q**= [C-P]/X= Quick or Acid Test Ratio= 1.01 times), (**M**= Margin of Contribution= 41,548 millions USD), (**U**= Utilized or Starting Capital= 64,720 millions USD), (**i**= I/S= Interest Portion= 1.00%),(**Q**= Quoted Longterm Liabilities= 35,351 millions USD), (**S**= Sales or Revenues= 51,294 millions USD), and (**I**= L/E= Leverage or Gearing Ratio= 99.00%), then it's (**F**= Fixed Cost planned), is:

$$F= U+M-Si-T-D-(Q+Vp/\{360[c-q]\})/I$$
$$= 64,720+41,548-51,294*0.0100$$
$$-3,385-2,370-(35,351$$
$$+9,746*240/\{360[1.2000$$
$$-1.0100]\})/0.9900$$
$$= 29,750 \text{ millions USD}$$

Corporate IFRS-GAAP (B/S-I/S), ISBN-13: **978-1720792789**, ISBN-10: **172079278X**

Law-7812:

If both (**p**= 360 P/V= Procured Inventory Days= 240 Days), (**V**= Variable Cost= 9,746 millions USD), (**D**= Dividend Paid= 2,370 millions USD), (**c**= C/X= Current Ratio= 1.20 times), (**T**= Tax Paid= 3,385 millions USD), (**q**= [C-P]/X= Quick or Acid Test Ratio= 1.01 times), (**M**= Margin of Contribution= 41,548 millions USD), (**U**= Utilized or Starting Capital= 64,720 millions USD), (**i**= I/S= Interest Portion= 1.00%),(**Q**= Quoted Longterm Liabilities= 35,351 millions USD), (**F**= Fixed Cost= 29,740 millions USD), and (**I**= L/E= Leverage or Gearing Ratio= 99.00%), then it's (**S**= Sales or Revenues planned), is:

$$S= [U+M-F-T-D-(Q+Vp/\{360[c-q]\})/I]/i$$
$$= [64,720+41,548-29,750-3,385$$
$$-2,370-(35,351+9,746$$
$$*240/\{360[1.2000$$
$$-1.0100]\})/0.9900]$$
$$/0.0100$$
$$= 51,294 \text{ millions USD}$$

Corporate IFRS-GAAP (B/S-I/S), ISBN-13: **978-1720792789**, ISBN-10: **172079278X**

<u>Law-7813</u>:

If both (**p**= 360 P/V= Procured Inventory Days= <u>240</u> Days), (**V**= Variable Cost= <u>9,746</u> millions USD), (**D**= Dividend Paid= <u>2,370</u> millions USD), (**c**= C/X= Current Ratio= <u>1.20</u> times), (**T**= Tax Paid= <u>3,385</u> millions USD), (**q**= [C-P]/X= Quick or Acid Test Ratio= <u>1.01</u> times), (**M**= Margin of Contribution= <u>41,548</u> millions USD), (**U**= Utilized or Starting Capital= <u>64,720</u> millions USD), (**$**= Sales or Revenues= <u>51,294</u> millions USD), (**Q**= Quoted Longterm Liabilities= <u>35,351</u> millions USD), (**F**= Fixed Cost= <u>29,740</u> millions USD), and (**/**= L/E= Leverage or Gearing Ratio= <u>99.00%</u>), then it's (**i**= Interest Portion planned), is:

$$i= [U+M-F-T-D-(Q+Vp/\{360[c-q]\})//]/\$$$

$$= [64,720+41,548-29,750-3,385$$
$$-2,370-(35,351+9,746$$
$$*240/\{360[1.2000$$
$$-1.0100]\})/0.9900]$$
$$/51,294$$

$$= \underline{1.00\%}$$

Corporate IFRS-GAAP (B/S-I/S), ISBN-13: **978-1720792789**, ISBN-10: **172079278X**

<u>Law-7814</u>:

If both (**p**= 360 P/V= Procured Inventory Days= <u>240</u> Days), (**V**= Variable Cost= <u>9,746</u> millions USD), (**D**= Dividend Paid= <u>2,370</u> millions USD), (**c**= C/X= Current Ratio= <u>1.20</u> times), (**i**= I/S= Interest Portion= <u>1.00%</u>), (**q**= [C-P]/X= Quick or Acid Test Ratio= <u>1.01</u> times), (**M**= Margin of Contribution= <u>41,548</u> millions USD), (**U**= Utilized or Starting Capital= <u>64,720</u> millions USD), (**S**= Sales or Revenues= <u>51,294</u> millions USD), (**Q**= Quoted Longterm Liabilities= <u>35,351</u> millions USD), (**F**= Fixed Cost= <u>29,740</u> millions USD), and (**F**= L/E= Leverage or Gearing Ratio= <u>99.00%</u>), then it's (**T**= Tax Paid planned), is:

$$T= U+M-F-Si-D-(Q+Vp/\{360[c-q]\})/I$$
$$= 64,720+41,548-29,750-51,294$$
$$*0.0100-2,370-(35,351$$
$$+9,746*240/\{360[1.2000$$
$$-1.0100]\})/0.9900$$
$$= \underline{3,385} \text{ millions USD}$$

Corporate IFRS-GAAP (B/S-I/S), ISBN-13: **978-1720792789**, ISBN-10: **172079278X**

<u>Law-7815</u>:

If both (**p**= 360 P/V= Procured Inventory Days= <u>240</u> Days), (**V**= Variable Cost= <u>9,746</u> millions USD), (**T**= Tax Paid= <u>3,385</u> millions USD), (**c**= C/X= Current Ratio= <u>1.20</u> times), (**i**= I/S= Interest Portion= <u>1.00%</u>), (**q**= [C-P]/X= Quick or Acid Test Ratio= <u>1.01</u> times), (**M**= Margin of Contribution= <u>41,548</u> millions USD), (**U**= Utilized or Starting Capital= <u>64,720</u> millions USD), (**$**= Sales or Revenues= <u>51,294</u> millions USD), (**Q**= Quoted Longterm Liabilities= <u>35,351</u> millions USD), (**F**= Fixed Cost= <u>29,740</u> millions USD), and (**I**= L/E= Leverage or Gearing Ratio= <u>99.00%</u>), then it's (**D**= Dividend Paid planned), is:

$$D= U+M-F-\$i-T-(Q+Vp/\{360[c-q]\})/I$$

$$= 64,720+41,548-29,750-51,294$$
$$*0.0100-3,385-(35,351$$
$$+9,746*240/\{360[1.2000$$
$$-1.0100]\})/0.9900$$

$$= \underline{2,370} \text{ millions USD}$$

Corporate IFRS-GAAP (B/S-I/S), ISBN-13: **978-1720792789**, ISBN-10: **172079278X**

Law-7816:

If both (**p**= 360 P/V= Procured Inventory Days= <u>240</u> Days), (**D**= Dividend Paid= <u>2,370</u> millions USD), (**T**= Tax Paid= <u>3,385</u> millions USD), (**c**= C/X= Current Ratio= <u>1.20</u> times), (**i**= I/S= Interest Portion= <u>1.00%</u>), (**q**= [C-P]/X= Quick or Acid Test Ratio= <u>1.01</u> times), (**M**= Margin of Contribution= <u>41,548</u> millions USD), (**U**= Utilized or Starting Capital= <u>64,720</u> millions USD), (**S**= Sales or Revenues= <u>51,294</u> millions USD), (**Q**= Quoted Longterm Liabilities= <u>35,351</u> millions USD), (**F**= Fixed Cost= <u>29,740</u> millions USD), and (**I**= L/E= Leverage or Gearing Ratio= <u>99.00%</u>), then it's (**V**= Variable Cost planned), is:

$$V= 360[c\text{-}q](I \{U+M\text{-}F\text{-}Si\text{-}T\text{-}D\}\text{-}Q)/p$$
$$= 360[1.2000\text{-}1.0100](0.9900$$
$$\{64,720+41,548\text{-}29,750$$
$$\text{-}51,294*0.0100\text{-}3,385$$
$$\text{-}2,370\}\text{-}35,351)/240$$
$$= \underline{9,746} \text{ millions USD}$$

Corporate IFRS-GAAP (B/S-I/S), ISBN-13: **978-1720792789**, ISBN-10: **172079278X**

Law-7817:

If both (**V**= Variable Cost= 9,746 millions USD), (**D**= Dividend Paid= 2,370 millions USD), (**T**= Tax Paid= 3,385 millions USD), (**c**= C/X= Current Ratio= 1.20 times), (**i**= I/S= Interest Portion= 1.00%), (**q**= Quick or Acid Test Ratio= [C-P]/X= 1.01 times), (**M**= Margin of Contribution= 41,548 millions USD), (**U**= Utilized or Starting Capital= 64,720 millions USD), (**S**= Sales or Revenues= 51,294 millions USD), (**Q**= Quoted Longterm Liabilities= 35,351 millions USD), (**F**= Fixed Cost= 29,740 millions USD), and (**I**= L/E= Leverage or Gearing Ratio= 99.00%), then it's (**p**= Procured Inventory Days planned), is:

$$p= 360[\text{c-q}](\textbf{\textit{I}} \{\textbf{U+M-F-Si-T-D}\}\text{-}\textbf{Q})/\textbf{V}$$
$$= 360[1.2000\text{-}1.0100](0.9900$$
$$\{64,720+41,548\text{-}29,750$$
$$-51,294*0.0100\text{-}3,385$$
$$-2,370\}\text{-}35,351)/9,746$$

$$= \underline{240} \text{ days}$$

Corporate IFRS-GAAP (B/S-I/S), ISBN-13: **978-1720792789**, ISBN-10: **172079278X**

<u>Law-7818</u>:

If both (**V**= Variable Cost= <u>9,746</u> millions USD), (**D**= Dividend Paid= <u>2,370</u> millions USD), (**T**= Tax Paid= <u>3,385</u> millions USD), (**p**= 360 P/V= Procured Inventory Days= <u>240</u> Days), (**i**= I/S= Interest Portion= <u>1.00%</u>), (**q**= [C-P]/X= Quick or Acid Test Ratio= <u>1.01</u> times), (**M**= Margin of Contribution= <u>41,548</u> millions USD), (**U**= Utilized or Starting Capital= <u>64,720</u> millions USD), (**$**= Sales or Revenues= <u>51,294</u> millions USD), (**Q**= Quoted Longterm Liabilities= <u>35,351</u> millions USD), (**F**= Fixed Cost= <u>29,740</u> millions USD), and (**I**= L/E= Leverage or Gearing Ratio= <u>99.00%</u>), then it's (**c**= Current Ratio planned), is:

$$c= q+Vp/[360(I\{U+M-F-Si-T-D\}-Q)]$$
$$= 1.0100+9,746*240/[360(0.9900$$
$$\{64,720+41,548-29,750$$
$$-51,294*0.0100-3,385$$
$$-2,370\}-35,351)]$$
$$= \underline{1.20} \text{ times}$$

Law-7819:

If both (**V**= Variable Cost= 9,746 millions USD), (**D**= Dividend Paid= 2,370 millions USD), (**T**= Tax Paid= 3,385 millions USD), (**p**= 360 P/V= Procured Inventory Days= 240 Days), (**i**= I/S= Interest Portion= 1.00%), (**c**= C/X= Current Ratio= 1.20 times), (**M**= Margin of Contribution= 41,548 millions USD), (**U**= Utilized or Starting Capital= 64,720 millions USD), (**S**= Sales or Revenues= 51,294 millions USD), (**Q**= Quoted Longterm Liabilities= 35,351 millions USD), (**F**= Fixed Cost= 29,740 millions USD), and (**I**= L/E= Leverage or Gearing Ratio= 99.00%), then it's (**q**= Quick or Acid Test Ratio planned), is:

$$q= c\text{-}Vp/[360(I\,\{U+M\text{-}F\text{-}Si\text{-}T\text{-}D\}\text{-}Q)]$$
$$= 1.2000\text{-}9,746*240/[360(0.9900$$
$$\{64,720+41,548\text{-}29,750$$
$$-51,294*0.0100\text{-}3,385$$
$$-2,370\}\text{-}35,351)]$$
$$= 1.01 \text{ times}$$

Corporate IFRS-GAAP (B/S-I/S), ISBN-13: **978-1720792789**, ISBN-10: **172079278X**

Law-7820:

If both (v= V/S= Variable Portion= 19.00%), (D= Dividend Paid= 2,370 millions USD), (T= Tax Paid= 3,385 millions USD), (q= [C-P]/X= Quick or Acid Test Ratio= 1.01 times), (p= 360 P/V= Procured Inventory Days= 240 Days), (i= I/S= Interest Portion= 1.00%), (c= C/X= Current Ratio= 1.20 times), (M= Margin of Contribution= 41,548 millions USD), (U= Utilized or Starting Capital= 64,720 millions USD), (S= Sales or Revenues= 51,294 millions USD), (F= Fixed Cost= 29,740 millions USD), and (I= L/E= Leverage or Gearing Ratio= 99.00%), then it's (Q= Quoted Longterm Liabilities planned), is:

$$Q= I\{U+M-F-Si-T-D\}-Svp/\{360[c-q]\}$$
$$= 0.9900\{64,720+41,548-29,750$$
$$-51,294*0.0100-3,385$$
$$-2,370\}-51,294*0.1900$$
$$*240/\{360[1.2000$$
$$-1.0100]\}$$
$$= 35,351 \text{ millions USD}$$

Corporate IFRS-GAAP (B/S-I/S), ISBN-13: **978-1720792789**, ISBN-10: **172079278X**

<u>Law-7821</u>:

If both (**v**= V/S= Variable Portion= <u>19.00%</u>), (**D**= Dividend Paid= <u>2,370</u> millions USD), (**T**= Tax Paid= <u>3,385</u> millions USD), (**q**= [C-P]/X= Quick or Acid Test Ratio= <u>1.01</u> times), (**p**= 360 P/V= Procured Inventory Days= <u>240</u> Days), (**i**= I/S= Interest Portion= <u>1.00%</u>), (**c**= C/X= Current Ratio= <u>1.20</u> times), (**M**= Margin of Contribution= <u>41,548</u> millions USD), (**U**= Utilized or Starting Capital= <u>64,720</u> millions USD), (**S**= Sales or Revenues= <u>51,294</u> millions USD), (**F**= Fixed Cost= <u>29,740</u> millions USD), and (**Q**= Quoted Longterm Liabilities= <u>35,351</u> millions USD),then it's (**I** = Leverage or Gearing Ratio planned), is:

$$I = (Q + Svp/\{360[c-q]\})/\{U+M-F-Si-T-D\}$$
$$= (35,351+51,294*0.1900*240$$
$$/\{360[1.2000-1.0100]\})$$
$$/\{64,720+41,548-29,750$$
$$-51,294*0.0100-3,385$$
$$-2,370\}$$
$$= \underline{99.00\%}$$

Corporate IFRS-GAAP (B/S-I/S), ISBN-13: **978-1720792789**, ISBN-10: **172079278X**

Law-7822:

If both (v= V/S= Variable Portion= 19.00%), (D= Dividend Paid= 2,370 millions USD), (T= Tax Paid= 3,385 millions USD), (q= [C-P]/X= Quick or Acid Test Ratio= 1.01 times), (p= 360 P/V= Procured Inventory Days= 240 Days), (i= I/S= Interest Portion= 1.00%), (c= C/X= Current Ratio= 1.20 times), (M= Margin of Contribution= 41,548 millions USD), (l= L/E= Leverage or Gearing Ratio= 99.00%), (S= Sales or Revenues= 51,294 millions USD), (F= Fixed Cost= 29,740 millions USD), and (Q= Quoted Longterm Liabilities= 35,351 millions USD),then it's (U= Utilized or Starting Capital planned), is:

$$U= F+Si+T+D-M+(Q+Svp/\{360[c-q]\})/l$$
$$= 29,750+51,294*0.0100+3,385$$
$$+2,370-41,548+(35,351$$
$$+51,294*0.1900*240$$
$$/\{360[1.2000-1.0100]\})$$
$$/0.9900$$
$$= 64,720 \text{ millions USD}$$

Corporate IFRS-GAAP (B/S-I/S), ISBN-13: **978-1720792789**, ISBN-10: **172079278X**

Law-7823:

If both (**v**= V/S= Variable Portion= 19.00%), (**D**= Dividend Paid= 2,370 millions USD), (**T**= Tax Paid= 3,385 millions USD), (**q**= [C-P]/X= Quick or Acid Test Ratio= 1.01 times), (**p**= 360 P/V= Procured Inventory Days= 240 Days), (**i**= I/S= Interest Portion= 1.00%), (**c**= C/X= Current Ratio= 1.20 times), (**U**= Utilized or Starting Capital= 64,720 millions USD), (**F**= Fixed Cost= 29,740 millions USD), (***I***= L/E= Leverage or Gearing Ratio= 99.00%), (**S**= Sales or Revenues= 51,294 millions USD), and (**Q**= Quoted Longterm Liabilities= 35,351 millions USD),then it's (**M**= Margin of Contribution planned), is:

$$M= F+Si+T+D-U+(Q+Svp/\{360[c-q]\})/I$$
$$= 29,750+51,294*0.0100+3,385$$
$$+2,370-64,720+(35,351$$
$$+51,294*0.1900*240$$
$$/\{360[1.2000-1.0100]\})$$
$$/0.9900$$
$$= 41,548 \text{ millions USD}$$

Corporate IFRS-GAAP (B/S-I/S), ISBN-13: **978-1720792789**, ISBN-10: **172079278X**

<u>Law-7824</u>:

If both (**v**= V/S= Variable Portion= <u>19.00%</u>), (**D**= Dividend Paid= <u>2,370</u> millions USD), (**T**= Tax Paid= <u>3,385</u> millions USD), (**q**= [C-P]/X= Quick or Acid Test Ratio= <u>1.01</u> times), (**p**= 360 P/V= Procured Inventory Days= <u>240</u> Days), (**i**= I/S= Interest Portion= <u>1.00%</u>), (**c**= C/X= Current Ratio= <u>1.20</u> times), (**U**= Utilized or Starting Capital= <u>64,720</u> millions USD), (**M**= Margin of Contribution= <u>41,548</u> millions USD), (**F**= L/E= Leverage or Gearing Ratio= <u>99.00%</u>), (**S**= Sales or Revenues= <u>51,294</u> millions USD), and (**Q**= Quoted Longterm Liabilities= <u>35,351</u> millions USD),then it's (**F**= Fixed Cost planned), is:

$$F= U+M\text{-}Si\text{-}T\text{-}D\text{-}(Q+Sup/\{360[c\text{-}q]\})/I$$

$$= 64{,}720+41{,}548\text{-}51{,}294*0.0100$$
$$-3{,}385\text{-}2{,}370\text{-}(35{,}351$$
$$+51{,}294*0.1900*240$$
$$/\{360[1.2000\text{-}1.0100]\})$$
$$/0.9900$$

$$= \underline{29{,}750} \text{ millions USD}$$

Corporate IFRS-GAAP (B/S-I/S), ISBN-13: **978-1720792789**, ISBN-10: **172079278X**

Law-7825:

If both (**v**= V/S= Variable Portion= 19.00%), (**D**= Dividend Paid= 2,370 millions USD), (**T**= Tax Paid= 3,385 millions USD), (**q**= [C-P]/X= Quick or Acid Test Ratio= 1.01 times), (**p**= 360 P/V= Procured Inventory Days= 240 Days), (**i**= I/S= Interest Portion= 1.00%), (**c**= C/X= Current Ratio= 1.20 times), (**U**= Utilized or Starting Capital= 64,720 millions USD), (**M**= Margin of Contribution= 41,548 millions USD), (**I**= L/E= Leverage or Gearing Ratio= 99.00%), (**F**= Fixed Cost= 29,740 millions USD), and (**Q**= Quoted Longterm Liabilities= 35,351 millions USD),then it's (**S**= Sales or Revenues planned), is:

$$S= [U+M-F-T-D-(Q+Svp/\{360[c-q]\})/I]/i$$
$$= [64,720+41,548-29,750-3,385$$
$$-2,370-(35,351+51,294$$
$$*0.1900*240/\{360[1.2000$$
$$-1.0100]\})/0.9900]/0.0100$$
$$= 51,294 \text{ millions USD}$$

Corporate IFRS-GAAP (B/S-I/S), ISBN-13: **978-1720792789**, ISBN-10: **172079278X**

Law-7826:

If both (**ʊ**= V/S= Variable Portion= 19.00%), (**D**= Dividend Paid= 2,370 millions USD), (**T**= Tax Paid= 3,385 millions USD), (**q**= [C-P]/X= Quick or Acid Test Ratio= 1.01 times), (**p**= 360 P/V= Procured Inventory Days= 240 Days), (**c**= C/X= Current Ratio= 1.20 times), (**U**= Utilized or Starting Capital= 64,720 millions USD), (**M**= Margin of Contribution= 41,548 millions USD), (**S**= Sales or Revenues= 51,294 millions USD), (**I**= L/E= Leverage or Gearing Ratio= 99.00%), (**F**= Fixed Cost= 29,740 millions USD), and (**Q**= Quoted Longterm Liabilities= 35,351 millions USD),then it's (**i**= Interest Portion planned), is:

$$i= [U+M-F-T-D-(Q+Sʊp/\{360[c-q]\})/I]/S$$
$$= [64,720+41,548-29,750-3,385$$
$$-2,370-(35,351+51,294$$
$$*0.1900*240/\{360[1.2000$$
$$-1.0100]\})/0.9900]/51,294$$
$$= 1.00\%$$

Corporate IFRS-GAAP (B/S-I/S), ISBN-13: **978-1720792789**, ISBN-10: **172079278X**

Law-7827:

If both (v= V/S= Variable Portion= 19.00%), (D= Dividend Paid= 2,370 millions USD), (i= I/S= Interest Portion= 1.00%), (q= [C-P]/X= Quick or Acid Test Ratio= 1.01 times), (p= 360 P/V= Procured Inventory Days= 240 Days), (c= C/X= Current Ratio= 1.20 times), (U= Utilized or Starting Capital= 64,720 millions USD), (M= Margin of Contribution= 41,548 millions USD), (S= Sales or Revenues= 51,294 millions USD), (I= L/E= Leverage or Gearing Ratio= 99.00%), (F= Fixed Cost= 29,740 millions USD), and (Q= Quoted Longterm Liabilities= 35,351 millions USD),then it's (T= Tax Paid planned), is:

T= U+M-F-Si-D-(Q+Svp/{360[c-q]})/I

$$= 64,720+41,548-29,750-51,294$$
$$*0.0100-2,370-(35,351$$
$$+51,294*0.1900*240$$
$$/\{360[1.2000-1.0100]\})$$
$$/0.9900$$

$$= 3,385 \text{ millions USD}$$

Corporate IFRS-GAAP (B/S-I/S), ISBN-13: **978-1720792789**, ISBN-10: **172079278X**

<u>Law-7828</u>:

If both (**ʋ**= V/S= Variable Portion= <u>19.00%</u>), (**T**= Tax Paid= <u>3,385</u> millions USD), (**i**= I/S= Interest Portion= <u>1.00%</u>), (**q**= [C-P]/X= Quick or Acid Test Ratio= <u>1.01</u> times), (**p**= 360 P/V= Procured Inventory Days= <u>240</u> Days), (**c**= C/X= Current Ratio= <u>1.20</u> times), (**U**= Utilized or Starting Capital= <u>64,720</u> millions USD), (**M**= Margin of Contribution= <u>41,548</u> millions USD), (**$**= Sales or Revenues= <u>51,294</u> millions USD), (**⌿**= L/E= Leverage or Gearing Ratio= <u>99.00%</u>), (**F**= Fixed Cost= <u>29,740</u> millions USD), and (**Q**= Quoted Longterm Liabilities= <u>35,351</u> millions USD),then it's (**D**= Dividend Paid planned), is:

$$D= U+M-F-\$i-T-(Q+\$ʋp/\{360[c-q]\})/⌿$$

$$= 64,720+41,548-29,750-51,294$$
$$*0.0100-3,385-(35,351$$
$$+51,294*0.1900*240$$
$$/\{360[1.2000-1.0100]\})$$
$$/0.9900$$

$$= \underline{2,370} \text{ millions USD}$$

Corporate IFRS-GAAP (B/S-I/S), ISBN-13: **978-1720792789**, ISBN-10: **172079278X**

<u>Law-7829</u>:

If both (**D**= Dividend Paid= <u>2,370</u> millions USD), (**T**= Tax Paid= <u>3,385</u> millions USD), (**i**= I/S= Interest Portion= <u>1.00%</u>), (**q**= [C-P]/X= Quick or Acid Test Ratio= <u>1.01</u> times), (**p**= 360 P/V= Procured Inventory Days= <u>240</u> Days), (**c**= C/X= Current Ratio= <u>1.20</u> times), (**U**= Utilized or Starting Capital= <u>64,720</u> millions USD), (**M**= Margin of Contribution= <u>41,548</u> millions USD), (**S**= Sales or Revenues= <u>51,294</u> millions USD), (**I**= L/E= Leverage or Gearing Ratio= <u>99.00%</u>), (**F**= Fixed Cost= <u>29,740</u> millions USD), and (**Q**= Quoted Longterm Liabilities= <u>35,351</u> millions USD),then it's (**v**= Variable Portion planned), is:

$$v= 360[c\text{-}q](I\{U+M\text{-}F\text{-}Si\text{-}T\text{-}D\}\text{-}Q)/[Sp]$$
$$= 360[1.2000\text{-}1.0100](0.9900\{64,720$$
$$+41,548\text{-}29,750\text{-}51,294$$
$$*0.0100\text{-}3,385\text{-}2,370\}$$
$$-35,351)/[51,294*240]$$
$$= \underline{19.00\%}$$

Corporate IFRS-GAAP (B/S-I/S), ISBN-13: **978-1720792789**, ISBN-10: **172079278X**

<u>Law-7830</u>:

If both (**D**= Dividend Paid= <u>2,370</u> millions USD),
(**T**= Tax Paid= <u>3,385</u> millions USD), (**i**= I/S= Interest
Portion= <u>1.00%</u>), (**q**= [C-P]/X= Quick or Acid Test
Ratio= <u>1.01</u> times), (**v**= V/S= Variable Portion=
<u>19.00%</u>), (**c**= C/X= Current Ratio= <u>1.20</u> times), (**U**=
Utilized or Starting Capital= <u>64,720</u> millions USD),
(**M**= Margin of Contribution= <u>41,548</u> millions USD),
(**S**= Sales or Revenues= <u>51,294</u> millions USD), (**l**=
L/E= Leverage or Gearing Ratio= <u>99.00%</u>), (**F**=
Fixed Cost= <u>29,740</u> millions USD), and (**Q**= Quoted
Longterm Liabilities= <u>35,351</u> millions USD),then it's
(**p**= Procured Inventory Days planned), is:

$$p= 360[c\text{-}q](l\{U+M\text{-}F\text{-}Si\text{-}T\text{-}D\}\text{-}Q)/[Sv]$$
$$= 360[1.2000\text{-}1.0100](0.9900\{64,720$$
$$+41,548\text{-}29,750\text{-}51,294$$
$$*0.0100\text{-}3,385\text{-}2,370\}$$
$$-35,351)/[51,294*0.1900]$$
$$= \underline{240} \text{ days}$$

Corporate IFRS-GAAP (B/S-I/S), ISBN-13: **978-1720792789**, ISBN-10: **172079278X**

Law-7831:

If both (**D**= Dividend Paid= 2,370 millions USD), (**T**= Tax Paid= 3,385 millions USD), (**i**= I/S= Interest Portion= 1.00%), (**q**= [C-P]/X= Quick or Acid Test Ratio= 1.01 times), (**v**= V/S= Variable Portion= 19.00%), (**p**= 360 P/V= Procured Inventory Days= 240 Days), (**U**= Utilized or Starting Capital= 64,720 millions USD), (**M**= Margin of Contribution= 41,548 millions USD), (**S**= Sales or Revenues= 51,294 millions USD), (**I**= L/E= Leverage or Gearing Ratio= 99.00%), (**F**= Fixed Cost= 29,740 millions USD), and (**Q**= Quoted Longterm Liabilities= 35,351 millions USD),then it's (**c**= Current Ratio planned), is:

$$c= q+Svp/[360(I \{U+M-F-Si-T-D\}-Q)]$$
$$= 1.0100+51,294*0.1900*240$$
$$/[360(0.9900\{64,720$$
$$+41,548-29,750-51,294$$
$$*0.0100-3,385-2,370\}$$
$$-35,351)]$$
$$= 1.20 \text{ times}$$

Corporate IFRS-GAAP (B/S-I/S), ISBN-13: **978-1720792789**, ISBN-10: **172079278X**

<u>Law-7832</u>:

If both (**D**= Dividend Paid= <u>2,370</u> millions USD), (**T**= Tax Paid= <u>3,385</u> millions USD), (**i**= I/S= Interest Portion= <u>1.00%</u>), (**c**= C/X= Current Ratio= <u>1.20</u> times), (**v**= V/S= Variable Portion= <u>19.00%</u>), (**p**= 360 P/V= Procured Inventory Days= <u>240</u> Days), (**U**= Utilized or Starting Capital= <u>64,720</u> millions USD), (**M**= Margin of Contribution= <u>41,548</u> millions USD), (**$**= Sales or Revenues= <u>51,294</u> millions USD), (**l**= L/E= Leverage or Gearing Ratio= <u>99.00%</u>), (**F**= Fixed Cost= <u>29,740</u> millions USD), and (**Q**= Quoted Longterm Liabilities= <u>35,351</u> millions USD),then it's (**q**= Quick or Acid Test Ratio planned), is:

$$q= c\text{-}\$vp/[360(\textbf{\textit{l}}\,\{U+M\text{-}F\text{-}\$i\text{-}T\text{-}D\}\text{-}Q)]$$
$$= 1.2000\text{-}51,294*0.1900*240$$
$$/[360(0.9900\{64,720$$
$$+41,548\text{-}29,750\text{-}51,294$$
$$*0.0100\text{-}3,385\text{-}2,370\}$$
$$-35,351)]$$
$$= \underline{1.01}\ \text{times}$$

Corporate IFRS-GAAP (B/S-I/S), ISBN-13: **978-1720792789**, ISBN-10: **172079278X**

Law-7833:

If both (**D**= Dividend Paid= 2,370 millions USD), (**T**= Tax Paid= 3,385 millions USD), (**i**= I/S= Interest Portion= 1.00%), (**S'**= Sales of Past Year= 48,851 millions USD), (**s**= [S/S']-1= Sales Growth= 5.00%), (**c**= C/X= Current Ratio= 1.20 times), (**v**= V/S= Variable Portion= 19.00%), (**p**= 360 P/V= Procured Inventory Days= 240 Days), (**U**= Utilized or Starting Capital= 64,720 millions USD), (**M**= Margin of Contribution= 41,548 millions USD), (**S**= Sales or Revenues= 51,294 millions USD), (**l**= L/E= Leverage or Gearing Ratio= 99.00%), (**F**= Fixed Cost= 29,740 millions USD), and (**q**= [C-P]/X= Quick or Acid Test Ratio= 1.01 times), then it's (**Q**= Quoted Longterm Liabilities planned), is:

$$Q = l\{U+M-F-Si-T-D\}-S'vp[1+s]/\{360[c-q]\}$$
$$= 0.9900\{64,720+41,548-29,750$$
$$-51,294*0.0100-3,385$$
$$-2,370\}-48,851*0.1900$$
$$*240[1+0.0500]/\{360$$
$$[1.2000-1.0100]\}$$
$$= 35,351 \text{ millions USD}$$

Corporate IFRS-GAAP (B/S-I/S), ISBN-13: **978-1720792789**, ISBN-10: **172079278X**

Law-7834:

If both (**D**= Dividend Paid= 2,370 millions USD),
(**T**= Tax Paid= 3,385 millions USD), (**i**= I/S= Interest
Portion= 1.00%), (**S'**= Sales of Past Year= 48,851
millions USD), (**s**= [S/S']-1= Sales Growth= 5.00%),
(**c**= C/X= Current Ratio= 1.20 times), (**v**= V/S=
Variable Portion= 19.00%), (**Q**= Quoted Longterm
Liabilities= 35,351 millions USD), (**p**= 360 P/V=
Procured Inventory Days= 240 Days), (**M**= Margin of
Contribution= 41,548 millions USD), (**S**= Sales or
Revenues= 51,294 millions USD), (**l**= L/E= Leverage
or Gearing Ratio= 99.00%), (**F**= Fixed Cost= 29,740
millions USD), and (**q**= [C-P]/X= Quick or Acid Test
Ratio= 1.01 times), then it's (**l** = Leverage or Gearing
Ratio planned), is:

$$l = (Q+S'vp[1+s]/\{360[c-q]\})$$
$$/\{U+M-F-Si-T-D\}$$
$$= (35,351+48,851*0.1900*240$$
$$[1+0.0500]/\{360[1.2000$$
$$-1.0100]\})/\{64,720+41,548$$
$$-29,750-51,294*0.0100$$
$$-3,385-2,370\}$$
$$= 99.00\%$$

Law-7835:

If both (**D**= Dividend Paid= 2,370 millions USD),
(**T**= Tax Paid= 3,385 millions USD), (**i**= I/S= Interest
Portion= 1.00%), (**S'**= Sales of Past Year= 48,851
millions USD), (**s**= [S/S']-1= Sales Growth= 5.00%),
(**c**= C/X= Current Ratio= 1.20 times), (**v**= V/S=
Variable Portion= 19.00%), (**p**= 360P/V= Procured
Inventory Days= 240 Days), (**Q**= Quoted Longterm
Liabilities= 35,351 millions USD), (**M**= Margin of
Contribution= 41,548 millions USD), (**S**= Sales or
Revenues= 51,294 millions USD), (**l**= L/E= Leverage
or Gearing Ratio= 99.00%), (**F**= Fixed Cost= 29,740
millions USD), and (**q**= [C-P]/X= Quick or Acid Test
Ratio= 1.01 times), then it's (**U**= Utilized or Starting
Capital planned), is :

$$U = F + Si + T + D - M + (Q + S'vp[1+s]$$
$$/\{360[c-q]\})/l$$
$$= 29,750 + 51,294*0.0100 + 3,385$$
$$+ 2,370 - 41,548 + (35,351$$
$$+ 48,851*0.1900*240$$
$$[1+0.0500]/\{360[1.2000$$
$$-1.0100]\})/0.9900$$
$$= 64,720 \text{ millions USD}$$

Corporate IFRS-GAAP (B/S-I/S), ISBN-13: **978-1720792789**, ISBN-10: **172079278X**

Law-7836:

If both (**D**= Dividend Paid= 2,370 millions USD), (**T**= Tax Paid= 3,385 millions USD), (**i**= I/S= Interest Portion= 1.00%), (**S'**= Sales of Past Year= 48,851 millions USD), (**s**= [S/S']-1= Sales Growth= 5.00%), (**c**= C/X= Current Ratio= 1.20 times), (**v**= V/S= Variable Portion= 19.00%), (**p**= 360 P/V= Procured Inventory Days= 240 Days), (**Q**= Quoted Longterm Liabilities= 35,351 millions USD), (**U**= Utilized or Starting Capital= 64,720 millions USD), (**S**= Sales or Revenues= 51,294 millions USD), (**l**= L/E= Leverage or Gearing Ratio= 99.00%), (**F**= Fixed Cost= 29,740 millions USD), and (**q**= [C-P]/X= Quick or Acid Test Ratio= 1.01 times), then it's (**M**= Margin of Contribution planned), is:

$$M= F+Si+T+D-U+(Q+S'vp[1+s]$$
$$/\{360[c-q]\})/l$$
$$= 29,750+51,294*0.0100+3,385$$
$$+2,370-64,720+(35,351$$
$$+48,851*0.1900*240$$
$$[1+0.0500]/\{360[1.2000$$
$$-1.0100]\})/0.9900$$
$$= 41,548 \text{ millions USD}$$

Corporate IFRS-GAAP (B/S-I/S), ISBN-13: **978-1720792789**, ISBN-10: **172079278X**

Law-7837:

If both (**D**= Dividend Paid= 2,370 millions USD),
(**T**= Tax Paid= 3,385 millions USD), (**i**= I/S= Interest
Portion= 1.00%), (**$'**= Sales of Past Year= 48,851
millions USD), (**s**= [S/S']-1= Sales Growth= 5.00%),
(**c**= C/X= Current Ratio= 1.20 times), (**v**= V/S=
Variable Portion= 19.00%), (**p**= 360 P/V= Procured
Inventory Days= 240 Days), (**Q**= Quoted Longterm
Liabilities= 35,351 millions USD), (**U**= Utilized or
Starting Capital= 64,720 millions USD), (**$**= Sales or
Revenues= 51,294 millions USD), (**/**= L/E= Leverage
or Gearing Ratio= 99.00%), (**M**= Margin of
Contribution= 41,548 millions USD), and (**q**= Quick
or Acid Test Ratio= [C-P]/X= 1.01 times), then it's
(**F**= Fixed Cost planned), is:

$$F= U+M-Si-T-D-(Q+S'vp[1+s]/\{360[c-q]\})/I$$

$$= 64,720+41,548-51,294*0.0100$$
$$-3,385-2,370-(35,351$$
$$+48,851*0.1900*240$$
$$[1+0.0500]/\{360[1.2000$$
$$-1.0100]\})/0.9900$$

$$= 29,750 \text{ millions USD}$$

Corporate IFRS-GAAP (B/S-I/S), ISBN-13: **978-1720792789**, ISBN-10: **172079278X**

Law-7838:

If both (**D**= Dividend Paid= 2,370 millions USD),
(**T**= Tax Paid= 3,385 millions USD), (**i**= I/S= Interest
Portion= 1.00%), (**S'**= Sales of Past Year= 48,851
millions USD), (**s**= [S/S']-1= Sales Growth= 5.00%),
(**c**= C/X= Current Ratio= 1.20 times), (**v**= V/S=
Variable Portion= 19.00%), (**p**= 360 P/V= Procured
Inventory Days= 240 Days), (**Q**= Quoted Longterm
Liabilities= 35,351 millions USD), (**U**= Utilized or
Starting Capital= 64,720 millions USD), (**F**= Fixed
Cost= 29,740 millions USD), (**l**= L/E= Leverage or
Gearing Ratio= 99.00%), (**M**= Margin of
Contribution= 41,548 millions USD), and (**q**= Quick
or Acid Test Ratio= [C-P]/X= 1.01 times), then it's
(**S**= Sales or Revenues planned), is:

$$S= [U+M-F-T-D-(Q+S'vp[1+s]$$
$$/\{360[c-q]\})/l\,]/i$$
$$= [64,720+41,548-29,750-3,385$$
$$-2,370-(35,351+48,851$$
$$*0.1900*240[1+0.0500]$$
$$/\{360[1.2000-1.0100]\})$$
$$/0.9900]/0.0100$$
$$= 51,294 \text{ millions USD}$$

Corporate IFRS-GAAP (B/S-I/S), ISBN-13: **978-1720792789**, ISBN-10: **172079278X**

Law-7839:

If both (**D**= Dividend Paid= 2,370 millions USD), (**T**= Tax Paid= 3,385 millions USD), (**S**= Sales or Revenues= 51,294 millions USD), (**S'**= Sales of Past Year= 48,851 millions USD), (**s**= [S/S']-1= Sales Growth= 5.00%), (**c**= C/X= Current Ratio= 1.20 times), (**v**= V/S= Variable Portion= 19.00%), (**p**= 360 P/V= Procured Inventory Days= 240 Days), (**Q**= Quoted Longterm Liabilities= 35,351 millions USD), (**U**= Utilized or Starting Capital= 64,720 millions USD), (**F**= Fixed Cost= 29,740 millions USD), (**I**= L/E= Leverage or Gearing Ratio= 99.00%), (**M**= Margin of Contribution= 41,548 millions USD), and (**q**= [C-P]/X= Quick or Acid Test Ratio= 1.01 times), then it's (**i**= Interest Portion planned), is:

$$\mathbf{i}= [\mathbf{U+M-F-T-D-(Q+S'vp}[1+s]$$
$$/\{360[\mathbf{c-q}]\})/\mathbf{I}]/\mathbf{S}$$
$$= [64,720+41,548-29,750-3,385$$
$$-2,370-(35,351+48,851$$
$$*0.1900*240[1+0.0500]$$
$$/\{360[1.2000-1.0100]\})$$
$$/0.9900]/51,294$$

$$= 1.00\%$$

Corporate IFRS-GAAP (B/S-I/S), ISBN-13: **978-1720792789**, ISBN-10: **172079278X**

Law-7840:

If both (**D**= Dividend Paid= 2,370 millions USD), (**i**= I/S= Interest Portion= 1.00%), (**$**= Sales or Revenues= 51,294 millions USD), (**$'**= Sales of Past Year= 48,851 millions USD), (**$**= [S/S']-1= Sales Growth= 5.00%), (**c**= C/X= Current Ratio= 1.20 times), (**v**= V/S= Variable Portion= 19.00%), (**p**= 360 P/V= Procured Inventory Days= 240 Days), (**Q**= Quoted Longterm Liabilities= 35,351 millions USD), (**U**= Utilized or Starting Capital= 64,720 millions USD), (**F**= Fixed Cost= 29,740 millions USD), (**l**= L/E= Leverage or Gearing Ratio= 99.00%), (**M**= Margin of Contribution= 41,548 millions USD), and (**q**= [C-P]/X= Quick or Acid Test Ratio= 1.01 times), then it's (**T**= Tax Paid planned), is:

$$T= U+M-F-Si-D-(Q+S'vp[1+s]$$
$$/\{360[c-q]\})/l$$
$$= 64,720+41,548-29,750-51,294$$
$$*0.0100-2,370-(35,351$$
$$+48,851*0.1900*240$$
$$[1+0.0500]/\{360[1.2000$$
$$-1.0100]\})/0.9900$$
$$= 3,385 \text{ millions USD}$$

Corporate IFRS-GAAP (B/S-I/S), ISBN-13: **978-1720792789**, ISBN-10: **172079278X**

<u>Law-7841</u>:

If both (**T**= Tax Paid= <u>3,385</u> millions USD), (**i**= I/S= Interest Portion= <u>1.00%</u>), (**S**= Sales or Revenues= <u>51,294</u> millions USD), (**S'**= Sales of Past Year= <u>48,851</u> millions USD), (**s**= [S/S']-1= Sales Growth= <u>5.00%</u>), (**c**= C/X= Current Ratio= <u>1.20</u> times), (**v**= V/S= Variable Portion= <u>19.00%</u>), (**p**= 360 P/V= Procured Inventory Days= <u>240</u> Days), (**Q**= Quoted Longterm Liabilities= <u>35,351</u> millions USD), (**U**= Utilized or Starting Capital= <u>64,720</u> millions USD), (**F**= Fixed Cost= <u>29,740</u> millions USD), (**F**= L/E= Leverage or Gearing Ratio= <u>99.00%</u>), (**M**= Margin of Contribution= <u>41,548</u> millions USD), and (**q**= Quick or Acid Test Ratio= [C-P]/X= <u>1.01</u> times), then it's (**D**= Dividend Paid planned), is:

$$D= U+M-F-Si-T-(Q+S'vp[1+s]/\{360[c-q]\})/I$$

$$\begin{aligned}
&= 64,720+41,548-29,750-51,294\\
&\quad *0.0100-3,385-(35,351\\
&\quad +48,851*0.1900*240\\
&\quad [1+0.0500]/\{360[1.2000\\
&\quad -1.0100]\})/0.9900\\
&= \underline{2,370} \text{ millions USD}
\end{aligned}$$

Corporate IFRS-GAAP (B/S-I/S), ISBN-13: **978-1720792789**, ISBN-10: **172079278X**

Law-7842:

If both (**T**= Tax Paid= 3,385 millions USD), (**i**= I/S= Interest Portion= 1.00%), (**S**= Sales or Revenues= 51,294 millions USD), (**D**= Dividend Paid= 2,370 millions USD), (**s**= [S/S']-1= Sales Growth= 5.00%), (**c**= C/X= Current Ratio= 1.20 times), (**v**= V/S= Variable Portion= 19.00%), (**p**= 360 P/V= Procured Inventory Days= 240 Days), (**Q**= Quoted Longterm Liabilities= 35,351 millions USD), (**U**= Utilized or Starting Capital= 64,720 millions USD), (**F**= Fixed Cost= 29,740 millions USD), (**l**= L/E= Leverage or Gearing Ratio= 99.00%), (**M**= Margin of Contribution= 41,548 millions USD), and (**q**= Quick or Acid Test Ratio= [C-P]/X= 1.01 times), then it's (**S'**= Sales Past), must be:

$$S'= 360[\textbf{c-q}](\textbf{\textit{l}} \{U+M-F-Si-T-D\}-Q)$$
$$/\{\textbf{vp}[1+\textbf{s}]\}$$
$$= 360[1.2000-1.0100](0.9900$$
$$\{64,720+41,548-29,750$$
$$-51,294*0.0100-3,385-$$
$$2,370\}-35,351)/\{0.1900$$
$$*240[1+0.0500]\}$$
$$= \underline{48,851} \text{ millions USD}$$

Corporate IFRS-GAAP (B/S-I/S), ISBN-13: **978-1720792789**, ISBN-10: **172079278X**

Law-7843:

If both (**T**= Tax Paid= 3,385 millions USD), (**i**= I/S= Interest Portion= 1.00%), (**S**= Sales or Revenues= 51,294 millions USD), (**D**= Dividend Paid= 2,370 millions USD), (**s**= [S/S']-1= Sales Growth= 5.00%), (**c**= C/X= Current Ratio= 1.20 times), (**S'**= Sales of Past Year= 48,851 millions USD), (**p**= 360 P/V= Procured Inventory Days= 240 Days), (**Q**= Quoted Longterm Liabilities= 35,351 millions USD), (**U**= Utilized or Starting Capital= 64,720 millions USD), (**F**= Fixed Cost= 29,740 millions USD), (**l**= L/E= Leverage or Gearing Ratio= 99.00%), (**M**= Margin of Contribution= 41,548 millions USD), and (**q**= Quick or Acid Test Ratio= [C-P]/X= 1.01 times), then it's (**v**= Variable Portion planned), is:

$$v= 360[c\text{-}q](l \{U\text{+}M\text{-}F\text{-}Si\text{-}T\text{-}D\}\text{-}Q) / \{S'p[1\text{+}s]\}$$

$$= 360[1.2000\text{-}1.0100](0.9900 \{64,720\text{+}41,548\text{-}29,750 -51,294*0.0100\text{-}3,385 -2,370\}\text{-}35,351)/\{48,851 *240[1\text{+}0.0500]\}$$

$$= 19.00\%$$

Corporate IFRS-GAAP (B/S-I/S), ISBN-13: **978-1720792789**, ISBN-10: **172079278X**

<u>Law-7844</u>:

If both (**T**= Tax Paid= <u>3,385</u> millions USD), (**i**= I/S= Interest Portion= <u>1.00%</u>), (**S**= Sales or Revenues= <u>51,294</u> millions USD), (**D**= Dividend Paid= <u>2,370</u> millions USD), (**s**= [S/S']-1= Sales Growth= <u>5.00%</u>), (**c**= C/X= Current Ratio= <u>1.20</u> times), (**S'**= Sales of Past Year= <u>48,851</u> millions USD), (**v**= V/S= Variable Portion= <u>19.00%</u>), (**Q**= Quoted Longterm Liabilities= <u>35,351</u> millions USD), (**U**= Utilized or Starting Capital= <u>64,720</u> millions USD), (**F**= Fixed Cost= <u>29,740</u> millions USD), (**l**= L/E= Leverage or Gearing Ratio= <u>99.00%</u>), (**M**= Margin of Contribution= <u>41,548</u> millions USD), and (**q**= [C-P]/X= Quick or Acid Test Ratio= <u>1.01</u> times), then it's (**p**= Procured Inventory Days planned), is:

$$p= 360[c-q](l\ \{U+M-F-Si-T-D\}-Q)$$
$$/\{S'v[1+s]\}$$

$$= 360[1.2000-1.0100](0.9900$$
$$\{64,720+41,548-29,750$$
$$-51,294*0.0100-3,385$$
$$-2,370\}-35,351)/\{48,851$$
$$*0.1900[1+0.0500]\}$$

$$= \underline{240}\ \text{days}$$

Corporate IFRS-GAAP (B/S-I/S), ISBN-13: **978-1720792789**, ISBN-10: **172079278X**

Law-7845:

If both (**T**= Tax Paid= 3,385 millions USD), (**i**= I/S= Interest Portion= 1.00%), (**S**= Sales or Revenues= 51,294 millions USD), (**D**= Dividend Paid= 2,370 millions USD), (**p**= 360 P/V= Procured Inventory Days= 240 Days), (**c**= C/X= Current Ratio= 1.20 times), (**S'**= Sales of Past Year= 48,851 millions USD), (**v**= V/S= Variable Portion= 19.00%), (**Q**= Quoted Longterm Liabilities= 35,351 millions USD), (**U**= Utilized or Starting Capital= 64,720 millions USD), (**F**= Fixed Cost= 29,740 millions USD), (**l**= L/E= Leverage or Gearing Ratio= 99.00%), (**M**= Margin of Contribution= 41,548 millions USD), and (**q**= [C-P]/X= Quick or Acid Test Ratio= 1.01 times), then it's (**s**= Sales Growth planned), is:

$$s= 360[c\text{-}q](l\,\{U+M\text{-}F\text{-}Si\text{-}T\text{-}D\}\text{-}Q)/[S'vp]-1$$
$$= 360[1.2000\text{-}1.0100](0.9900$$
$$\{64,720+41,548\text{-}29,750$$
$$-51,294*0.0100\text{-}3,385$$
$$-2,370\}\text{-}35,351)/[48,851$$
$$*0.1900*240]-1$$
$$= 5.00\%$$

Corporate IFRS-GAAP (B/S-I/S), ISBN-13: **978-1720792789**, ISBN-10: **172079278X**

Law-7846:

If both (**T**= Tax Paid= 3,385 millions USD), (**i**= I/S= Interest Portion= 1.00%), (**S**= Sales or Revenues= 51,294 millions USD), (**D**= Dividend Paid= 2,370 millions USD), (**p**= 360 P/V= Procured Inventory Days= 240 Days), (**s**= [S/S']-1= Sales Growth= 5.00%), (**S'**= Sales of Past Year= 48,851 millions USD), (**v**= V/S= Variable Portion= 19.00%), (**Q**= Quoted Longterm Liabilities= 35,351 millions USD), (**U**= Utilized or Starting Capital= 64,720 millions USD), (**F**= Fixed Cost= 29,740 millions USD), (**I**= L/E= Leverage or Gearing Ratio= 99.00%), (**M**= Margin of Contribution= 41,548 millions USD), and (**q**= [C-P]/X= Quick or Acid Test Ratio= 1.01 times), then it's (**c**= Current Ratio planned), is:

$$c= q+S'vp[1+s]/[360(I \{U+M-F-Si-T-D\}-Q)]$$
$$= 1.0100+48.851*0.1900*240$$
$$[1+0.0500]/[360(0.9900$$
$$\{64,720+41,548-29,750$$
$$-51,294*0.0100-3,385$$
$$-2,370\}-35,351)]$$
$$= 1.20 \text{ times}$$

Corporate IFRS-GAAP (B/S-I/S), ISBN-13: **978-1720792789**, ISBN-10: **172079278X**

<u>Law-7847</u>:

If both (**T**= Tax Paid= <u>3,385</u> millions USD), (**i**= I/S= Interest Portion= <u>1.00%</u>), (**S**= Sales or Revenues= <u>51,294</u> millions USD), (**D**= Dividend Paid= <u>2,370</u> millions USD), (**p**= 360 P/V= Procured Inventory Days= <u>240</u> Days), (**s**= [S/S']-1= Sales Growth= <u>5.00%</u>), (**S'**= Sales of Past Year= <u>48,851</u> millions USD), (**v**= V/S= Variable Portion= <u>19.00%</u>), (**Q**= Quoted Longterm Liabilities= <u>35,351</u> millions USD), (**U**= Utilized or Starting Capital= <u>64,720</u> millions USD), (**F**= Fixed Cost= <u>29,740</u> millions USD), (**l**= L/E= Leverage or Gearing Ratio= <u>99.00%</u>), (**M**= Margin of Contribution= <u>41,548</u> millions USD), and (**c**= C/X= Current Ratio= <u>1.20</u> times),then it's (**q**= Quick or Acid Test Ratio planned), is:

$$q= c\text{-}S'vp[1+s]/[360(l\ \{U+M\text{-}F\text{-}Si\text{-}T\text{-}D\}\text{-}Q)]$$
$$= 1.2000\text{-}48,851*0.1900*240$$
$$[1+0.0500]/[360(0.9900$$
$$\{64,720+41,548\text{-}29,750$$
$$-51,294*0.0100\text{-}3,385$$
$$-2,370\}\text{-}35,351)]$$

$$= \underline{1.01}\ \text{times}$$

Corporate IFRS-GAAP (B/S-I/S), ISBN-13: **978-1720792789**, ISBN-10: **172079278X**

Law-7848:

If both (**T**= Tax Paid= 3,385 millions USD), (**i**= I/S= Interest Portion= 1.00%), (**S**= Sales or Revenues= 51,294 millions USD), (**d**= D/A= Dividend Portion or Payout= 30.00%), (**U**= Utilized or Starting Capital= 64,720 millions USD), (**F**= Fixed Cost= 29,740 millions USD), (**l**= L/E= Leverage or Gearing Ratio= 99.00%), (**M**= Margin of Contribution= 41,548 millions USD), and (**X**= Xpress or Current Debt= 34,196 millions USD), then it's (**Q**= Quoted Longterm Liabilities planned), is:

$$Q= l\{U+[M-F-Si-T][1-d]\}-X$$
$$= 0.9900\{64,720+[41,548-29,750$$
$$-51,294*0.0100-3,385]$$
$$[1-0.3000]\}-34,196$$
$$= 35,351 \text{ millions USD}$$

Corporate IFRS-GAAP (B/S-I/S), ISBN-13: **978-1720792789**, ISBN-10: **172079278X**

<u>Law-7849</u>:

If both (**T**= Tax Paid= <u>3,385</u> millions USD), (**i**= I/S= Interest Portion= <u>1.00%</u>), (**S**= Sales or Revenues= <u>51,294</u> millions USD), (**d**= D/A= Dividend Portion or Payout= <u>30.00%</u>), (**U**= Utilized or Starting Capital= <u>64,720</u> millions USD), (**F**= Fixed Cost= <u>29,740</u> millions USD), (**Q**= Quoted Longterm Liabilities= <u>35,351</u> millions USD), (**M**= Margin of Contribution= <u>41,548</u> millions USD), and (**X**= Xpress or Current Debt= <u>34,196</u> millions USD), then it's (**I** = Leverage or Gearing Ratio planned), is:

$$I = [\mathbf{Q}+\mathbf{X}]/\{\mathbf{U}+[\mathbf{M}-\mathbf{F}-\mathbf{Si}-\mathbf{T}][1-\mathbf{d}]\}$$
$$= [35,351+34,196]/\{64,720+[41,548$$
$$-29,750-51.294*0.0100$$
$$-3,385][1-0.3000]\}$$
$$= \underline{99.00\%}$$

Corporate IFRS-GAAP (B/S-I/S), ISBN-13: **978-1720792789**, ISBN-10: **172079278X**

Law-7850:

If both (**T**= Tax Paid= 3,385 millions USD), (**i**= I/S= Interest Portion= 1.00%), (**S**= Sales or Revenues= 51,294 millions USD), (**d**= D/A= Dividend Portion or Payout= 30.00%), (**I**= L/E= Leverage or Gearing Ratio= 99.00%), (**F**= Fixed Cost= 29,740 millions USD), (**Q**= Quoted Longterm Liabilities= 35,351 millions USD), (**M**= Margin of Contribution= 41,548 millions USD), and (**X**= Xpress or Current Debt= 34,196 millions USD), then it's (**U**= Utilized or Starting Capital planned), is:

$$U= [Q+X]/I -[M-F-Si-T][1-d]$$
$$= [35,351+34,196]/0.9900-[41,548$$
$$-29,750-51,294*0.0100$$
$$-3,385][1-0.3000]$$
$$= 64,720 \text{ millions USD}$$

Corporate IFRS-GAAP (B/S-I/S), ISBN-13: **978-1720792789**, ISBN-10: **172079278X**

Law-7851:

If both (**T**= Tax Paid= 3,385 millions USD), (**i**= I/S= Interest Portion= 1.00%), (**S**= Sales or Revenues= 51,294 millions USD), (**d**= D/A= Dividend Portion or Payout= 30.00%), (**I**= L/E= Leverage or Gearing Ratio= 99.00%), (**F**= Fixed Cost= 29,740 millions USD), (**Q**= Quoted Longterm Liabilities= 35,351 millions USD), (**U**= Utilized or Starting Capital= 64,720 millions USD), and (**X**= Xpress or Current Debt= 34,196 millions USD), then it's (**M**= Margin of Contribution planned), is:

$$M= F+Si+T+\{[Q+X]/I -U\}/[1-d]$$
$$= 29,750+51,294*0.0100+3,385$$
$$+\{[35,351+34,196]/0.9900$$
$$-64,720\}/[1-0.3000]$$
$$= 41,548 \text{ millions USD}$$

Corporate IFRS-GAAP (B/S-I/S), ISBN-13: **978-1720792789**, ISBN-10: **172079278X**

Law-7852:

If both (**T**= Tax Paid= 3,385 millions USD), (**i**= I/S= Interest Portion= 1.00%), (**S**= Sales or Revenues= 51,294 millions USD), (**d**= D/A= Dividend Portion or Payout= 30.00%), (**I**= L/E= Leverage or Gearing Ratio= 99.00%), (**M**= Margin of Contribution= 41,548 millions USD), (**Q**= Quoted Longterm Liabilities= 35,351 millions USD), (**U**= Utilized or Starting Capital= 64,720 millions USD), and (**X**= Xpress or Current Debt= 34,196 millions USD), then it's (**F**= Fixed Cost planned), is:

$$F= M\text{-}Si\text{-}T\text{-}\{[Q+X]/I\text{-}U\}/[1\text{-}d]$$
$$= 41,548\text{-}51,294*0.0100\text{-}3,385$$
$$-\{[35,351+34,196]/0.9900$$
$$-64,720\}/[1\text{-}0.3000]$$
$$= \underline{29,750} \text{ millions USD}$$

Corporate IFRS-GAAP (B/S-I/S), ISBN-13: **978-1720792789**, ISBN-10: **172079278X**

Law-7853:

If both (**T**= Tax Paid= 3,385 millions USD), (**i**= I/S=
Interest Portion= 1.00%), (**F**= Fixed Cost= 29,740
millions USD), (**d**= D/A= Dividend Portion or
Payout= 30.00%), (**I**= L/E= Leverage or Gearing
Ratio= 99.00%), (**M**= Margin of Contribution=
41,548 millions USD), (**Q**= Quoted Longterm
Liabilities= 35,351 millions USD), (**U**= Utilized or
Starting Capital= 64,720 millions USD), and (**X**=
Xpress or Current Debt= 34,196 millions USD), then
it's (**S**= Sales or Revenues planned), is:

$$S= (M-F-T-\{[Q+X]/I-U\}/[1-d])/i$$
$$= (41,548-29,750-3,385-\{[35,351$$
$$+34,196]/0.9900-64,720\}$$
$$/[1-0.3000])/ 0.0100$$
$$= 51,294 \text{ millions USD}$$

Corporate IFRS-GAAP (B/S-I/S), ISBN-13: **978-1720792789**, ISBN-10: **172079278X**

Law-7854:

If both (**T**= Tax Paid= 3,385 millions USD), (**S**= Sales or Revenues= 51,294 millions USD), (**F**= Fixed Cost= 29,740 millions USD), (**d**= D/A= Dividend Portion or Payout= 30.00%), (**I**= L/E= Leverage or Gearing Ratio= 99.00%), (**M**= Margin of Contribution= 41,548 millions USD), (**Q**= Quoted Longterm Liabilities= 35,351 millions USD), (**U**= Utilized or Starting Capital= 64,720 millions USD), and (**X**= Xpress or Current Debt= 34,196 millions USD), then it's (**i**= Interest Portion planned), is:

$$i= (M-F-T-\{[Q+X]/I -U\}/[1-d])/S$$
$$= (41,548-29,750-3,385-\{[35,351$$
$$+34,196]/0.9900-64,720\}$$
$$/[1-0.3000])/51,294$$
$$= 1.00\%$$

Corporate IFRS-GAAP (B/S-I/S), ISBN-13: **978-1720792789**, ISBN-10: **172079278X**

Law-7855:

If both (**i**= I/S= Interest Portion= 1.00%), (**$**= Sales or Revenues= 51,294 millions USD), (**F**= Fixed Cost= 29,740 millions USD), (**d**= D/A= Dividend Portion or Payout= 30.00%), (**I**= L/E= Leverage or Gearing Ratio= 99.00%), (**M**= Margin of Contribution= 41,548 millions USD), (**Q**= Quoted Longterm Liabilities= 35,351 millions USD), (**U**= Utilized or Starting Capital= 64,720 millions USD), and (**X**= Xpress or Current Debt= 34,196 millions USD), then it's (**T**= Tax Paid planned), is:

$$T= M\text{-}F\text{-}\$i\text{-}\{[Q+X]/I\text{-}U\}/[1\text{-}d]$$
$$= 41,548\text{-}29,750\text{-}51,294*0.0100$$
$$-\{[35,351+34,196]/0.9900$$
$$-64,720\}/[1\text{-}0.3000]$$
$$= 3,385 \text{ millions USD}$$

Corporate IFRS-GAAP (B/S-I/S), ISBN-13: **978-1720792789**, ISBN-10: **172079278X**

Law-7856:

If both (**i**= I/S= Interest Portion= 1.00%), (**S**= Sales or Revenues= 51,294 millions USD), (**F**= Fixed Cost= 29,740 millions USD), (**T**= Tax Paid= 3,385 millions USD), (**I**= L/E= Leverage or Gearing Ratio= 99.00%), (**M**= Margin of Contribution= 41,548 millions USD), (**Q**= Quoted Longterm Liabilities= 35,351 millions USD), (**U**= Utilized or Starting Capital= 64,720 millions USD), and (**X**= Xpress or Current Debt= 34,196 millions USD), then it's (**d**= Dividend Portion or Payout planned), is :

$$d= 1-\{[\mathbf{Q+X}]/\mathbf{I}-\mathbf{U}\}/[\mathbf{M-F-Si-T}]$$
$$= 1-\{[35,351+34,196]/0.9900-64,720\}$$
$$/[41,548-29,750-51,294$$
$$*0.0100-3,385]$$
$$= 30.00\%$$

Corporate IFRS-GAAP (B/S-I/S), ISBN-13: **978-1720792789**, ISBN-10: **172079278X**

Law-7857:

If both (**i**= I/S= Interest Portion= 1.00%), (**S**= Sales or Revenues= 51,294 millions USD), (**F**= Fixed Cost= 29,740 millions USD), (**T**= Tax Paid= 3,385 millions USD), (**l**= L/E= Leverage or Gearing Ratio= 99.00%), (**M**= Margin of Contribution= 41,548 millions USD), (**Q**= Quoted Longterm Liabilities= 35,351 millions USD), (**U**= Utilized or Starting Capital= 64,720 millions USD), and (**d**= D/A= Dividend Portion or Payout= 30.00%), then it's (**X**= Xpress or Current Debt planned), is:

$$\mathbf{X= l}\{\mathbf{U}+[\mathbf{M\text{-}F\text{-}Si\text{-}T}][1\text{-}\mathbf{d}]\}\text{-}\mathbf{Q}$$

$$= 0.9900\{64,720+[41,548\text{-}29,750$$
$$-51,294*0.0100\text{-}3,385]$$
$$[1\text{-}0.3000]\}\text{-}35,351$$

$$= \underline{34,196} \text{ millions USD}$$

Corporate IFRS-GAAP (B/S-I/S), ISBN-13: **978-1720792789**, ISBN-10: **172079278X**

<u>Law-7858</u>:

If both (**i**= I/S= Interest Portion= <u>1.00%</u>), (**c**= C/X= Current Ratio= <u>1.20</u> times), (**q**= [C-P]/X= Quick or Acid Test Ratio= <u>1.01</u> times), (**P**= Procured Inventories= <u>6,497</u> millions USD), (**S**= Sales or Revenues= <u>51,294</u> millions USD), (**F**= Fixed Cost= <u>29,740</u> millions USD), (**T**= Tax Paid= <u>3,385</u> millions USD), (**l**= L/E= Leverage or Gearing Ratio= <u>99.00%</u>), (**M**= Margin of Contribution= <u>41,548</u> millions USD), (**U**= Utilized or Starting Capital= <u>64,720</u> millions USD), and (**d**= D/A= Dividend Portion or Payout= <u>30.00%</u>), then it's (**Q**= Quoted Longterm Liabilities planned), is:

$$Q = l\{U+[M-F-Si-T][1-d]\}-P/[c-q]$$

$$= 0.9900\{64,720+[41,548-29,750$$
$$-51,294*0.0100-3,385]$$
$$[1-0.3000]\}-6,497/[1.2000$$
$$-1.0100]$$

$$= \underline{35,351} \text{ millions USD}$$

Corporate IFRS-GAAP (B/S-I/S), ISBN-13: **978-1720792789**, ISBN-10: **172079278X**

Law-7859:

If both (**i**= I/S= Interest Portion= 1.00%), (**c**= C/X= Current Ratio= 1.20 times), (**q**= [C-P]/X= Quick or Acid Test Ratio= 1.01 times), (**P**= Procured Inventories= 6,497 millions USD), (**S**= Sales or Revenues= 51,294 millions USD), (**F**= Fixed Cost= 29,740 millions USD), (**T**= Tax Paid= 3,385 millions USD), (**Q**= Quoted Longterm Liabilities= 35,351 millions USD), (**M**= Margin of Contribution= 41,548 millions USD), (**U**= Utilized or Starting Capital= 64,720 millions USD), and (**d**= D/A= Dividend Portion or Payout= 30.00%), then it's (**I** = Leverage or Gearing Ratio planned), is:

$$I = \{Q+P/[c-q]\}/\{U+[M-F-Si-T][1-d]\}$$
$$= \{35,351+6,497/[1.2000-1.0100]\}$$
$$/\{64,720+[41,548-29,750$$
$$-51,294*0.0100-3,385]$$
$$[1-0.3000]\}$$
$$= 99.00\%$$

Corporate IFRS-GAAP (B/S-I/S), ISBN-13: **978-1720792789**, ISBN-10: **172079278X**

Law-7860:

If both (**i**= I/S= Interest Portion= 1.00%), (**c**= C/X= Current Ratio= 1.20 times), (**q**= [C-P]/X= Quick or Acid Test Ratio= 1.01 times), (**P**= Procured Inventories= 6,497 millions USD), (**S**= Sales or Revenues= 51,294 millions USD), (**F**= Fixed Cost= 29,740 millions USD), (**T**= Tax Paid= 3,385 millions USD), (**Q**= Quoted Longterm Liabilities= 35,351 millions USD), (**M**= Margin of Contribution= 41,548 millions USD), (**I**= L/E= Leverage or Gearing Ratio= 99.00%), and (**d**= D/A= Dividend Portion or Payout= 30.00%), then it's (**U**= Utilized or Starting Capital planned), is:

$$U= \{Q+P/[c-q]\}/I -[M-F-Si-T][1-d]$$
$$= \{35,351+6,497/[1.2000-1.0100]\}$$
$$/0.9900-[41,548-29,750$$
$$-51,294*0.0100-3,385]$$
$$[1-0.3000]$$
$$= \underline{64,720} \text{ millions USD}$$

Corporate IFRS-GAAP (B/S-I/S), ISBN-13: **978-1720792789**, ISBN-10: **172079278X**

Law-7861:

If both (**i**= I/S= Interest Portion= 1.00%), (**c**= C/X= Current Ratio= 1.20 times), (**q**= [C-P]/X= Quick or Acid Test Ratio= 1.01 times), (**P**= Procured Inventories= 6,497 millions USD), (**S**= Sales or Revenues= 51,294 millions USD), (**F**= Fixed Cost= 29,740 millions USD), (**T**= Tax Paid= 3,385 millions USD), (**Q**= Quoted Longterm Liabilities= 35,351 millions USD), (**U**= Utilized or Starting Capital= 64,720 millions USD), (**l**= L/E= Leverage or Gearing Ratio= 99.00%), and (**d**= D/A= Dividend Portion or Payout= 30.00%), then it's (**M**= Margin of Contribution planned), is:

$$M= F+Si+T+(\{Q+P/[c-q]\}/l-U)/[1-d]$$
$$= 29,750+51,294*0.0100+3,385$$
$$+(\{35,351+6,497/[1.2000$$
$$-1.0100]\}/0.9900-64,720)$$
$$/[1-0.3000]$$
$$= 41,548 \text{ millions USD}$$

Corporate IFRS-GAAP (B/S-I/S), ISBN-13: **978-1720792789**, ISBN-10: **172079278X**

<u>Law-7862</u>:

If both (**i**= I/S= Interest Portion= <u>1.00%</u>), (**c**= C/X= Current Ratio= <u>1.20</u> times), (**q**= [C-P]/X= Quick or Acid Test Ratio= <u>1.01</u> times), (**P**= Procured Inventories= <u>6,497</u> millions USD), (**S**= Sales or Revenues= <u>51,294</u> millions USD), (**M**= Margin of Contribution= <u>41,548</u> millions USD), (**T**= Tax Paid= <u>3,385</u> millions USD), (**Q**= Quoted Longterm Liabilities= <u>35,351</u> millions USD), (**U**= Utilized or Starting Capital= <u>64,720</u> millions USD), (**l**= L/E= Leverage or Gearing Ratio= <u>99.00%</u>), and (**d**= D/A= Dividend Portion or Payout= <u>30.00%</u>), then it's (**F**= Fixed Cost planned), is:

$$F = M\text{-}Si\text{-}T\text{-}(\{Q+P/[c\text{-}q]\}/l\text{-}U)/[1\text{-}d]$$

$$= 41{,}548\text{-}51{,}294*0.0100\text{-}3{,}385$$
$$-(\{35{,}351+6{,}497/[1.2000$$
$$-1.0100]\}/0.9900\text{-}64{,}720)$$
$$/[1\text{-}0.3000]$$

$$= \underline{29{,}750} \text{ millions USD}$$

Corporate IFRS-GAAP (B/S-I/S), ISBN-13: **978-1720792789**, ISBN-10: **172079278X**

<u>Law-7863</u>:

If both (**i**= I/S= Interest Portion= <u>1.00%</u>), (**c**= C/X= Current Ratio= <u>1.20</u> times), (**q**= [C-P]/X= Quick or Acid Test Ratio= <u>1.01</u> times), (**P**= Procured Inventories= <u>6,497</u> millions USD), (**F**= Fixed Cost= <u>29,740</u> millions USD), (**M**= Margin of Contribution= <u>41,548</u> millions USD), (**T**= Tax Paid= <u>3,385</u> millions USD), (**Q**= Quoted Longterm Liabilities= <u>35,351</u> millions USD), (**U**= Utilized or Starting Capital= <u>64,720</u> millions USD), (**l**= L/E= Leverage or Gearing Ratio= <u>99.00%</u>), and (**d**= D/A= Dividend Portion or Payout= <u>30.00%</u>), then it's (**S**= Sales or Revenues planned), is:

$$S= [M-F-T-(\{Q+P/[c-q]\}/l-U)/[1-d]]/i$$
$$= [41,548-29,750-3,385-(\{35,351$$
$$+6,497/[1.2000-1.0100]\}$$
$$/0.9900-64,720)$$
$$/[1-0.3000]]/0.0100$$
$$= \underline{51,294} \text{ millions USD}$$

Corporate IFRS-GAAP (B/S-I/S), ISBN-13: **978-1720792789**, ISBN-10: **172079278X**

Law-7864:

If both ($= Sales or Revenues= 51,294 millions
USD), (**c**= C/X= Current Ratio= 1.20 times), (**q**= [C-
P]/X= Quick or Acid Test Ratio= 1.01 times), (**P**=
Procured Inventories= 6,497 millions USD), (**F**=
Fixed Cost= 29,740 millions USD), (**M**= Margin of
Contribution= 41,548 millions USD), (**T**= Tax Paid=
3,385 millions USD), (**Q**= Quoted Longterm
Liabilities= 35,351 millions USD), (**U**= Utilized or
Starting Capital= 64,720 millions USD), (**I**= L/E=
Leverage or Gearing Ratio= 99.00%), and (**d**= D/A=
Dividend Portion or Payout= 30.00%), then it's (**i**=
Interest Portion planned), is:

$$i= [M\text{-}F\text{-}T\text{-}(\{Q+P/[c\text{-}q]\}/I\text{-}U)/[1\text{-}d]]/\$$$

$$= [41,548\text{-}29,750\text{-}3,385\text{-}(\{35,351$$
$$+6,497/[1.2000\text{-}1.0100]\}$$
$$/0.9900\text{-}64,720)/[1\text{-}0.3000]]$$
$$/51,294$$

$$= 1.00\%$$

Corporate IFRS-GAAP (B/S-I/S), ISBN-13: **978-1720792789**, ISBN-10: **172079278X**

Law-7865:

If both ($\textbf{S}$= Sales or Revenues= 51,294 millions USD), ($\textbf{c}$= C/X= Current Ratio= 1.20 times), ($\textbf{q}$= [C-P]/X= Quick or Acid Test Ratio= 1.01 times), ($\textbf{P}$= Procured Inventories= 6,497 millions USD), ($\textbf{F}$= Fixed Cost= 29,740 millions USD), ($\textbf{M}$= Margin of Contribution= 41,548 millions USD), ($\textbf{i}$= I/S= Interest Portion= 1.00%), ($\textbf{Q}$= Quoted Longterm Liabilities= 35,351 millions USD), ($\textbf{U}$= Utilized or Starting Capital= 64,720 millions USD), ($\textbf{l}$= L/E= Leverage or Gearing Ratio= 99.00%), and ($\textbf{d}$= D/A= Dividend Portion or Payout= 30.00%), then it's ($\textbf{T}$= Tax Paid planned), is:

$$\textbf{T}= \textbf{M-F-Si-}(\{\textbf{Q}+\textbf{P}/[\textbf{c-q}]\}/\textbf{l}-\textbf{U})/[1-\textbf{d}]$$
$$= 41,548-29,750-51,294*0.0100$$
$$-(\{35,351+6,497/[1.2000$$
$$-1.0100]\}/0.9900-64,720)$$
$$/[1-0.3000]$$
$$= 3,385 \text{ millions USD}$$

Corporate IFRS-GAAP (B/S-I/S), ISBN-13: **978-1720792789**, ISBN-10: **172079278X**

Law-7866:

If both ($= Sales or Revenues= 51,294 millions
USD), (c= C/X= Current Ratio= 1.20 times), (q= [C-
P]/X= Quick or Acid Test Ratio= 1.01 times), (P=
Procured Inventories= 6,497 millions USD), (F=
Fixed Cost= 29,740 millions USD), (M= Margin of
Contribution= 41,548 millions USD), (i= I/S=
Interest Portion= 1.00%), (Q= Quoted Longterm
Liabilities= 35,351 millions USD), (U= Utilized or
Starting Capital= 64,720 millions USD), (F= L/E=
Leverage or Gearing Ratio= 99.00%), and (T= Tax
Paid= 3,385 millions USD), then it's (d= Dividend
Portion or Payout planned), is:

$$d= 1-(\{Q+P/[c-q]\}/I-U)/[M-F-Si-T]$$
$$= 1-(\{35,351+6,497/[1.2000$$
$$-1.0100]\}/0.9900-64,720)$$
$$/[41,548-29,750-51,294$$
$$*0.0100-3,385]$$
$$= 30.00\%$$

Corporate IFRS-GAAP (B/S-I/S), ISBN-13: **978-1720792789**, ISBN-10: **172079278X**

<u>Law-7867</u>:

If both (**$**= Sales or Revenues= <u>51,294</u> millions USD), (**c**= C/X= Current Ratio= <u>1.20</u> times), (**q**= [C-P]/X= Quick or Acid Test Ratio= <u>1.01</u> times), (**d**= D/A= Dividend Portion or Payout= <u>30.00%</u>), (**F**= Fixed Cost= <u>29,740</u> millions USD), (**M**= Margin of Contribution= <u>41,548</u> millions USD), (**i**= I/S= Interest Portion= <u>1.00%</u>), (**Q**= Quoted Longterm Liabilities= <u>35,351</u> millions USD), (**U**= Utilized or Starting Capital= <u>64,720</u> millions USD), (**l**= L/E= Leverage or Gearing Ratio= <u>99.00%</u>), and (**T**= Tax Paid= <u>3,385</u> millions USD), then it's (**P**= Procured Inventories planned), is:

$$P= [c\text{-}q](l\,\{U+[M\text{-}F\text{-}Si\text{-}T][1\text{-}d]\}\text{-}Q)$$
$$= [1.2000\text{-}1.0100](0.9900\{64,720$$
$$+[41,548\text{-}29,750\text{-}51,294$$
$$*0.0100\text{-}3,385][1\text{-}0.3000]\}$$
$$-35,351)$$
$$= \underline{6,497} \text{ millions USD}$$

Corporate IFRS-GAAP (B/S-I/S), ISBN-13: **978-1720792789**, ISBN-10: **172079278X**

Law-7868:

    If both (**$**= Sales or Revenues= 51,294 millions
USD), (**P**= Procured Inventories= 6,497 millions
USD), (**q**= [C-P]/X= Quick or Acid Test Ratio= 1.01
times), (**d**= D/A= Dividend Portion or Payout=
30.00%), (**F**= Fixed Cost= 29,740 millions USD),
(**M**= Margin of Contribution= 41,548 millions USD),
(**i**= I/S= Interest Portion= 1.00%), (**Q**= Quoted
Longterm Liabilities= 35,351 millions USD), (**U**=
Utilized or Starting Capital= 64,720 millions USD),
(**l**= L/E= Leverage or Gearing Ratio= 99.00%), and
(**T**= Tax Paid= 3,385 millions USD), then it's (**c**=
Current Ratio planned), is:

$$c = q + P/(l\{U + [M-F-Si-T][1-d]\}-Q)$$

$$= 1.0100 + 6,497/(0.9900\{64,720$$
$$+[41,548-29,750-51,294$$
$$*0.0100-3,385][1-0.3000]\}$$
$$-35,351)$$

$$= \underline{1.20} \text{ times}$$

Corporate IFRS-GAAP (B/S-I/S), ISBN-13: **978-1720792789**, ISBN-10: **172079278X**

<u>Law-7869</u>:

If both (**$**= Sales or Revenues= <u>51,294</u> millions USD), (**P**= Procured Inventories= <u>6,497</u> millions USD), (**c**= C/X= Current Ratio= <u>1.20</u> times), (**d**= D/A= Dividend Portion or Payout= <u>30.00%</u>), (**F**= Fixed Cost= <u>29,740</u> millions USD), (**M**= Margin of Contribution= <u>41,548</u> millions USD), (**i**= I/S= Interest Portion= <u>1.00%</u>), (**Q**= Quoted Longterm Liabilities= <u>35,351</u> millions USD), (**U**= Utilized or Starting Capital= <u>64,720</u> millions USD), (**l**= L/E= Leverage or Gearing Ratio= <u>99.00%</u>), and (**T**= Tax Paid= <u>3,385</u> millions USD), then it's (**q**= Quick or Acid Test Ratio planned), is:

$$q = c\text{-}P/(l\,\{U+[M\text{-}F\text{-}Si\text{-}T][1\text{-}d]\}\text{-}Q)$$
$$= 1.2000\text{-}6,497/(0.9900\{64,720$$
$$+[41,548\text{-}29,750\text{-}51,294$$
$$*0.0100\text{-}3,385][1\text{-}0.3000]\}$$
$$-35,351)$$
$$= \underline{1.01}\ \text{times}$$

Corporate IFRS-GAAP (B/S-I/S), ISBN-13: **978-1720792789**, ISBN-10: **172079278X**

<u>Law-7870</u>:

If both (**$**= Sales or Revenues= <u>51,294</u> millions USD), (**p**= 360 P/V= Procured Inventory Days= <u>240</u> Days), (**V**= Variable Cost= <u>9,746</u> millions USD), (**c**= C/X= Current Ratio= <u>1.20</u> times), (**d**= D/A= Dividend Portion or Payout= <u>30.00%</u>), (**F**= Fixed Cost= <u>29,740</u> millions USD), (**M**= Margin of Contribution= <u>41,548</u> millions USD), (**i**= I/S= Interest Portion= <u>1.00%</u>), (**q**= [C-P]/X= Quick or Acid Test Ratio= <u>1.01</u> times), (**U**= Utilized or Starting Capital= <u>64,720</u> millions USD), (**ℓ**= L/E= Leverage or Gearing Ratio= <u>99.00%</u>), and (**T**= Tax Paid= <u>3,385</u> millions USD), then it's (**Q**= Quoted Longterm Liabilities planned), is:

$$Q = \ell\{U+[M-F-\$i-T][1-d]\}-Vp/\{360[c-q]\}$$
$$= 0.9900\{64,720+[41,548-29,750$$
$$-51,294*0.0100-3,385]$$
$$[1-0.3000]\}-9,746*240$$
$$/\{360[1.2000-1.0100]\}$$
$$= \underline{35,351} \text{ millions USD}$$

Corporate IFRS-GAAP (B/S-I/S), ISBN-13: **978-1720792789**, ISBN-10: **172079278X**

<u>Law-7871</u>:

If both (**$**= Sales or Revenues= <u>51,294</u> millions USD), (**p**= 360 P/V= Procured Inventory Days= <u>240</u> Days), (**V**= Variable Cost= <u>9,746</u> millions USD), (**c**= C/X= Current Ratio= <u>1.20</u> times), (**d**= D/A= Dividend Portion or Payout= <u>30.00%</u>), (**F**= Fixed Cost= <u>29,740</u> millions USD), (**M**= Margin of Contribution= <u>41,548</u> millions USD), (**i**= I/S= Interest Portion= <u>1.00%</u>), (**q**= [C-P]/X= Quick or Acid Test Ratio= <u>1.01</u> times), (**U**= Utilized or Starting Capital= <u>64,720</u> millions USD), (**Q**= Quoted Longterm Liabilities= <u>35,351</u> millions USD), and (**T**= Tax Paid= <u>3,385</u> millions USD), then it's (**I** = Leverage or Gearing Ratio planned), is:

$$I = (Q+Vp/\{360[c-q]\})/\{U+[M-F-Si-T][1-d]\}$$
$$= (35,351+9,746*240/\{360[1.2000$$
$$-1.0100]\})/\{64,720+[41,548$$
$$-29,750-51,294*0.0100$$
$$-3,385][1-0.3000]\}$$
$$= \underline{99.00\%}$$

Corporate IFRS-GAAP (B/S-I/S), ISBN-13: **978-1720792789**, ISBN-10: **172079278X**

Law-7872:

If both ($=$ Sales or Revenues= 51,294 millions USD), ($p=$ 360 P/V= Procured Inventory Days= 240 Days), ($V=$ Variable Cost= 9,746 millions USD), ($c=$ C/X= Current Ratio= 1.20 times), ($d=$ D/A= Dividend Portion or Payout= 30.00%), ($F=$ Fixed Cost= 29,740 millions USD), ($M=$ Margin of Contribution= 41,548 millions USD), ($i=$ I/S= Interest Portion= 1.00%), ($Q=$ Quoted Longterm Liabilities= 35,351 millions USD), ($q=$ [C-P]/X= Quick or Acid Test Ratio= 1.01 times), ($l=$ L/E= Leverage or Gearing Ratio= 99.00%), and ($T=$ Tax Paid= 3,385 millions USD), then it's ($U=$ Utilized or Starting Capital planned), is:

$$U = (Q + Vp/\{360[c-q]\})/l - [M-F-Si-T][1-d]$$
$$= (35,351 + 9,746*240/\{360[1.2000$$
$$-1.0100]\})/0.9900 - [41,548$$
$$-29,750 - 51,294*0.0100$$
$$-3,385][1-0.3000]$$
$$= 64,720 \text{ millions USD}$$

Corporate IFRS-GAAP (B/S-I/S), ISBN-13: **978-1720792789**, ISBN-10: **172079278X**

Law-7873:

If both ($= Sales or Revenues= 51,294 millions
USD), (**p**= 360 P/V= Procured Inventory Days=
240 Days), (**V**= Variable Cost= 9,746 millions USD),
(**c**= C/X= Current Ratio= 1.20 times), (**d**= D/A=
Dividend Portion or Payout= 30.00%), (**F**= Fixed
Cost= 29,740 millions USD), (**U**= Utilized or Starting
Capital= 64,720 millions USD), (**i**= I/S= Interest
Portion= 1.00%), (**Q**= Quoted Longterm Liabilities=
35,351 millions USD), (**q**= [C-P]/X= Quick or Acid
Test Ratio= 1.01 times), (**I**= L/E= Leverage or
Gearing Ratio= 99.00%), and (**T**= Tax Paid= 3,385
millions USD), then it's (**M**= Margin of Contribution
planned), is:

$$M= F+Si+T+[(Q+Vp/\{360[c-q]\})/I-U]/[1-d]$$
$$= 29,750+51,294*0.0100+3,385$$
$$+[(35,351+9,746*240/\{360$$
$$[1.2000-1.0100]\})/0.9900$$
$$-64,720]/[1-0.3000]$$
$$= 41,548 \text{ millions USD}$$

Corporate IFRS-GAAP (B/S-I/S), ISBN-13: **978-1720792789**, ISBN-10: **172079278X**

<u>Law-7874</u>:

If both (**$**= Sales or Revenues= <u>51,294</u> millions USD), (**p**= 360 P/V= Procured Inventory Days= <u>240</u> Days), (**V**= Variable Cost= <u>9,746</u> millions USD), (**c**= C/X= Current Ratio= <u>1.20</u> times), (**d**= D/A= Dividend Portion or Payout= <u>30.00%</u>), (**M**= Margin of Contribution= <u>41,548</u> millions USD), (**U**= Utilized or Starting Capital= <u>64,720</u> millions USD), (**i**= I/S= Interest Portion= <u>1.00%</u>), (**Q**= Quoted Longterm Liabilities= <u>35,351</u> millions USD), (**q**= [C-P]/X= Quick or Acid Test Ratio= <u>1.01</u> times), (***l***= L/E= Leverage or Gearing Ratio= <u>99.00%</u>), and (**T**= Tax Paid= <u>3,385</u> millions USD), then it's (**F**= Fixed Cost planned), is:

$$F= M\text{-}\$i\text{-}T\text{-}[(Q+Vp/\{360[c\text{-}q]\})/\textit{l}\text{ -}U]/[1\text{-}d]$$

$$= 41,548\text{-}51,294*0.0100\text{-}3,385$$
$$-[(35,351+9,746*240/\{360$$
$$[1.2000\text{-}1.0100]\})/0.9900$$
$$-64,720]/[1\text{-}0.3000]$$

$$= \underline{29,750}\text{ millions USD}$$

Corporate IFRS-GAAP (B/S-I/S), ISBN-13: **978-1720792789**, ISBN-10: **172079278X**

Law-7875:

If both (**F**= Fixed Cost= 29,740 millions USD), (**p**= 360 P/V= Procured Inventory Days= 240 Days), (**V**= Variable Cost= 9,746 millions USD), (**c**= C/X= Current Ratio= 1.20 times), (**d**= D/A= Dividend Portion or Payout= 30.00%), (**M**= Margin of Contribution= 41,548 millions USD), (**U**= Utilized or Starting Capital= 64,720 millions USD), (**i**= I/S= Interest Portion= 1.00%), (**Q**= Quoted Longterm Liabilities= 35,351 millions USD), (**q**= [C-P]/X= Quick or Acid Test Ratio= 1.01 times), (**ℓ**= L/E= Leverage or Gearing Ratio= 99.00%), and (**T**= Tax Paid= 3,385 millions USD), then it's (**S**= Sales or Revenues planned), is:

$$S= \{M\text{-}F\text{-}T\text{-}[(Q+Vp/\{360[c\text{-}q]\})/\ell\text{-}U]/[1\text{-}d]\}/i$$
$$= \{41,548\text{-}29,750\text{-}3.385\text{-}[(35,351$$
$$+9,746*240/\{360[1.2000$$
$$-1.0100]\}\ )/0.9900\text{-}64,720]$$
$$/[1\text{-}0.3000]\}/0.0100$$
$$= 51,294 \text{ millions USD}$$

Corporate IFRS-GAAP (B/S-I/S), ISBN-13: **978-1720792789**, ISBN-10: **172079278X**

Law-7876:

If both (**F**= Fixed Cost= <u>29,740</u> millions USD), (**p**= 360 P/V= Procured Inventory Days= <u>240</u> Days), (**V**= Variable Cost= <u>9,746</u> millions USD), (**c**= C/X= Current Ratio= <u>1.20</u> times), (**d**= D/A= Dividend Portion or Payout= <u>30.00%</u>), (**M**= Margin of Contribution= <u>41,548</u> millions USD), (**U**= Utilized or Starting Capital= <u>64,720</u> millions USD), (**S**= Sales or Revenues= <u>51,294</u> millions USD), (**Q**= Quoted Longterm Liabilities= <u>35,351</u> millions USD), (**q**= Quick or Acid Test Ratio= [C-P]/X= <u>1.01</u> times), (**l**= L/E= Leverage or Gearing Ratio= <u>99.00%</u>), and (**T**= Tax Paid= <u>3,385</u> millions USD), then it's (**i**= Interest Portion planned), is:

$$\mathbf{i} = \{\mathbf{M} \text{-} \mathbf{F} \text{-} \mathbf{T} \text{-} [(\mathbf{Q} + \mathbf{Vp}/\{360[\mathbf{c} \text{-} \mathbf{q}]\})/\mathbf{l} \text{-} \mathbf{U}]$$
$$/[1 \text{-} \mathbf{d}]\}/\mathbf{S}$$
$$= \{41,548 \text{-} 29,750 \text{-} 3,385 \text{-} [(35,351$$
$$+9,746*240/\{360[1.2000$$
$$-1.0100]\})/0.9900 \text{-} 64,720]$$
$$/[1 \text{-} 0.3000]\}/51,294$$
$$= \underline{1.00\%}$$

Corporate IFRS-GAAP (B/S-I/S), ISBN-13: **978-1720792789**, ISBN-10: **172079278X**

<u>Law-7877</u>:

If both (**F**= Fixed Cost= <u>29,740</u> millions USD), (**p**= 360 P/V= Procured Inventory Days= <u>240</u> Days), (**V**= Variable Cost= <u>9,746</u> millions USD), (**c**= C/X= Current Ratio= <u>1.20</u> times), (**d**= D/A= Dividend Portion or Payout= <u>30.00%</u>), (**M**= Margin of Contribution= <u>41,548</u> millions USD), (**U**= Utilized or Starting Capital= <u>64,720</u> millions USD), (**S**= Sales or Revenues= <u>51,294</u> millions USD), (**Q**= Quoted Longterm Liabilities= <u>35,351</u> millions USD), (**q**= Quick or Acid Test Ratio= [C-P]/X= <u>1.01</u> times), (**I**= L/E= Leverage or Gearing Ratio= <u>99.00%</u>), and (**i**= I/S= Interest Portion= <u>1.00%</u>), then it's (**T**= Tax Paid planned), is:

$$T= M\text{-}F\text{-}Si\text{-}[(Q+Vp/\{360[c\text{-}q]\})/I\text{-}U]/[1\text{-}d]$$
$$= 41,548\text{-}29,750\text{-}51,294*0.0100$$
$$-[(35,351+9,746*240/\{360$$
$$[1.2000\text{-}1.0100]\})/0.9900$$
$$-64,720]/[1\text{-}0.3000]$$
$$= \underline{3,385} \text{ millions USD}$$

Corporate IFRS-GAAP (B/S-I/S), ISBN-13: **978-1720792789**, ISBN-10: **172079278X**

<u>Law-7878</u>:

If both (**F**= Fixed Cost= <u>29,740</u> millions USD), (**p**= 360 P/V= Procured Inventory Days= <u>240</u> Days), (**V**= Variable Cost= <u>9,746</u> millions USD), (**c**= C/X= Current Ratio= <u>1.20</u> times), (**T**= Tax Paid= <u>3,385</u> millions USD), (**M**= Margin of Contribution= <u>41,548</u> millions USD), (**U**= Utilized or Starting Capital= <u>64,720</u> millions USD), (**S**= Sales or Revenues= <u>51,294</u> millions USD), (**Q**= Quoted Longterm Liabilities= <u>35,351</u> millions USD), (**q**= [C-P]/X= Quick or Acid Test Ratio= <u>1.01</u> times), (**I**= L/E= Leverage or Gearing Ratio= <u>99.00%</u>), and (**i**= I/S= Interest Portion= <u>1.00%</u>), then it's (**d**= Dividend Portion or Payout planned), is:

$$\mathbf{d} = 1 - \mathbf{[(Q + Vp}/\{360[\mathbf{c\text{-}q}]\})/\mathbf{I}\,\mathbf{\text{-}U]}$$
$$/\mathbf{[M\text{-}F\text{-}Si\text{-}T]}$$
$$= 1 - \mathbf{[}(35{,}351 + 9{,}746*240/\{360[1.2000}$$
$$-1.0100]\})/0.9900 - 64{,}720]$$
$$/[41{,}548 - 29{,}750 - 51{,}294$$
$$*0.0100 - 3{,}385\mathbf{]}$$
$$= \underline{30.00\%}$$

Corporate IFRS-GAAP (B/S-I/S), ISBN-13: **978-1720792789**, ISBN-10: **172079278X**

Law-7879:

If both (**F**= Fixed Cost= <u>29,740</u> millions USD), (**p**= 360 P/V= Procured Inventory Days= <u>240</u> Days), (**d**= D/A= Dividend Portion or Payout= <u>30.00%</u>), (**c**= C/X= Current Ratio= <u>1.20</u> times), (**T**= Tax Paid= <u>3,385</u> millions USD), (**M**= Margin of Contribution= <u>41,548</u> millions USD), (**U**= Utilized or Starting Capital= <u>64,720</u> millions USD), (**S**= Sales or Revenues= <u>51,294</u> millions USD), (**Q**= Quoted Longterm Liabilities= <u>35,351</u> millions USD), (**q**= Quick or Acid Test Ratio= [C-P]/X= <u>1.01</u> times), (**l**= L/E= Leverage or Gearing Ratio= <u>99.00%</u>), and (**i**= I/S= Interest Portion= <u>1.00%</u>), then it's (**V**= Variable Cost planned), is:

$$\mathbf{V} = 360[\mathbf{c}\text{-}\mathbf{q}](\mathbf{l}\ \{\mathbf{U}+[\mathbf{M}\text{-}\mathbf{F}\text{-}\mathbf{Si}\text{-}\mathbf{T}][1\text{-}\mathbf{d}]\}\text{-}\mathbf{Q})/\mathbf{p}$$

$$= 360[1.2000\text{-}1.0100]\})/0.9900$$
$$\{64,720+[41,548\text{-}29,750$$
$$-51,294*0.0100\text{-}3,385]$$
$$[1\text{-}0.3000]\}\text{-}35,351)/240$$

$$= \underline{9,746}\ \text{millions USD}$$

Corporate IFRS-GAAP (B/S-I/S), ISBN-13: **978-1720792789**, ISBN-10: **172079278X**

Law-7880:

If both (**F**= Fixed Cost= 29,740 millions USD), (**V**=
Variable Cost= 9,746 millions USD), (**d**= D/A=
Dividend Portion or Payout= 30.00%), (**c**= C/X=
Current Ratio= 1.20 times), (**T**= Tax Paid= 3,385
millions USD), (**M**= Margin of Contribution= 41,548
millions USD), (**U**= Utilized or Starting Capital=
64,720 millions USD), (**S**= Sales or Revenues=
51,294 millions USD), (**Q**= Quoted Longterm
Liabilities= 35,351 millions USD), (**q**= [C-P]/X=
Quick or Acid Test Ratio= 1.01 times), (**I**= L/E=
Leverage or Gearing Ratio= 99.00%), and (**i**= I/S=
Interest Portion= 1.00%), then it's (**p**= Procured
Inventory Days planned), is:

$$p= 360[\text{c-q}](I \{U+[M-F-Si-T]$$
$$[1-d]\}-Q)/V$$
$$= 360[1.2000-1.0100]\})/0.9900$$
$$\{64,720+[41,548-29,750$$
$$-51,294*0.0100-3,385]$$
$$[1-0.3000]\}-35,351)/9,746$$
$$= \underline{240} \text{ days}$$

Corporate IFRS-GAAP (B/S-I/S), ISBN-13: **978-1720792789**, ISBN-10: **172079278X**

Law-7881:

If both (**F**= Fixed Cost= 29,740 millions USD), (**V**= Variable Cost= 9,746 millions USD), (**d**= D/A= Dividend Portion or Payout= 30.00%), (**p**= 360 P/V= Procured Inventory Days= 240 Days), (**T**= Tax Paid= 3,385 millions USD), (**M**= Margin of Contribution= 41,548 millions USD), (**U**= Utilized or Starting Capital= 64,720 millions USD), (**$**= Sales or Revenues= 51,294 millions USD), (**Q**= Quoted Longterm Liabilities= 35,351 millions USD), (**q**= Quick or Acid Test Ratio= [C-P]/X= 1.01 times), (**/**= L/E= Leverage or Gearing Ratio= 99.00%), and (**i**= I/S= Interest Portion= 1.00%), then it's (**c**= Current Ratio planned), is:

$$c= q+Vp/[360(/ \{U+[M-F-$i-T][1-d]\}-Q)]$$
$$= 1.0100+9,746*240/[360(0.9900$$
$$\{64,720+[41,548-29,750$$
$$-51,294*0.0100-3,385]$$
$$[1-0.3000]\}-35,351)]$$
$$= 1.20 \text{ times}$$

Corporate IFRS-GAAP (B/S-I/S), ISBN-13: **978-1720792789**, ISBN-10: **172079278X**

Law-7882:

If both (**F**= Fixed Cost= 29,740 millions USD), (**V**= Variable Cost= 9,746 millions USD), (**d**= D/A= Dividend Portion or Payout= 30.00%), (**p**= 360 P/V= Procured Inventory Days= 240 Days), (**T**= Tax Paid= 3,385 millions USD), (**M**= Margin of Contribution= 41,548 millions USD), (**U**= Utilized or Starting Capital= 64,720 millions USD), (**S**= Sales or Revenues= 51,294 millions USD), (**Q**= Quoted Longterm Liabilities= 35,351 millions USD), (**c**= C/X= Current Ratio= 1.20 times), (**l**= L/E= Leverage or Gearing Ratio= 99.00%), and (**i**= I/S= Interest Portion= 1.00%), then it's (**q**= Quick or Acid Test Ratio planned), is:

$$q = c\text{-}Vp/[360(\textbf{\textit{l}}\{U+[M\text{-}F\text{-}Si\text{-}T][1\text{-}d]\}\text{-}Q)]$$
$$= 1.2000\text{-}9,746*240/[360(0.9900$$
$$\{64,720+[41,548\text{-}29,750$$
$$-51,294*0.0100\text{-}3,385]$$
$$[1\text{-}0.3000]\}\text{-}35,351)]$$
$$= \underline{1.01} \text{ times}$$

Corporate IFRS-GAAP (B/S-I/S), ISBN-13: **978-1720792789**, ISBN-10: **172079278X**

<u>Law-7883</u>:

If both (**F**= Fixed Cost= <u>29,740</u> millions USD), (**v**= V/S= Variable Portion= <u>19.00%</u>), (**d**= D/A= Dividend Portion or Payout= <u>30.00%</u>), (**p**= 360 P/V= Procured Inventory Days= <u>240</u> Days), (**T**= Tax Paid= <u>3,385</u> millions USD), (**M**= Margin of Contribution= <u>41,548</u> millions USD), (**U**= Utilized or Starting Capital= <u>64,720</u> millions USD), (**S**= Sales or Revenues= <u>51,294</u> millions USD), (**q**= [C-P]/X= Quick or Acid Test Ratio= <u>1.01</u> times), (**c**= C/X= Current Ratio= <u>1.20</u> times), (**l**= L/E= Leverage or Gearing Ratio= <u>99.00%</u>), and (**i**= I/S= Interest Portion= <u>1.00%</u>), then it's (**Q**= Quoted Longterm Liabilities planned), is:

$$Q = l\{U+[M\text{-}F\text{-}Si\text{-}T][1\text{-}d]\}$$
$$-Svp/\{360[c\text{-}q]\}$$
$$= 0.9900\{64,720+[41,548\text{-}29,750$$
$$-51,294*0.0100\text{-}3,385]$$
$$[1\text{-}0.3000]\}\text{-}51,294$$
$$*0.1900*240/\{360[1.2000$$
$$-1.0100]\}$$
$$= \underline{35,351} \text{ millions USD}$$

Corporate IFRS-GAAP (B/S-I/S), ISBN-13: **978-1720792789**, ISBN-10: **172079278X**

<u>Law-7884</u>:

If both (**F**= Fixed Cost= <u>29,740</u> millions USD), (**v**= V/S= Variable Portion= <u>19.00%</u>), (**d**= D/A= Dividend Portion or Payout= <u>30.00%</u>), (**p**= 360 P/V= Procured Inventory Days= <u>240</u> Days), (**T**= Tax Paid= <u>3,385</u> millions USD), (**M**= Margin of Contribution= <u>41,548</u> millions USD), (**U**= Utilized or Starting Capital= <u>64,720</u> millions USD), (**S**= Sales or Revenues= <u>51,294</u> millions USD), (**q**= [C-P]/X= Quick or Acid Test Ratio= <u>1.01</u> times), (**c**= C/X= Current Ratio= <u>1.20</u> times), (**Q**= Quoted Longterm Liabilities= <u>35,351</u> millions USD), and (**i**= I/S= Interest Portion= <u>1.00%</u>), then it's (**I** = Leverage or Gearing Ratio planned), is:

$$I = (\mathbf{Q} + \mathbf{Svp}/\{360[\mathbf{c-q}]\})$$
$$/\{\mathbf{U} + [\mathbf{M-F-Si-T}][1-\mathbf{d}]\}$$
$$= (35,351 + 51,294*0.1900*240/\{360$$
$$[1.2000 - 1.0100]\})/\{64,720$$
$$+ [41,548 - 29,750 - 51,294$$
$$*0.0100 - 3,385][1 - 0.3000]\}$$
$$= \underline{99.00\%}$$

Corporate IFRS-GAAP (B/S-I/S), ISBN-13: **978-1720792789**, ISBN-10: **172079278X**

Law-7885:

If both (**F**= Fixed Cost= 29,740 millions USD), (**v**=
V/S= Variable Portion= 19.00%), (**d**= D/A=
Dividend Portion or Payout= 30.00%), (**p**= 360 P/V=
Procured Inventory Days= 240 Days), (**T**= Tax Paid=
3,385 millions USD), (**M**= Margin of Contribution=
41,548 millions USD), (**l**= L/E= Leverage or Gearing
Ratio= 99.00%), (**S**= Sales or Revenues= 51,294
millions USD), (**q**= [C-P]/X= Quick or Acid Test
Ratio= 1.01 times), (**c**= C/X= Current Ratio= 1.20
times), (**Q**= Quoted Longterm Liabilities= 35,351
millions USD), and (**i**= I/S= Interest Portion= 1.00%),
then it's (**U**= Utilized or Starting Capital planned), is:

$$U= (Q+Svp/\{360[c-q]\})/l -[M-F-Si-T][1-d]$$
$$= (35,351+51,294*0.1900*240/\{360$$
$$[1.2000-1.0100]\})/0.9900$$
$$-[41,548-29,750-51,294$$
$$*0.0100-3,385][1-0.3000]$$
$$= \underline{64,720} \text{ millions USD}$$

<u>Law-7886</u>:

If both (**F**= Fixed Cost= <u>29,740</u> millions USD), (**v**= V/S= Variable Portion= <u>19.00%</u>), (**d**= D/A= Dividend Portion or Payout= <u>30.00%</u>), (**p**= 360 P/V= Procured Inventory Days= <u>240</u> Days), (**T**= Tax Paid= <u>3,385</u> millions USD), (**U**= Utilized or Starting Capital= <u>64,720</u> millions USD), (**I**= L/E= Leverage or Gearing Ratio= <u>99.00%</u>), (**S**= Sales or Revenues= <u>51,294</u> millions USD), (**q**= [C-P]/X= Quick or Acid Test Ratio= <u>1.01</u> times), (**c**= C/X= Current Ratio= <u>1.20</u> times), (**Q**= Quoted Longterm Liabilities= <u>35,351</u> millions USD), and (**i**= I/S= Interest Portion= <u>1.00%</u>), then it's (**M**= Margin of Contribution planned), is:

$$M= F+Si+T+[(Q+Svp/\{360[c-q]\})/I -U]/[1-d]$$

$$= 29,750+51,294*0.0100+3,385$$
$$+[(35,351+51,294*0.1900$$
$$*240/\{360[1.2000-1.0100]\})$$
$$/0.9900-64,720]/[1-0.3000]$$

$$= \underline{41,548} \text{ millions USD}$$

Corporate IFRS-GAAP (B/S-I/S), ISBN-13: **978-1720792789**, ISBN-10: **172079278X**

Law-7887:

If both (**M**= Margin of Contribution= <u>41,548</u> millions USD), (**v**= V/S= Variable Portion= <u>19.00%</u>), (**d**= D/A= Dividend Portion or Payout= <u>30.00%</u>), (**p**= 360 P/V= Procured Inventory Days= <u>240</u> Days), (**T**= Tax Paid= <u>3,385</u> millions USD), (**U**= Utilized or Starting Capital= <u>64,720</u> millions USD), (**I**= L/E= Leverage or Gearing Ratio= <u>99.00%</u>), (**S**= Sales or Revenues= <u>51,294</u> millions USD), (**q**= [C-P]/X= Quick or Acid Test Ratio= <u>1.01</u> times), (**c**= C/X= Current Ratio= <u>1.20</u> times), (**Q**= Quoted Longterm Liabilities= <u>35,351</u> millions USD), and (**i**= I/S= Interest Portion= <u>1.00%</u>), then it's (**F** = Fixed Cost planned), is:

$$F= M\text{-}Si\text{-}T\text{-}[(Q+Svp/\{360[c\text{-}q]\})/I\text{-}U]$$
$$/[1\text{-}d]$$
$$= 41,548\text{-}51,294*0.0100\text{-}3,385$$
$$-[(35,351+51,294*0.1900$$
$$*240/\{360[1.2000\text{-}1.0100]\})$$
$$/0.9900\text{-}64,720]/[1\text{-}0.3000]$$
$$= \underline{29,750} \text{ millions USD}$$

Corporate IFRS-GAAP (B/S-I/S), ISBN-13: **978-1720792789**, ISBN-10: **172079278X**

Law-7888:

If both (**M**= Margin of Contribution= 41,548 millions USD), (**v**= V/S= Variable Portion= 19.00%), (**d**= D/A= Dividend Portion or Payout= 30.00%), (**p**= 360 P/V= Procured Inventory Days= 240 Days), (**T**= Tax Paid= 3,385 millions USD), (**U**= Utilized or Starting Capital= 64,720 millions USD), (**I**= L/E= Leverage or Gearing Ratio= 99.00%), (**F**= Fixed Cost= 29,740 millions USD), (**q**= [C-P]/X= Quick or Acid Test Ratio= 1.01 times), (**c**= C/X= Current Ratio= 1.20 times), (**Q**= Quoted Longterm Liabilities= 35,351 millions USD), and (**i**= I/S= Interest Portion= 1.00%), then it's (**S** = Sales or Revenues planned), is:

$$S= \{M\text{-}F\text{-}T\text{-}[(Q+Svp/\{360[c\text{-}q]\})/I\text{-}U]$$
$$/[1\text{-}d]\}/i$$
$$= \{41,548\text{-}29,750\text{-}3,385\text{-}[(35,351$$
$$+51,294*0.1900*240/\{360$$
$$[1.2000\text{-}1.0100]\})/0.9900$$
$$-64,720]/[1\text{-}0.3000]\}$$
$$/0.0100$$
$$= 51,294 \text{ millions USD}$$

Corporate IFRS-GAAP (B/S-I/S), ISBN-13: **978-1720792789**, ISBN-10: **172079278X**

<u>Law-7889</u>:

If both (**M**= Margin of Contribution= <u>41,548</u> millions USD), (**v**= V/S= Variable Portion= <u>19.00%</u>), (**d**= D/A= Dividend Portion or Payout= <u>30.00%</u>), (**p**= 360 P/V= Procured Inventory Days= <u>240</u> Days), (**T**= Tax Paid= <u>3,385</u> millions USD), (**U**= Utilized or Starting Capital= <u>64,720</u> millions USD), (**I**= L/E= Leverage or Gearing Ratio= <u>99.00%</u>), (**F**= Fixed Cost= <u>29,740</u> millions USD), (**q**= [C-P]/X= Quick or Acid Test Ratio= <u>1.01</u> times), (**c**= C/X= Current Ratio= <u>1.20</u> times), (**Q**= Quoted Longterm Liabilities= <u>35,351</u> millions USD), and (**$**= Sales or Revenues= <u>51,294</u> millions USD), then it's (**i** = Interest Portion planned), is:

$$i= \{M\text{-}F\text{-}T\text{-}[(Q+\$vp/\{360[c\text{-}q]\})/I\text{-}U]$$
$$/[1\text{-}d]\}/\$$$
$$= \{41,548\text{-}29,750\text{-}3,385\text{-}[(35,351$$
$$+51,294*0.1900*240/\{360$$
$$[1.2000\text{-}1.0100]\})/0.9900$$
$$-64,720]/[1\text{-}0.3000]\}/51,294$$
$$= \underline{1.00\%}$$

Corporate IFRS-GAAP (B/S-I/S), ISBN-13: **978-1720792789**, ISBN-10: **172079278X**

Law-7890:

If both (**M**= Margin of Contribution= 41,548 millions USD), (**v**= V/S= Variable Portion= 19.00%), (**d**= D/A= Dividend Portion or Payout= 30.00%), (**p**= 360 P/V= Procured Inventory Days= 240 Days), (**i**= I/S= Interest Portion= 1.00%), (**U**= Utilized or Starting Capital= 64,720 millions USD), (**l**= L/E= Leverage or Gearing Ratio= 99.00%), (**F**= Fixed Cost= 29,740 millions USD), (**q**= [C-P]/X= Quick or Acid Test Ratio= 1.01 times), (**c**= C/X= Current Ratio= 1.20 times), (**Q**= Quoted Longterm Liabilities= 35,351 millions USD), and (**S**= Sales or Revenues= 51,294 millions USD), then it's (**T** = Tax Paid planned), is:

$$T= M\text{-}F\text{-}Si\text{-}[(Q+Svp/\{360[c\text{-}q]\})/l\text{-}U]$$
$$/[1\text{-}d]$$

$$= 41,548\text{-}29,750\text{-}51,294*0.0100$$
$$-[(35,351+51,294*0.1900$$
$$*240/\{360[1.2000\text{-}1.0100]\})$$
$$/0.9900\text{-}64,720]/[1\text{-}0.3000]$$

$$= \underline{3,385} \text{ millions USD}$$

Corporate IFRS-GAAP (B/S-I/S), ISBN-13: **978-1720792789**, ISBN-10: **172079278X**

Law-7891:

If both (**M**= Margin of Contribution= 41,548 millions USD), (**v**= V/S= Variable Portion= 19.00%), (**T**= Tax Paid= 3,385 millions USD), (**p**= 360 P/V= Procured Inventory Days= 240 Days), (**i**= I/S= Interest Portion= 1.00%), (**U**= Utilized or Starting Capital= 64,720 millions USD), (**F**= L/E= Leverage or Gearing Ratio= 99.00%), (**F**= Fixed Cost= 29,740 millions USD), (**q**= [C-P]/X= Quick or Acid Test Ratio= 1.01 times), (**c**= C/X= Current Ratio= 1.20 times), (**Q**= Quoted Longterm Liabilities= 35,351 millions USD), and (**S**= Sales or Revenues= 51,294 millions USD), then it's (**d** = Dividend Portion or Payout planned), is:

$$\mathbf{d}= 1-[(\mathbf{Q+Svp}/\{360[\mathbf{c-q}]\})/\mathbf{I-U}]$$
$$/[\mathbf{M-F-Si-T}]$$

$$= 1-[(35,351+51,294*0.1900*240$$
$$/\{360[1.2000-1.0100]\})$$
$$/0.9900-64,720]/[41,548$$
$$-29,750-51,294*0.0100$$
$$-3,385]$$

$$= 30.00\%$$

Corporate IFRS-GAAP (B/S-I/S), ISBN-13: **978-1720792789**, ISBN-10: **172079278X**

<u>Law-7892</u>:

If both (**M**= Margin of Contribution= <u>41,548</u> millions USD), (**d**= D/A= Dividend Portion or Payout= <u>30.00%</u>), (**T**= Tax Paid= <u>3,385</u> millions USD), (**p**= 360 P/V= Procured Inventory Days= <u>240</u> Days), (**i**= I/S= Interest Portion= <u>1.00%</u>), (**U**= Utilized or Starting Capital= <u>64,720</u> millions USD), (**l**= L/E= Leverage or Gearing Ratio= <u>99.00%</u>), (**F**= Fixed Cost= <u>29,740</u> millions USD), (**q**= [C-P]/X= Quick or Acid Test Ratio= <u>1.01</u> times), (**c**= C/X= Current Ratio= <u>1.20</u> times), (**Q**= Quoted Longterm Liabilities= <u>35,351</u> millions USD), and (**S**= Sales or Revenues= <u>51,294</u> millions USD), then it's (**v**= Variable Portion planned), is:

$$v= 360[c-q](l\{U+[M-F-Si-T][1-d]\}-Q)/[Sp]$$

$$= 360[1.2000-1.0100](0.9900$$
$$\{64,720]+[41,548-29,750$$
$$-51,294*0.0100-3,385]$$
$$[1-0.3000]\}-35,351)$$
$$/[51,294*240]$$

$$= \underline{19.00\%}$$

Corporate IFRS-GAAP (B/S-I/S), ISBN-13: **978-1720792789**, ISBN-10: **172079278X**

Law-7893:

If both (**M**= Margin of Contribution= 41,548 millions USD), (**v**= V/S= Variable Portion= 19.00%), (**T**= Tax Paid= 3,385 millions USD), (**d**= D/A= Dividend Portion or Payout= 30.00%), (**i**= I/S= Interest Portion= 1.00%), (**U**= Utilized or Starting Capital= 64,720 millions USD), (**l**= L/E= Leverage or Gearing Ratio= 99.00%), (**F**= Fixed Cost= 29,740 millions USD), (**q**= [C-P]/X= Quick or Acid Test Ratio= 1.01 times), (**c**= C/X= Current Ratio= 1.20 times), (**Q**= Quoted Longterm Liabilities= 35,351 millions USD), and (**\$**= Sales or Revenues= 51,294 millions USD), then it's (**p** = Procured Inventory Days planned), is:

$$p= 360[\textbf{c-q}](\textbf{l} \{\textbf{U}+[\textbf{M-F-\$i-T}][1-\textbf{d}]\}-\textbf{Q})$$
$$/[\textbf{\$v}]$$

$$= 360[1.2000-1.0100](0.9900$$
$$\{64,720\}+[41,548-29,750$$
$$-51,294*0.0100-3,385]$$
$$[1-0.3000]\}-35,351)/[51,294$$
$$*0.1900]$$

$$= \underline{240} \text{ days}$$

Corporate IFRS-GAAP (B/S-I/S), ISBN-13: **978-1720792789**, ISBN-10: **172079278X**

Law-7894:

If both (**M**= Margin of Contribution= 41,548 millions USD), (**v**= V/S= Variable Portion= 19.00%), (**T**= Tax Paid= 3,385 millions USD), (**d**= D/A= Dividend Portion or Payout= 30.00%), (**i**= I/S= Interest Portion= 1.00%), (**U**= Utilized or Starting Capital= 64,720 millions USD), (**l**= L/E= Leverage or Gearing Ratio= 99.00%), (**F**= Fixed Cost= 29,740 millions USD), (**q**= [C-P]/X= Quick or Acid Test Ratio= 1.01 times), (**p**= 360 P/V= Procured Inventory Days= 240 Days), (**Q**= Quoted Longterm Liabilities= 35,351 millions USD), and (**S**= Sales or Revenues= 51,294 millions USD), then it's (**c** = Current Ratio planned), is:

$$c= q+Svp/[360(l\{U+[M-F-Si-T][1-d]\}-Q)]$$
$$= 1.0100+51,294*0.1900*240/[360$$
$$(0.9900\{64,720]+[41,548$$
$$-29,750-51,294*0.0100$$
$$-3,385][1-0.3000]\}-35,351)]$$
$$= 1.20 \text{ times}$$

Corporate IFRS-GAAP (B/S-I/S), ISBN-13: **978-1720792789**, ISBN-10: **172079278X**

Law-7895:

If both (**M**= Margin of Contribution= 41,548 millions USD), (**v**= V/S= Variable Portion= 19.00%), (**T**= Tax Paid= 3,385 millions USD), (**d**= D/A= Dividend Portion or Payout= 30.00%), (**i**= I/S= Interest Portion= 1.00%), (**U**= Utilized or Starting Capital= 64,720 millions USD), (**⌐**= L/E= Leverage or Gearing Ratio= 99.00%), (**F**= Fixed Cost= 29,740 millions USD), (**c**= C/X= Current Ratio= 1.20 times), (**p**= 360 P/V= Procured Inventory Days= 240 Days), (**Q**= Quoted Longterm Liabilities= 35,351 millions USD), and (**S**= Sales or Revenues= 51,294 millions USD), then it's (**q** = Quick or Acid Test Ratio planned), is:

$$q= c\text{-}Svp/[360(\text{⌐}\{U+[M\text{-}F\text{-}Si\text{-}T][1\text{-}d]\}\text{-}Q)]$$
$$= 1.2000\text{-}51,294*0.1900*240/[360$$
$$(0.9900\{64,720]+[41,548$$
$$-29,750\text{-}51,294*0.0100$$
$$-3,385][1\text{-}0.3000]\}$$
$$-35,351)]$$
$$= 1.01 \text{ times}$$

Law-7896:

If both (**M**= Margin of Contribution= 41,548 millions USD), (**S'**= Sales of Past Year= 48,851 millions USD), (**s**= [S/S']-1= Sales Growth= 5.00%), (**v**= V/S= Variable Portion= 19.00%), (**T**= Tax Paid= 3,385 millions USD), (**d**= D/A= Dividend Portion or Payout= 30.00%), (**i**= I/S= Interest Portion= 1.00%), (**U**= Utilized or Starting Capital= 64,720 millions USD), (**l**= L/E= Leverage or Gearing Ratio= 99.00%), (**F**= Fixed Cost= 29,740 millions USD), (**c**= C/X= Current Ratio= 1.20 times), (**p**= 360 P/V= Procured Inventory Days= 240 Days), (**q**= [C-P]/X= Quick or Acid Test Ratio= 1.01 times), and (**S**= Sales or Revenues= 51,294 millions USD), then it's (**Q**= Quoted Longterm Liabilities planned), is:

$$Q = l\{U+[M-F-Si-T][1-d]\}$$
$$-S'vp[1+s]/\{360[c-q]\}$$
$$= 0.9900\{64,720]+[41,548-29,750$$
$$-51,294*0.0100-3,385]$$
$$[1-0.3000]\}-48,851*0.1900$$
$$*240/\{360[1.2000-1.0100]\}$$
$$= 35,351 \text{ millions USD}$$

Corporate IFRS-GAAP (B/S-I/S), ISBN-13: **978-1720792789**, ISBN-10: **172079278X**

Law-7897:

If both (**M**= Margin of Contribution= 41,548 millions
USD), (**S'**= Sales of Past Year= 48,851 millions
USD), (**s**= [S/S']-1= Sales Growth= 5.00%), (**v**=
V/S= Variable Portion= 19.00%), (**T**= Tax Paid=
3,385 millions USD), (**d**= D/A= Dividend Portion or
Payout= 30.00%), (**i**= I/S= Interest Portion= 1.00%),
(**U**= Utilized or Starting Capital= 64,720 millions
USD), (**Q**= Quoted Longterm Liabilities= 35,351
millions USD), (**F**= Fixed Cost= 29,740 millions
USD), (**c**= C/X= Current Ratio= 1.20 times), (**p**= 360
P/V= Procured Inventory Days= 240 Days), (**q**=
Quick or Acid Test Ratio= [C-P]/X= 1.01 times), and
(**S**= Sales or Revenues= 51,294 millions USD), then
it's (**I** = Leverage or Gearing Ratio planned), is:

$$I = (\mathbf{Q+S'vp}[1+\mathbf{s}]/\{360[\mathbf{c\text{-}q}]\})$$
$$/\{\mathbf{U}+[\mathbf{M\text{-}F\text{-}Si\text{-}T}][1\text{-}\mathbf{d}]\}$$
$$= (35,351+48,851*0.1900*240$$
$$/\{360[1.2000\text{-}1.0100]\})$$
$$/\{64,720]+[41,548\text{-}29,750$$
$$-51,294*0.0100\text{-}3,385]$$
$$[1\text{-}0.3000]\}$$
$$= 99.00\%$$

Corporate IFRS-GAAP (B/S-I/S), ISBN-13: **978-1720792789**, ISBN-10: **172079278X**

<u>Law-7898</u>:

If both (**M**= Margin of Contribution= <u>41,548</u> millions USD), (**S'**= Sales of Past Year= <u>48,851</u> millions USD), (**s**= [S/S']-1= Sales Growth= <u>5.00%</u>), (**v**= V/S= Variable Portion= <u>19.00%</u>), (**T**= Tax Paid= <u>3,385</u> millions USD), (**d**= D/A= Dividend Portion or Payout= <u>30.00%</u>), (**i**= I/S= Interest Portion= <u>1.00%</u>), (**Q**= Quoted Longterm Liabilities= <u>35,351</u> millions USD), (**I**= L/E= Leverage or Gearing Ratio= <u>99.00%</u>), (**F**= Fixed Cost= <u>29,740</u> millions USD), (**c**= C/X= Current Ratio= <u>1.20</u> times), (**p**= 360 P/V= Procured Inventory Days= <u>240</u> Days), (**q**= [C-P]/X= Quick or Acid Test Ratio= <u>1.01</u> times), and (**S**= Sales or Revenues= <u>51,294</u> millions USD), then it's (**U**= Utilized or Starting Capital planned), is:

$$U= (Q+S'vp[1+s]/\{360[c-q]\})/I$$
$$-[M-F-Si-T][1-d]$$

$$= (35,351+48,851*0.1900*240$$
$$/\{360[1.2000-1.0100]\})$$
$$/0.9900-[41,548-29,750$$
$$-51,294*0.0100-3,385]$$
$$[1-0.3000]$$

$$= \underline{64,720} \text{ millions USD}$$

Corporate IFRS-GAAP (B/S-I/S), ISBN-13: **978-1720792789**, ISBN-10: **172079278X**

Law-7899:

If both (**U**= Utilized or Starting Capital= 64,720 millions USD), (**S'**= Sales of Past Year= 48,851 millions USD), (**s**= [S/S']-1= Sales Growth= 5.00%), (**v**= V/S= Variable Portion= 19.00%), (**T**= Tax Paid= 3,385 millions USD), (**d**= D/A= Dividend Portion or Payout= 30.00%), (**i**= I/S= Interest Portion= 1.00%), (**Q**= Quoted Longterm Liabilities= 35,351 millions USD), (**l**= L/E= Leverage or Gearing Ratio= 99.00%), (**F**= Fixed Cost= 29,740 millions USD), (**c**= C/X= Current Ratio= 1.20 times), (**p**= 360 P/V= Procured Inventory Days= 240 Days), (**q**= [C-P]/X= Quick or Acid Test Ratio= 1.01 times), and (**S**= Sales or Revenues= 51,294 millions USD), then it's (**M**= Margin of Contribution planned), is:

$$M= F+Si+T+[(Q+S'vp[1+s]$$
$$/\{360[c-q]\})/l-U]/[1-d]$$
$$= 29,750+51,294*0.0100+3,385$$
$$+[(35,351+48,851*0.1900$$
$$*240/\{360[1.2000$$
$$-1.0100]\})/0.9900$$
$$-64,720]/[1-0.3000]$$
$$= 41,548 \text{ millions USD}$$

Corporate IFRS-GAAP (B/S-I/S), ISBN-13: **978-1720792789**, ISBN-10: **172079278X**

<u>Law-7900</u>:

If both (**U**= Utilized or Starting Capital= <u>64,720</u> millions USD), (**S'**= Sales of Past Year= <u>48,851</u> millions USD), (**s**= [S/S']-1= Sales Growth= <u>5.00%</u>), (**v**= V/S= Variable Portion= <u>19.00%</u>), (**T**= Tax Paid= <u>3,385</u> millions USD), (**d**= D/A= Dividend Portion or Payout= <u>30.00%</u>), (**i**= I/S= Interest Portion= <u>1.00%</u>), (**Q**= Quoted Longterm Liabilities= <u>35,351</u> millions USD), (**l**= L/E= Leverage or Gearing Ratio= <u>99.00%</u>), (**M**= Margin of Contribution= <u>41,548</u> millions USD), (**c**= C/X= Current Ratio= <u>1.20</u> times), (**p**= 360 P/V= Procured Inventory Days= <u>240</u> Days), (**q**= [C-P]/X= Quick or Acid Test Ratio= <u>1.01</u> times), and (**S**= Sales or Revenues= <u>51,294</u> millions USD), then it's (**F**= Fixed Cost planned), is:

$$F= M\text{-}Si\text{-}T\text{-}[(Q+S'vp[1+s]$$
$$/\{360[c\text{-}q]\})/l\text{-}U]/[1\text{-}d]$$
$$= 41{,}548\text{-}51{,}294*0.0100\text{-}3{,}385$$
$$\text{-}[(35{,}351+48{,}851*0.1900$$
$$*240/\{360[1.2000$$
$$\text{-}1.0100]\})/0.9900$$
$$\text{-}64{,}720]/[1\text{-}0.3000]$$
$$= 29{,}750 \text{ millions USD}$$

Corporate IFRS-GAAP (B/S-I/S), ISBN-13: **978-1720792789**, ISBN-10: **172079278X**

Law-7901:

If both (**U**= Utilized or Starting Capital= 64,720 millions USD), (**S'**= Sales of Past Year= 48,851 millions USD), (**s**= [S/S']-1= Sales Growth= 5.00%), (**v**= V/S= Variable Portion= 19.00%), (**T**= Tax Paid= 3,385 millions USD), (**d**= D/A= Dividend Portion or Payout= 30.00%), (**i**= I/S= Interest Portion= 1.00%), (**Q**= Quoted Longterm Liabilities= 35,351 millions USD), (**/**= L/E= Leverage or Gearing Ratio= 99.00%), (**M**= Margin of Contribution= 41,548 millions USD), (**c**= C/X= Current Ratio= 1.20 times), (**p**= 360 P/V= Procured Inventory Days= 240 Days), (**q**= [C-P]/X= Quick or Acid Test Ratio= 1.01 times), and (**F**= Fixed Cost= 29,740 millions USD), then it's (**S** = Sales or Revenues planned), is:

$$S= \{M\text{-}F\text{-}T\text{-}[(Q+S'vp[1+s]$$
$$/\{360[c\text{-}q]\})//\text{-}U]/[1\text{-}d]\}/i$$
$$= \{41,548\text{-}29,750\text{-}3,385\text{-}[(35,351$$
$$+48,851*0.1900*240/\{360$$
$$[1.2000\text{-}1.0100]\})/0.9900$$
$$-64,720]/[1\text{-}0.3000]\}/0.0100$$
$$= \underline{51,294} \text{ millions USD}$$

Corporate IFRS-GAAP (B/S-I/S), ISBN-13: **978-1720792789**, ISBN-10: **172079278X**

Law-7902:

If both (**U**= Utilized or Starting Capital= 64,720
millions USD), (**S'**= Sales of Past Year= 48,851
millions USD), (**s**= [S/S']-1= Sales Growth= 5.00%),
(**v**= V/S= Variable Portion= 19.00%), (**T**= Tax Paid=
3,385 millions USD), (**d**= D/A= Dividend Portion or
Payout= 30.00%), (**S**= Sales or Revenues= 51,294
millions USD), (**Q**= Quoted Longterm Liabilities=
35,351 millions USD), (**l**= L/E= Leverage or Gearing
Ratio= 99.00%), (**M**= Margin of Contribution=
41,548 millions USD), (**c**= C/X= Current Ratio= 1.20
times), (**p**= 360 P/V= Procured Inventory Days= 240
Days), (**q**= [C-P]/X= Quick or Acid Test Ratio= 1.01
times), and (**F**= Fixed Cost= 29,740 millions USD),
then it's (**i** = Interest Portion planned), is :

$$S= \{M\text{-}F\text{-}T\text{-}[(Q+S'vp[1+s]$$
$$/\{360[c\text{-}q]\})/l\text{-}U]/[1\text{-}d]\}/S$$
$$= \{41,548\text{-}29,750\text{-}3,385\text{-}[(35,351$$
$$+48,851*0.1900*240/\{360$$
$$[1.2000\text{-}1.0100]\})/0.9900$$
$$-64,720]/[1\text{-}0.3000]\}/51,294$$
$$= 1.00\%$$

Corporate IFRS-GAAP (B/S-I/S), ISBN-13: **978-1720792789**, ISBN-10: **172079278X**

<u>Law-7903</u>:

If both (**U**= Utilized or Starting Capital= <u>64,720</u> millions USD), (**S'**= Sales of Past Year= <u>48,851</u> millions USD), (**s**= [S/S']-1= Sales Growth= <u>5.00%</u>), (**v**= V/S= Variable Portion= <u>19.00%</u>), (**i**= I/S= Interest Portion= <u>1.00%</u>), (**d**= D/A= Dividend Portion or Payout= <u>30.00%</u>), (**S**= Sales or Revenues= <u>51,294</u> millions USD), (**Q**= Quoted Longterm Liabilities= <u>35,351</u> millions USD), (**I**= L/E= Leverage or Gearing Ratio= <u>99.00%</u>), (**M**= Margin of Contribution= <u>41,548</u> millions USD), (**c**= C/X= Current Ratio= <u>1.20</u> times), (**p**= 360 P/V= Procured Inventory Days= <u>240</u> Days), (**q**= [C-P]/X= Quick or Acid Test Ratio= <u>1.01</u> times), and (**F**= Fixed Cost= <u>29,740</u> millions USD), then it's (**T** = Tax Paid planned), is:

$$T= M\text{-}F\text{-}Si\text{-}[(Q+S'vp[1+s]$$
$$/\{360[c\text{-}q]\})/I\text{-}U]/[1\text{-}d]$$
$$= 41{,}548\text{-}29{,}750\text{-}51{,}294*0.0100$$
$$\text{-}[(35{,}351+48{,}851*0.1900$$
$$*240/\{360[1.2000\text{-}1.0100]\})$$
$$/0.9900\text{-}64{,}720]/[1\text{-}0.3000]$$
$$= \underline{3{,}385} \text{ millions USD}$$

Corporate IFRS-GAAP (B/S-I/S), ISBN-13: **978-1720792789**, ISBN-10: **172079278X**

<u>Law-7904</u>:

If both (**U**= Utilized or Starting Capital= <u>64,720</u> millions USD), (**$'**= Sales of Past Year= <u>48,851</u> millions USD), (**s**= [S/S']-1= Sales Growth= <u>5.00%</u>), (**v**= V/S= Variable Portion= <u>19.00%</u>), (**i**= I/S= Interest Portion= <u>1.00%</u>), (**T**= Tax Paid= <u>3,385</u> millions USD), (**$**= Sales or Revenues= <u>51,294</u> millions USD), (**Q**= Quoted Longterm Liabilities= <u>35,351</u> millions USD), (**l**= L/E= Leverage or Gearing Ratio= <u>99.00%</u>), (**M**= Margin of Contribution= <u>41,548</u> millions USD), (**c**= C/X= Current Ratio= <u>1.20</u> times), (**p**= 360 P/V= Procured Inventory Days= <u>240</u> Days), (**q**= [C-P]/X= Quick or Acid Test Ratio= <u>1.01</u> times), and (**F**= Fixed Cost= <u>29,740</u> millions USD), then it's (**d**= Dividend Portion or Payout planned), is:

$$d= 1-[(Q+\$'vp[1+s]/\{360[c-q]\})/l-U]$$
$$/[M-F-\$i-T]$$

$$= 1-[(35,351+48,851*0.1900*240$$
$$/\{360[1.2000-1.0100]\})$$
$$/0.9900-64,720]/[41,548$$
$$-29,750-51,294*0.0100$$
$$-3,385]$$

$$= \underline{30.00\%}$$

Corporate IFRS-GAAP (B/S-I/S), ISBN-13: **978-1720792789**, ISBN-10: **172079278X**

Law-7905:

If both (**U**= Utilized or Starting Capital= 64,720
millions USD), (**d**= D/A= Dividend Portion or
Payout= 30.00%), (**s**= [S/S']-1= Sales Growth=
5.00%), (**v**= V/S= Variable Portion= 19.00%), (**i**=
I/S= Interest Portion= 1.00%), (**T**= Tax Paid= 3,385
millions USD), (**S**= Sales or Revenues= 51,294
millions USD), (**Q**= Quoted Longterm Liabilities=
35,351 millions USD), (**I**= L/E= Leverage or Gearing
Ratio= 99.00%), (**M**= Margin of Contribution=
41,548 millions USD), (**c**= C/X= Current Ratio= 1.20
times), (**p**= 360 P/V= Procured Inventory Days= 240
Days), (**q**= [C-P]/X= Quick or Acid Test Ratio= 1.01
times), and (**F**= Fixed Cost= 29,740 millions USD),
then it's (**S'** = Sales Past), must be:

$$S' = 360[c\text{-}q](I \{U+[M\text{-}F\text{-}Si\text{-}T][1\text{-}d]\}\text{-}Q)$$
$$/\{vp[1+s]\}$$

$$= 360[1.2000\text{-}1.0100](0.9900\{64,720$$
$$+[41,548\text{-}29,750\text{-}51,294$$
$$*0.0100\text{-}3,385][1\text{-}0.3000]\}$$
$$- 35,351)/\{0.1900*240$$
$$[1+0.0500]\}$$

$$= \underline{48,851} \text{ millions USD}$$

Corporate IFRS-GAAP (B/S-I/S), ISBN-13: **978-1720792789**, ISBN-10: **172079278X**

<u>Law-7906</u>:

If both (**U**= Utilized or Starting Capital= <u>64,720</u> millions USD), (**d**= D/A= Dividend Portion or Payout= <u>30.00%</u>), (**s**= [S/S']-1= Sales Growth= <u>5.00%</u>), (**S'**= Sales of Past Year= <u>48,851</u> millions USD), (**i**= I/S= Interest Portion= <u>1.00%</u>), (**T**= Tax Paid= <u>3,385</u> millions USD), (**S**= Sales or Revenues= <u>51,294</u> millions USD), (**Q**= Quoted Longterm Liabilities= <u>35,351</u> millions USD), (**F**= L/E= Leverage or Gearing Ratio= <u>99.00%</u>), (**M**= Margin of Contribution= <u>41,548</u> millions USD), (**c**= C/X= Current Ratio= <u>1.20</u> times), (**p**= 360 P/V= Procured Inventory Days= <u>240</u> Days), (**q**= [C-P]/X= Quick or Acid Test Ratio= <u>1.01</u> times), and (**F**= Fixed Cost= <u>29,740</u> millions USD), then it's (**v**= Variable Portion planned), is :

$$v= 360[\textbf{c-q}](\textbf{\textit{I}} \{U+[\textbf{M-F-Si-T}][1-\textbf{d}]\}-\textbf{Q})$$
$$/\{\textbf{S'p}[1+\textbf{s}]\}$$
$$= 360[1.2000-1.0100](0.9900\{64,720$$
$$+[41,548-29,750-51,294$$
$$*0.0100-3,385][1-0.3000]\}$$
$$-35,351)/\{48,851*240$$
$$[1+0.0500]\}$$
$$= \underline{19.00\%}$$

Corporate IFRS-GAAP (B/S-I/S), ISBN-13: **978-1720792789**, ISBN-10: **172079278X**

<u>Law-7907</u>:

If both (**U**= Utilized or Starting Capital= <u>64,720</u> millions USD), (**d**= D/A= Dividend Portion or Payout= <u>30.00%</u>), (**s**= [S/S']-1= Sales Growth= <u>5.00%</u>), (**S'**= Sales of Past Year= <u>48,851</u> millions USD), (**i**= I/S= Interest Portion= <u>1.00%</u>), (**T**= Tax Paid= <u>3,385</u> millions USD), (**S**= Sales or Revenues= <u>51,294</u> millions USD), (**Q**= Quoted Longterm Liabilities= <u>35,351</u> millions USD), (**l**= L/E= Leverage or Gearing Ratio= <u>99.00%</u>), (**M**= Margin of Contribution= <u>41,548</u> millions USD), (**c**= C/X= Current Ratio= <u>1.20</u> times), (**v**= V/S= Variable Portion= <u>19.00%</u>), (**q**= [C-P]/X= Quick or Acid Test Ratio= <u>1.01</u> times), and (**F**= Fixed Cost= <u>29,740</u> millions USD), then it's (**p**= Procured Inventory Days planned), is:

$$p= 360[\textbf{c-q}](\textbf{l} \{\textbf{U}+[\textbf{M-F-Si-T}][1-\textbf{d}]\}-\textbf{Q})$$
$$/\{\textbf{S'v}[1+\textbf{s}]\}$$
$$= 360[1.2000-1.0100](0.9900$$
$$\{64,720+[41,548-29,750$$
$$-51,294*0.0100-3,385]$$
$$[1-0.3000]\}-35,351)$$
$$/\{48,851*0.1900$$
$$[1+0.0500]\}$$
$$= \underline{240} \text{ days}$$

Corporate IFRS-GAAP (B/S-I/S), ISBN-13: **978-1720792789**, ISBN-10: **172079278X**

Law-7908:

If both (**U**= Utilized or Starting Capital= 64,720 millions USD), (**d**= D/A= Dividend Portion or Payout= 30.00%), (**p**= 360 P/V= Procured Inventory Days= 240 Days), (**S'**= Sales of Past Year= 48,851 millions USD), (**i**= I/S= Interest Portion= 1.00%), (**T**= Tax Paid= 3,385 millions USD), (**S**= Sales or Revenues= 51,294 millions USD), (**Q**= Quoted Longterm Liabilities= 35,351 millions USD), (**l**= L/E= Leverage or Gearing Ratio= 99.00%), (**M**= Margin of Contribution= 41,548 millions USD), (**c**= C/X= Current Ratio= 1.20 times), (**v**= V/S= Variable Portion= 19.00%), (**q**= [C-P]/X= Quick or Acid Test Ratio= 1.01 times), and (**F**= Fixed Cost= 29,740 millions USD), then it's (**s**= Sales Growth planned), is :

$$s= 360[\mathbf{c}\text{-}\mathbf{q}](\mathbf{l} \{\mathbf{U}+[\mathbf{M}\text{-}\mathbf{F}\text{-}\mathbf{Si}\text{-}\mathbf{T}][1\text{-}\mathbf{d}]\}\text{-}\mathbf{Q})$$
$$/[\mathbf{S'vp}]\text{-}1$$
$$= 360[1.2000\text{-}1.0100](0.9900$$
$$\{64,720+[41,548\text{-}29,750$$
$$\text{-}51,294*0.0100\text{-}3,385]$$
$$[1\text{-}0.3000]\}\text{-}35,351)$$
$$/[48,851*0.1900*240]\text{-}1$$
$$= \underline{5.00\%}$$

Corporate IFRS-GAAP (B/S-I/S), ISBN-13: **978-1720792789**, ISBN-10: **172079278X**

<u>Law-7909</u>:

If both (**U**= Utilized or Starting Capital= <u>64,720</u> millions USD), (**d**= D/A= Dividend Portion or Payout= <u>30.00%</u>), (**p**= 360 P/V= Procured Inventory Days= <u>240</u> Days), (**S'**= Sales of Past Year= <u>48,851</u> millions USD), (**i**= I/S= Interest Portion= <u>1.00%</u>), (**T**= Tax Paid= <u>3,385</u> millions USD), (**S**= Sales or Revenues= <u>51,294</u> millions USD), (**Q**= Quoted Longterm Liabilities= <u>35,351</u> millions USD), (**l**= L/E= Leverage or Gearing Ratio= <u>99.00%</u>), (**M**= Margin of Contribution= <u>41,548</u> millions USD), (**s**= [S/S']-1= Sales Growth= <u>5.00%</u>), (**v**= V/S= Variable Portion= <u>19.00%</u>), (**q**= [C-P]/X= Quick or Acid Test Ratio= <u>1.01</u> times), and (**F**= Fixed Cost= <u>29,740</u> millions USD), then it's (**c**= Current Ratio planned), is :

$$c= q+S'vp[1+s]/[360(l \{U+[M-F-Si-T] [1-d]\}-Q)]$$

$$= 1.0100+48,851*0.1900*240$$
$$[1+0.0500]/[360(0.9900$$
$$\{64,720+[41,548-29,750$$
$$-51,294*0.0100-3,385]$$
$$[1-0.3000]\}-35,351)]$$

$$= \underline{1.20} \text{ times}$$

Corporate IFRS-GAAP (B/S-I/S), ISBN-13: **978-1720792789**, ISBN-10: **172079278X**

<u>Law-7910</u>:

If both (**U**= Utilized or Starting Capital= <u>64,720</u> millions USD), (**d**= D/A= Dividend Portion or Payout= <u>30.00%</u>), (**p**= 360 P/V= Procured Inventory Days= <u>240</u> Days), (**S'**= Sales of Past Year= <u>48,851</u> millions USD), (**i**= I/S= Interest Portion= <u>1.00%</u>), (**T**= Tax Paid= <u>3,385</u> millions USD), (**S**= Sales or Revenues= <u>51,294</u> millions USD), (**Q**= Quoted Longterm Liabilities= <u>35,351</u> millions USD), (**l**= L/E= Leverage or Gearing Ratio= <u>99.00%</u>), (**M**= Margin of Contribution= <u>41,548</u> millions USD), (**s**= [S/S']-1= Sales Growth= <u>5.00%</u>), (**v**= V/S= Variable Portion= <u>19.00%</u>), (**c**= C/X= Current Ratio= <u>1.20</u> times), and (**F**= Fixed Cost= <u>29,740</u> millions USD), then it's (**q**= Quick or Acid Test Ratio planned), is:

$$q= c\text{-}S'vp[1+s]/[360(l\ \{U+[M\text{-}F\text{-}Si\text{-}T]\ [1\text{-}d]\}\text{-}Q)]$$

$$= 1.2000\text{-}48,851*0.1900*240$$
$$[1+0.0500]/[360(0.9900$$
$$\{64,720+[41,548\text{-}29,750$$
$$\text{-}51,294*0.0100\text{-}3,385]$$
$$[1\text{-}0.3000]\}\text{-}35,351)]$$

$$= \underline{1.01}\ \text{times}$$

Corporate IFRS-GAAP (B/S-I/S), ISBN-13: **978-1720792789**, ISBN-10: **172079278X**

Law-7911:

If both (**U**= Utilized or Starting Capital= <u>64,720</u> millions USD), (**d**= D/A= Dividend Portion or Payout= <u>30.00%</u>), (**A**= After Tax Income= <u>7,899</u> millions USD), (**i**= I/S= Interest Portion= <u>1.00%</u>), (**T**= Tax Paid= <u>3,385</u> millions USD), (**$**= Sales or Revenues= <u>51,294</u> millions USD), (**F**= Fixed Cost= <u>29,740</u> millions USD), (**X**= Xpress or Current Debt= <u>34,196</u> millions USD), (**$\digamma$**= L/E= Leverage or Gearing Ratio= <u>99.00%</u>), and (**M**= Margin of Contribution= <u>41,548</u> millions USD), then it's (**Q**= Quoted Longterm Liabilities planned), is:

$$Q = \digamma [U+M-F-Si-T-Ad]-X$$
$$= 0.9900[64,720+41,548-29,750$$
$$-51,294*0.0100-3,385$$
$$-7,899*0.3000]-34,196$$
$$= \underline{35,351} \text{ millions USD}$$

Corporate IFRS-GAAP (B/S-I/S), ISBN-13: **978-1720792789**, ISBN-10: **172079278X**

Law-7912:

If both (**U**= Utilized or Starting Capital= 64,720 millions USD), (**d**= D/A= Dividend Portion or Payout= 30.00%), (**A**= After Tax Income= 7,899 millions USD), (**i**= I/S= Interest Portion= 1.00%), (**T**= Tax Paid= 3,385 millions USD), (**$**= Sales or Revenues= 51,294 millions USD), (**F**= Fixed Cost= 29,740 millions USD), (**X**= Xpress or Current Debt= 34,196 millions USD), (**Q**= Quoted Longterm Liabilities= 35,351 millions USD), and (**M**= Margin of Contribution= 41,548 millions USD), then it's (**I**= Leverage or Gearing Ratio planned), is:

$$I = [Q+X]/[U+M-F-Si-T-Ad]$$
$$= [35,351+34,196]/[64,720+41,548$$
$$-29,750-51,294*0.0100$$
$$-3,385-7,899*0.3000]$$
$$= \underline{99.00\%}$$

Corporate IFRS-GAAP (B/S-I/S), ISBN-13: **978-1720792789**, ISBN-10: **172079278X**

Law-7913:

If both (**I**= L/E= Leverage or Gearing Ratio= 99.00%), (**d**= D/A= Dividend Portion or Payout= 30.00%), (**A**= After Tax Income= 7,899 millions USD), (**i**= I/S= Interest Portion= 1.00%), (**T**= Tax Paid= 3,385 millions USD), (**S**= Sales or Revenues= 51,294 millions USD), (**F**= Fixed Cost= 29,740 millions USD), (**X**= Xpress or Current Debt= 34,196 millions USD), (**Q**= Quoted Longterm Liabilities= 35,351 millions USD), and (**M**= Margin of Contribution= 41,548 millions USD), then it's (**U**= Utilized or Starting Capital planned), is:

$$U= F+Si+T+Ad-M+[Q+X]/I$$
$$= 29,750+51,294*0.0100+3,385$$
$$+7,899*0.3000-41,548$$
$$+[35,351+34,196]/0.9900$$
$$= 64,720 \text{ millions USD}$$

Corporate IFRS-GAAP (B/S-I/S), ISBN-13: **978-1720792789**, ISBN-10: **172079278X**

Law-7914:

If both (**I**= L/E= Leverage or Gearing Ratio= 99.00%), (**d**= D/A= Dividend Portion or Payout= 30.00%), (**A**= After Tax Income= 7,899 millions USD), (**i**= I/S= Interest Portion= 1.00%), (**T**= Tax Paid= 3,385 millions USD), (**S**= Sales or Revenues= 51,294 millions USD), (**F**= Fixed Cost= 29,740 millions USD), (**X**= Xpress or Current Debt= 34,196 millions USD), (**Q**= Quoted Longterm Liabilities= 35,351 millions USD), and (**U**= Utilized or Starting Capital= 64,720 millions USD), then it's (**M**= Margin of Contribution planned), is:

$$M= F+Si+T+Ad-U+[Q+X]/I$$
$$= 29,750+51,294*0.0100+3,385$$
$$+7,899*0.3000-64,720$$
$$+[35,351+34,196]/0.9900$$
$$= 41,548 \text{ millions USD}$$

Corporate IFRS-GAAP (B/S-I/S), ISBN-13: **978-1720792789**, ISBN-10: **172079278X**

Law-7915:

If both (**F**= L/E= Leverage or Gearing Ratio= 99.00%), (**d**= D/A= Dividend Portion or Payout= 30.00%), (**A**= After Tax Income= 7,899 millions USD), (**i**= I/S= Interest Portion= 1.00%), (**T**= Tax Paid= 3,385 millions USD), (**$**= Sales or Revenues= 51,294 millions USD), (**M**= Margin of Contribution= 41,548 millions USD), (**X**= Xpress or Current Debt= 34,196 millions USD), (**Q**= Quoted Longterm Liabilities= 35,351 millions USD), and (**U**= Utilized or Starting Capital= 64,720 millions USD), then it's (**F**= Fixed Cost planned), is:

$$F= U+M-\$i-T-Ad-[Q+X]/I$$
$$= 64,720+41,548-51,294*0.0100$$
$$-3,385-7,899*0.3000$$
$$+[35,351+34,196]/0.9900$$
$$= \underline{29,750} \text{ millions USD}$$

Corporate IFRS-GAAP (B/S-I/S), ISBN-13: **978-1720792789**, ISBN-10: **172079278X**

<u>Law-7916</u>:

If both (**I**= L/E= Leverage or Gearing Ratio= <u>99.00%</u>), (**d**= D/A= Dividend Portion or Payout= <u>30.00%</u>), (**A**= After Tax Income= <u>7,899</u> millions USD), (**i**= I/S= Interest Portion= <u>1.00%</u>), (**T**= Tax Paid= <u>3,385</u> millions USD), (**F**= Fixed Cost= <u>29,740</u> millions USD), (**M**= Margin of Contribution= <u>41,548</u> millions USD), (**X**= Xpress or Current Debt= <u>34,196</u> millions USD), (**Q**= Quoted Longterm Liabilities= <u>35,351</u> millions USD),and (**U**= Utilized or Starting Capital= <u>64,720</u> millions USD), then it's (**S**= Sales or Revenues planned), is:

$$S= \{U+M-F-T-Ad-[Q+X]/I\}/i$$
$$= \{64,720+41,548-29,750-3,385$$
$$-7,899*0.3000-[35,351$$
$$+34,196]/0.9900\}/0.0100$$
$$= \underline{51,294} \text{ millions USD}$$

Corporate IFRS-GAAP (B/S-I/S), ISBN-13: **978-1720792789**, ISBN-10: **172079278X**

Law-7917:

If both (**I**= L/E= Leverage or Gearing Ratio= 99.00%), (**d**= D/A= Dividend Portion or Payout= 30.00%), (**A**= After Tax Income= 7,899 millions USD), (**S**= Sales or Revenues= 51,294 millions USD), (**T**= Tax Paid= 3,385 millions USD), (**F**= Fixed Cost= 29,740 millions USD), (**M**= Margin of Contribution= 41,548 millions USD), (**X**= Xpress or Current Debt= 34,196 millions USD), (**Q**= Quoted Longterm Liabilities= 35,351 millions USD), and (**U**= Utilized or Starting Capital= 64,720 millions USD), then it's (**i**= Interest Portion planned), is:

$$i= \{U+M-F-T-Ad-[Q+X]/I\}/S$$

$$= \{64,720+41,548-29,750-3,385$$
$$-7,899*0.3000-[35,351$$
$$+34,196]/0.9900\}/51,294$$

$$= 1.00\%$$

Corporate IFRS-GAAP (B/S-I/S), ISBN-13: **978-1720792789**, ISBN-10: **172079278X**

Law-7918:

If both (**I**= L/E= Leverage or Gearing Ratio= 99.00%), (**d**= D/A= Dividend Portion or Payout= 30.00%), (**A**= After Tax Income= 7,899 millions USD), (**S**= Sales or Revenues= 51,294 millions USD), (**i**= I/S= Interest Portion= 1.00%), (**F**= Fixed Cost= 29,740 millions USD), (**M**= Margin of Contribution= 41,548 millions USD), (**X**= Xpress or Current Debt= 34,196 millions USD), (**Q**= Quoted Longterm Liabilities= 35,351 millions USD),and (**U**= Utilized or Starting Capital= 64,720 millions USD), then it's (**T**= Tax Paid planned), is:

$$T= U+M-F-Si-Ad-[Q+X]/I$$
$$= 64,720+41,548-29,750-51,294$$
$$*0.0100-7,899*0.3000$$
$$-[35,351+34,196]/0.9900$$
$$= 3,385 \text{ millions USD}$$

Corporate IFRS-GAAP (B/S-I/S), ISBN-13: **978-1720792789**, ISBN-10: **172079278X**

Law-7919:

If both (**I**= L/E= Leverage or Gearing Ratio=
99.00%), (**d**= D/A= Dividend Portion or Payout=
30.00%), (**T**= Tax Paid= 3,385 millions USD), (**S**=
Sales or Revenues= 51,294 millions USD), (**i**= I/S=
Interest Portion= 1.00%), (**F**= Fixed Cost= 29,740
millions USD), (**M**= Margin of Contribution= 41,548
millions USD), (**X**= Xpress or Current Debt= 34,196
millions USD), (**Q**= Quoted Longterm Liabilities=
35,351 millions USD),and (**U**= Utilized or Starting
Capital= 64,720 millions USD), then it's (**A**= After
Tax Income planned), is:

$$A= \{U+M-F-Si-T-[Q+X]/I \}/d$$
$$= \{64,720+41,548-29,750-51,294$$
$$*0.0100-3,385-[35,351$$
$$+34,196]/0.9900\}/0.3000$$
$$= 7,899 \text{ millions USD}$$

Corporate IFRS-GAAP (B/S-I/S), ISBN-13: **978-1720792789**, ISBN-10: **172079278X**

<u>Law-7920</u>:

If both (**I**= L/E= Leverage or Gearing Ratio= <u>99.00%</u>), (**A**= After Tax Income= <u>7,899</u> millions USD), (**T**= Tax Paid= <u>3,385</u> millions USD), (**S**= Sales or Revenues= <u>51,294</u> millions USD), (**i**= I/S= Interest Portion= <u>1.00%</u>), (**F**= Fixed Cost= <u>29,740</u> millions USD), (**M**= Margin of Contribution= <u>41,548</u> millions USD), (**X**= Xpress or Current Debt= <u>34,196</u> millions USD), (**Q**= Quoted Longterm Liabilities= <u>35,351</u> millions USD),and (**U**= Utilized or Starting Capital= <u>64,720</u> millions USD), then it's (**d**= Dividend Portion or Payout planned), is:

$$d= \{U+M-F-Si-T-[Q+X]/I\}/A$$
$$= \{64,720+41,548-29,750-51,294$$
$$*0.0100-3,385-[35,351$$
$$+34,196]/0.9900\}/7,899$$
$$= \underline{30.00\%}$$

Corporate IFRS-GAAP (B/S-I/S), ISBN-13: **978-1720792789**, ISBN-10: **172079278X**

<u>Law-7921</u>:

If both (**I**= L/E= Leverage or Gearing Ratio=
<u>99.00%</u>), (**A**= After Tax Income= <u>7,899</u> millions
USD), (**T**= Tax Paid= <u>3,385</u> millions USD), (**S**= Sales
or Revenues= <u>51,294</u> millions USD), (**i**= I/S= Interest
Portion= <u>1.00%</u>), (**F**= Fixed Cost= <u>29,740</u> millions
USD), (**M**= Margin of Contribution= <u>41,548</u> millions
USD), (**d**= D/A= Dividend Portion or Payout=
<u>30.00%</u>), (**Q**= Quoted Longterm Liabilities= <u>35,351</u>
millions USD),and (**U**= Utilized or Starting Capital=
<u>64,720</u> millions USD), then it's (**X**= Xpress or
Current Debt planned), is:

$$X= I[U+M-F-Si-T-Ad]-Q$$
$$= 0.9900[64,720+41,548-29,750$$
$$-51,294*0.0100-3,385$$
$$-7,899*0.3000]-35,351$$
$$= \underline{34,196} \text{ millions USD}$$

Corporate IFRS-GAAP (B/S-I/S), ISBN-13: **978-1720792789**, ISBN-10: **172079278X**

Law-7922:

If both (**f**= L/E= Leverage or Gearing Ratio= 99.00%), (**A**= After Tax Income= 7,899 millions USD), (**T**= Tax Paid= 3,385 millions USD), (**S**= Sales or Revenues= 51,294 millions USD), (**i**= I/S= Interest Portion= 1.00%), (**F**= Fixed Cost= 29,740 millions USD), (**M**= Margin of Contribution= 41,548 millions USD), (**d**= D/A= Dividend Portion or Payout= 30.00%), (**c**= C/X= Current Ratio= 1.20 times), (**q**= [C-P]/X= Quick or Acid Test Ratio= 1.01 times), (**P**= Procured Inventories= 6,497 millions USD), and (**U**= Utilized or Starting Capital= 64,720 millions USD), then it's (**Q**= Quoted Longterm Liabilities planned), is:

$$Q= f[U+M-F-Si-T-Ad]-P/[c-q]$$
$$= 0.9900[64,720+41,548-29,750$$
$$-51,294*0.0100-3,385$$
$$-7,899*0.3000]-6,497$$
$$/[1.2000-1.0100]$$
$$= \underline{35,351} \text{ millions USD}$$

Corporate IFRS-GAAP (B/S-I/S), ISBN-13: **978-1720792789**, ISBN-10: **172079278X**

Law-7923:

If both (Q= Quoted Longterm Liabilities= 35,351 millions USD), (A= After Tax Income= 7,899 millions USD), (T= Tax Paid= 3,385 millions USD), ($\$$= Sales or Revenues= 51,294 millions USD), (i= I/S= Interest Portion= 1.00%), (F= Fixed Cost= 29,740 millions USD), (M= Margin of Contribution= 41,548 millions USD), (d= D/A= Dividend Portion or Payout= 30.00%), (c= C/X= Current Ratio= 1.20 times), (q= [C-P]/X= Quick or Acid Test Ratio= 1.01 times), (P= Procured Inventories= 6,497 millions USD), and (U= Utilized or Starting Capital= 64,720 millions USD), then it's (I = Leverage or Gearing Ratio planned), is:

$$I = \{Q+P/[c-q]\}/[U+M-F-Si-T-Ad]$$
$$= \{35,351+6,497/[1.2000-1.0100]\}$$
$$/[64,720+41,548-29,750$$
$$-51,294*0.0100-3,385$$
$$-7,899*0.3000]$$
$$= 99.00\%$$

Corporate IFRS-GAAP (B/S-I/S), ISBN-13: **978-1720792789**, ISBN-10: **172079278X**

<u>Law-7924</u>:

If both (**Q**= Quoted Longterm Liabilities= <u>35,351</u> millions USD), (**A**= After Tax Income= <u>7,899</u> millions USD), (**T**= Tax Paid= <u>3,385</u> millions USD), (**$**= Sales or Revenues= <u>51,294</u> millions USD), (**i**= I/S= Interest Portion= <u>1.00%</u>), (**F**= Fixed Cost= <u>29,740</u> millions USD), (**M**= Margin of Contribution= <u>41,548</u> millions USD), (**d**= D/A= Dividend Portion or Payout= <u>30.00%</u>), (**c**= C/X= Current Ratio= <u>1.20</u> times), (**q**= [C-P]/X= Quick or Acid Test Ratio= <u>1.01</u> times), (**P**= Procured Inventories= <u>6,497</u> millions USD), and (**I**= L/E= Leverage or Gearing Ratio= <u>99.00%</u>), then it's (**U** = Utilized or Starting Capital planned), is:

$$U= F+Si+T+Ad-M+\{Q+P/[c-q]\}/I$$
$$= 29,750+51,294*0.0100+3,385+7,899$$
$$*0.3000]-41,548+\{35,351$$
$$+6,497/[1.2000-1.0100]\}$$
$$/0.9900$$
$$= \underline{64,720} \text{ millions USD}$$

Corporate IFRS-GAAP (B/S-I/S), ISBN-13: **978-1720792789**, ISBN-10: **172079278X**

Law-7925:

If both (**Q**= Quoted Longterm Liabilities= 35,351 millions USD), (**A**= After Tax Income= 7,899 millions USD), (**T**= Tax Paid= 3,385 millions USD), (**$**= Sales or Revenues= 51,294 millions USD), (**i**= I/S= Interest Portion= 1.00%), (**F**= Fixed Cost= 29,740 millions USD), (**U**= Utilized or Starting Capital= 64,720 millions USD), (**d**= D/A= Dividend Portion or Payout= 30.00%), (**c**= C/X= Current Ratio= 1.20 times), (**q**= [C-P]/X= Quick or Acid Test Ratio= 1.01 times), (**P**= Procured Inventories= 6,497 millions USD), and (**l**= L/E= Leverage or Gearing Ratio= 99.00%), then it's (**M**= Margin of Contribution planned), is:

$$M= F+Si+T+Ad-U+\{Q+P/[c-q]\}/l$$
$$= 29,750+51,294*0.0100+3,385$$
$$+7,899*0.3000]-64,720$$
$$+\{35,351+6,497/[1.2000$$
$$-1.0100]\}/0.9900$$
$$= \underline{41,548} \text{ millions USD}$$

Corporate IFRS-GAAP (B/S-I/S), ISBN-13: **978-1720792789**, ISBN-10: **172079278X**

Law-7926:

If both (**Q**= Quoted Longterm Liabilities= 35,351 millions USD), (**A**= After Tax Income= 7,899 millions USD), (**T**= Tax Paid= 3,385 millions USD), (**$**= Sales or Revenues= 51,294 millions USD), (**i**= I/S= Interest Portion= 1.00%), (**M**= Margin of Contribution= 41,548 millions USD), (**U**= Utilized or Starting Capital= 64,720 millions USD), (**d**= D/A= Dividend Portion or Payout= 30.00%), (**c**= C/X= Current Ratio= 1.20 times), (**q**= [C-P]/X= Quick or Acid Test Ratio= 1.01 times), (**P**= Procured Inventories= 6,497 millions USD), and (**l**= L/E= Leverage or Gearing Ratio= 99.00%), then it's (**F** = Margin of Contribution planned), is :

$$F= U+M\text{-}Si\text{-}T\text{-}Ad\text{-}\{Q+P/[c\text{-}q]\}/l$$

$$= 64{,}720+41{,}548\text{-}51{,}294*0.0100$$
$$-3{,}385\text{-}7{,}899*0.3000]$$
$$-\{35{,}351+6{,}497/[1.2000$$
$$-1.0100]\}/0.9900$$

$$= 29{,}750 \text{ millions USD}$$

Corporate IFRS-GAAP (B/S-I/S), ISBN-13: **978-1720792789**, ISBN-10: **172079278X**

<u>Law-7927</u>:

If both (**Q**= Quoted Longterm Liabilities= <u>35,351</u> millions USD), (**A**= After Tax Income= <u>7,899</u> millions USD), (**T**= Tax Paid= <u>3,385</u> millions USD), (**F**= Fixed Cost= <u>29,740</u> millions USD), (**i**= I/S= Interest Portion= <u>1.00%</u>), (**M**= Margin of Contribution= <u>41,548</u> millions USD), (**U**= Utilized or Starting Capital= <u>64,720</u> millions USD), (**d**= D/A= Dividend Portion or Payout= <u>30.00%</u>), (**c**= C/X= Current Ratio= <u>1.20</u> times), (**q**= [C-P]/X= Quick or Acid Test Ratio= <u>1.01</u> times), (**P**= Procured Inventories= <u>6,497</u> millions USD), and (**l**= L/E= Leverage or Gearing Ratio= <u>99.00%</u>), then it's (**S**= Sales or Revenues planned), is:

$$S= (U+M-F-T-Ad-\{Q+P/[c-q]\}/l)/i$$
$$= (64,720+41,548-29,750-3,385$$
$$-7,899*0.3000]-\{35,351$$
$$+6,497/[1.2000-1.0100]\}$$
$$/0.9900)/0.0100$$
$$= \underline{51,294} \text{ millions USD}$$

Corporate IFRS-GAAP (B/S-I/S), ISBN-13: **978-1720792789**, ISBN-10: **172079278X**

<u>Law-7928</u>:

If both (**Q**= Quoted Longterm Liabilities= <u>35,351</u> millions USD), (**A**= After Tax Income= <u>7,899</u> millions USD), (**T**= Tax Paid= <u>3,385</u> millions USD), (**F**= Fixed Cost= <u>29,740</u> millions USD), (**S**= Sales or Revenues= <u>51,294</u> millions USD), (**M**= Margin of Contribution= <u>41,548</u> millions USD), (**U**= Utilized or Starting Capital= <u>64,720</u> millions USD), (**d**= D/A= Dividend Portion or Payout= <u>30.00%</u>), (**c**= C/X= Current Ratio= <u>1.20</u> times), (**q**= [C-P]/X= Quick or Acid Test Ratio= <u>1.01</u> times), (**P**= Procured Inventories= <u>6,497</u> millions USD), and (***l***= L/E= Leverage or Gearing Ratio= <u>99.00%</u>), then it's (**i**= Interest Portion planned), is:

$$i = (U+M-F-T-Ad-\{Q+P/[c-q]\}/l)/S$$

$$= (64,720+41,548-29,750-3,385$$
$$-7,899*0.3000]-\{35,351$$
$$+6,497/[1.2000-1.0100]\}$$
$$/0.9900)/51,294$$

$$= \underline{1.00\%}$$

Corporate IFRS-GAAP (B/S-I/S), ISBN-13: **978-1720792789**, ISBN-10: **172079278X**

Law-7929:

If both (**Q**= Quoted Longterm Liabilities= 35,351 millions USD), (**A**= After Tax Income= 7,899 millions USD), (**i**= I/S= Interest Portion= 1.00%), (**F**= Fixed Cost= 29,740 millions USD), (**S**= Sales or Revenues= 51,294 millions USD), (**M**= Margin of Contribution= 41,548 millions USD), (**U**= Utilized or Starting Capital= 64,720 millions USD), (**d**= D/A= Dividend Portion or Payout= 30.00%), (**c**= C/X= Current Ratio= 1.20 times), (**q**= [C-P]/X= Quick or Acid Test Ratio= 1.01 times), (**P**= Procured Inventories= 6,497 millions USD), and (**l**= L/E= Leverage or Gearing Ratio= 99.00%), then it's (**T**= Tax Paid planned), is:

$$T= U+M-F-Si-Ad-\{Q+P/[c-q]\}/l$$
$$= 64,720+41,548-29,750-51,294$$
$$*0.0100-7,899*0.3000]$$
$$-\{35,351+6,497/[1.2000$$
$$-1.0100]\}/0.9900$$
$$= \underline{3,385} \text{ millions USD}$$

Corporate IFRS-GAAP (B/S-I/S), ISBN-13: **978-1720792789**, ISBN-10: **172079278X**

Law-7930:

If both (**Q**= Quoted Longterm Liabilities= 35,351 millions USD), (**T**= Tax Paid= 3,385 millions USD), (**i**= I/S= Interest Portion= 1.00%), (**F**= Fixed Cost= 29,740 millions USD), (**S**= Sales or Revenues= 51,294 millions USD), (**M**= Margin of Contribution= 41,548 millions USD), (**U**= Utilized or Starting Capital= 64,720 millions USD), (**d**= D/A= Dividend Portion or Payout= 30.00%), (**c**= C/X= Current Ratio= 1.20 times), (**q**= [C-P]/X= Quick or Acid Test Ratio= 1.01 times), (**P**= Procured Inventories= 6,497 millions USD), and (**I**= L/E= Leverage or Gearing Ratio= 99.00%), then it's (**A**= After Tax Income planned), is:

$$A= (U+M-F-Si-T-\{Q+P/[c-q]\}/I\,)/d$$
$$= (64,720+41,548-29,750-51,294$$
$$*0.0100-3,385-\{35,351$$
$$+6,497/[1.2000-1.0100]\}$$
$$/0.9900)/0.3000$$
$$= \underline{7,899} \text{ millions USD}$$

Corporate IFRS-GAAP (B/S-I/S), ISBN-13: **978-1720792789**, ISBN-10: **172079278X**

Law-7931:

If both (**Q**= Quoted Longterm Liabilities= 35,351 millions USD), (**T**= Tax Paid= 3,385 millions USD), (**i**= I/S= Interest Portion= 1.00%), (**F**= Fixed Cost= 29,740 millions USD), (**S**= Sales or Revenues= 51,294 millions USD), (**M**= Margin of Contribution= 41,548 millions USD), (**U**= Utilized or Starting Capital= 64,720 millions USD), (**A**= After Tax Income= 7,899 millions USD), (**c**= C/X= Current Ratio= 1.20 times), (**q**= [C-P]/X= Quick or Acid Test Ratio= 1.01 times), (**P**= Procured Inventories= 6,497 millions USD), and (**l**= L/E= Leverage or Gearing Ratio= 99.00%), then it's (**d**= Dividend Portion or Payout planned), is:

$$d = (U+M-F-Si-T-\{Q+P/[c-q]\}/l)/A$$
$$= (64,720+41,548-29,750-51,294$$
$$*0.0100-3,385-\{35,351$$
$$+6,497/[1.2000-1.0100]\}$$
$$/0.9900)/7,899$$

$$= 30.00\%$$

Corporate IFRS-GAAP (B/S-I/S), ISBN-13: **978-1720792789**, ISBN-10: **172079278X**

<u>Law-7932</u>:

If both (**Q**= Quoted Longterm Liabilities= <u>35,351</u> millions USD), (**T**= Tax Paid= <u>3,385</u> millions USD), (**i**= I/S= Interest Portion= <u>1.00%</u>), (**F**= Fixed Cost= <u>29,740</u> millions USD), (**S**= Sales or Revenues= <u>51,294</u> millions USD), (**M**= Margin of Contribution= <u>41,548</u> millions USD), (**U**= Utilized or Starting Capital= <u>64,720</u> millions USD), (**A**= After Tax Income= <u>7,899</u> millions USD), (**c**= C/X= Current Ratio= <u>1.20</u> times), (**q**= [C-P]/X= Quick or Acid Test Ratio= <u>1.01</u> times), (**d**= D/A= Dividend Portion or Payout= <u>30.00%</u>), and (**l**= L/E= Leverage or Gearing Ratio= <u>99.00%</u>), then it's (**P**= Procured Inventories planned), is:

$$P= [c\text{-}q]\{l\,[U+M\text{-}F\text{-}Si\text{-}T\text{-}Ad]\text{-}Q\}$$
$$= [1.2000\text{-}1.0100]\{0.9900\{64,720$$
$$+41,548\text{-}29,750\text{-}51,294$$
$$*0.0100\text{-}3,385\text{-}7,899$$
$$*0.3000]\text{-}35,351\}$$
$$= \underline{6,497} \text{ millions USD}$$

Corporate IFRS-GAAP (B/S-I/S), ISBN-13: **978-1720792789**, ISBN-10: **172079278X**

Law-7933:

If both (**Q**= Quoted Longterm Liabilities= 35,351 millions USD), (**T**= Tax Paid= 3,385 millions USD), (**i**= I/S= Interest Portion= 1.00%), (**F**= Fixed Cost= 29,740 millions USD), (**S**= Sales or Revenues= 51,294 millions USD), (**M**= Margin of Contribution= 41,548 millions USD), (**U**= Utilized or Starting Capital= 64,720 millions USD), (**A**= After Tax Income= 7,899 millions USD), (**P**= Procured Inventories= 6,497 millions USD), (**q**= [C-P]/X= Quick or Acid Test Ratio= 1.01 times), (**d**= D/A= Dividend Portion or Payout= 30.00%), and (**l**= L/E= Leverage or Gearing Ratio= 99.00%), then it's (**c**= Current Ratio planned), is:

$$c= q+P/\{l\,[U+M-F-Si-T-Ad]-Q\}$$
$$= 1.0100+6,497/\{0.9900\{64,720$$
$$+41,548-29,750-51,294$$
$$*0.0100-3,385-7,899$$
$$*0.3000]-35,351\}$$
$$= \underline{1.20} \text{ times}$$

Corporate IFRS-GAAP (B/S-I/S), ISBN-13: **978-1720792789**, ISBN-10: **172079278X**

<u>Law-7934</u>:

If both (**Q**= Quoted Longterm Liabilities= <u>35,351</u> millions USD), (**T**= Tax Paid= <u>3,385</u> millions USD), (**i**= I/S= Interest Portion= <u>1.00%</u>), (**F**= Fixed Cost= <u>29,740</u> millions USD), (**S**= Sales or Revenues= <u>51,294</u> millions USD), (**M**= Margin of Contribution= <u>41,548</u> millions USD), (**U**= Utilized or Starting Capital= <u>64,720</u> millions USD), (**A**= After Tax Income= <u>7,899</u> millions USD), (**P**= Procured Inventories= <u>6,497</u> millions USD), (**c**= C/X= Current Ratio= <u>1.20</u> times), (**d**= D/A= Dividend Portion or Payout= <u>30.00%</u>), and (**l**= L/E= Leverage or Gearing Ratio= <u>99.00%</u>), then it's (**q**= Quick or Acid Test Ratio planned), is:

$$q= c\text{-}P/\{l\,[U+M\text{-}F\text{-}Si\text{-}T\text{-}Ad]\text{-}Q\}$$
$$= 1.2000\text{-}6,497/\{0.9900\{64,720$$
$$+41,548\text{-}29,750\text{-}51,294$$
$$*0.0100\text{-}3,385\text{-}7,899$$
$$*0.3000]\text{-}35,351\}$$
$$= \underline{1.01}\ \text{times}$$

Corporate IFRS-GAAP (B/S-I/S), ISBN-13: **978-1720792789**, ISBN-10: **172079278X**

Law-7935:

   If both (**q**= [C-P]/X= Quick or Acid Test Ratio= <u>1.01</u> times), (**T**= Tax Paid= <u>3,385</u> millions USD), (**i**= I/S= Interest Portion= <u>1.00%</u>), (**F**= Fixed Cost= <u>29,740</u> millions USD), (**$**= Sales or Revenues= <u>51,294</u> millions USD), (**M**= Margin of Contribution= <u>41,548</u> millions USD), (**U**= Utilized or Starting Capital= <u>64,720</u> millions USD), (**A**= After Tax Income= <u>7,899</u> millions USD), (**p**= 360 P/V= Procured Inventory Days= <u>240</u> Days), (**V**= Variable Cost= <u>9,746</u> millions USD), (**c**= C/X= Current Ratio= <u>1.20</u> times), (**d**= D/A= Dividend Portion or Payout= <u>30.00%</u>), and (**l**= L/E= Leverage or Gearing Ratio= <u>99.00%</u>), then it's (**Q** = Quoted Longterm Liabilities planned), is:

$$Q= l\,[U+M-F-Si-T-Ad]-Vp/\{360[c-q]\}$$
$$= 0.9900\{64,720+41,548-29,750$$
$$-51,294*0.0100-3,385-7,899$$
$$*0.3000]-9,746*240/\{360$$
$$[1.2000-1.0100]\}$$
$$= \underline{35,351} \text{ millions USD}$$

Corporate IFRS-GAAP (B/S-I/S), ISBN-13: **978-1720792789**, ISBN-10: **172079278X**

<u>Law-7936</u>:

If both (**q**= [C-P]/X= Quick or Acid Test Ratio= <u>1.01</u> times), (**T**= Tax Paid= <u>3,385</u> millions USD), (**i**= I/S= Interest Portion= <u>1.00%</u>), (**F**= Fixed Cost= <u>29,740</u> millions USD), (**S**= Sales or Revenues= <u>51,294</u> millions USD), (**M**= Margin of Contribution= <u>41,548</u> millions USD), (**U**= Utilized or Starting Capital= <u>64,720</u> millions USD), (**A**= After Tax Income= <u>7,899</u> millions USD), (**p**= 360 P/V= Procured Inventory Days= <u>240</u> Days), (**V**= Variable Cost= <u>9,746</u> millions USD), (**c**= C/X= Current Ratio= <u>1.20</u> times), (**d**= D/A= Dividend Portion or Payout= <u>30.00%</u>), and (**Q**= Quoted Longterm Liabilities= <u>35,351</u> millions USD), then it's (**I** = Leverage or Gearing Ratio planned), is:

$$I = (Q+Vp/\{360[c\text{-}q]\})/[U+M\text{-}F\text{-}Si\text{-}T\text{-}Ad]$$
$$= (35,351+9,746*240/\{360[1.2000$$
$$-1.0100]\})/\{64,720+41,548$$
$$-29,750-51,294*0.0100$$
$$-3,385-7,899 *0.3000]$$
$$= \underline{99.00\%}$$

Corporate IFRS-GAAP (B/S-I/S), ISBN-13: **978-1720792789**, ISBN-10: **172079278X**

<u>Law-7937</u>:

If both (**I**= L/E= Leverage or Gearing Ratio= <u>99.00%</u>), (**T**= Tax Paid= <u>3,385</u> millions USD), (**i**= I/S= Interest Portion= <u>1.00%</u>), (**q**= [C-P]/X= Quick or Acid Test Ratio= <u>1.01</u> times), (**F**= Fixed Cost= <u>29,740</u> millions USD), (**S**= Sales or Revenues= <u>51,294</u> millions USD), (**M**= Margin of Contribution= <u>41,548</u> millions USD), (**A**= After Tax Income= <u>7,899</u> millions USD), (**p**= 360 P/V= Procured Inventory Days= <u>240</u> Days), (**V**= Variable Cost= <u>9,746</u> millions USD), (**c**= C/X= Current Ratio= <u>1.20</u> times), (**d**= D/A= Dividend Portion or Payout= <u>30.00%</u>), and (**Q**= Quoted Longterm Liabilities= <u>35,351</u> millions USD), then it's (**U**= Utilized or Starting Capital planned), is:

$$U= F+Si+T+Ad-M+(Q+Vp/\{360[c-q]\})/I$$

$$= 29,750+51,294*0.0100+3,385$$
$$+7,899*0.3000-41,548$$
$$+(35,351+9,746*240$$
$$/\{360[1.2000-1.0100]\})$$
$$/0.9900$$

$$= \underline{64,720} \text{ millions USD}$$

Corporate IFRS-GAAP (B/S-I/S), ISBN-13: **978-1720792789**, ISBN-10: **172079278X**

Law-7938:

If both (**I**= L/E= Leverage or Gearing Ratio= 99.00%), (**T**= Tax Paid= 3,385 millions USD), (**i**= I/S= Interest Portion= 1.00%), (**q**= [C-P]/X= Quick or Acid Test Ratio= 1.01 times), (**F**= Fixed Cost= 29,740 millions USD), (**S**= Sales or Revenues= 51,294 millions USD), (**U**= Utilized or Starting Capital= 64,720 millions USD), (**A**= After Tax Income= 7,899 millions USD), (**p**= 360 P/V= Procured Inventory Days= 240 Days), (**V**= Variable Cost= 9,746 millions USD), (**c**= C/X= Current Ratio= 1.20 times), (**d**= D/A= Dividend Portion or Payout= 30.00%), and (**Q**= Quoted Longterm Liabilities= 35,351 millions USD), then it's (**M**= Margin of Contribution planned), is:

$$M= F+Si+T+Ad-U+(Q+Vp/\{360[c-q]\})/I$$
$$= 29,750+51,294*0.0100+3,385$$
$$+7,899 *0.3000-64,720$$
$$+(35,351+9,746*240$$
$$/\{360[1.2000-1.0100]\})$$
$$/0.9900$$
$$= 41,548 \text{ millions USD}$$

Corporate IFRS-GAAP (B/S-I/S), ISBN-13: **978-1720792789**, ISBN-10: **172079278X**

Law-7939:

If both (I= L/E= Leverage or Gearing Ratio= 99.00%), (T= Tax Paid= 3,385 millions USD), (i= I/S= Interest Portion= 1.00%), (q= [C-P]/X= Quick or Acid Test Ratio= 1.01 times), (M= Margin of Contribution= 41,548 millions USD), (S= Sales or Revenues= 51,294 millions USD), (U= Utilized or Starting Capital= 64,720 millions USD), (A= After Tax Income= 7,899 millions USD), (p= 360 P/V= Procured Inventory Days= 240 Days), (V= Variable Cost= 9,746 millions USD), (c= C/X= Current Ratio= 1.20 times), (d= D/A= Dividend Portion or Payout= 30.00%), and (Q= Quoted Longterm Liabilities= 35,351 millions USD), then it's (F= Fixed Cost planned), is:

$$F = U+M-Si-T-Ad-(Q+Vp/\{360[c-q]\})/I$$
$$= 64,720+41,548-51,294*0.0100$$
$$-3,385-7,899*0.3000$$
$$-(35,351+9,746*240$$
$$/\{360[1.2000-1.0100]\})$$
$$/0.9900$$
$$= 29,750 \text{ millions USD}$$

Corporate IFRS-GAAP (B/S-I/S), ISBN-13: **978-1720792789**, ISBN-10: **172079278X**

<u>Law-7940</u>:

If both (I= L/E= Leverage or Gearing Ratio= 99.00%), (T= Tax Paid= <u>3,385</u> millions USD), (i= I/S= Interest Portion= <u>1.00%</u>), (q= [C-P]/X= Quick or Acid Test Ratio= <u>1.01</u> times), (M= Margin of Contribution= <u>41,548</u> millions USD), (F= Fixed Cost= <u>29,740</u> millions USD), (U= Utilized or Starting Capital= <u>64,720</u> millions USD), (A= After Tax Income= <u>7,899</u> millions USD), (p= 360 P/V= Procured Inventory Days= <u>240</u> Days), (V= Variable Cost= <u>9,746</u> millions USD), (c= C/X= Current Ratio= <u>1.20</u> times), (d= D/A= Dividend Portion or Payout= <u>30.00%</u>), and (Q= Quoted Longterm Liabilities= <u>35,351</u> millions USD), then it's (S= Sales or Revenues planned), is:

$$S= [U+M-F-T-Ad-(Q+Vp/\{360[c-q]\})/I]/i$$
$$= [64,720+41,548-29,750-3,385$$
$$-7,899*0.3000-(35,351$$
$$+9,746*240/\{360[1.2000$$
$$-1.0100]\})/0.9900]/0.0100$$
$$= \underline{51,294} \text{ millions USD}$$

Corporate IFRS-GAAP (B/S-I/S), ISBN-13: **978-1720792789**, ISBN-10: **172079278X**

Law-7941:

If both (I= L/E= Leverage or Gearing Ratio= 99.00%), (**T**= Tax Paid= 3,385 millions USD), (S= Sales or Revenues= 51,294 millions USD), (**q**= Quick or Acid Test Ratio= [C-P]/X= 1.01 times), (**M**= Margin of Contribution= 41,548 millions USD), (**F**= Fixed Cost= 29,740 millions USD), (**U**= Utilized or Starting Capital= 64,720 millions USD), (**A**= After Tax Income= 7,899 millions USD), (**p**= 360 P/V= Procured Inventory Days= 240 Days), (**V**= Variable Cost= 9,746 millions USD), (**c**= C/X= Current Ratio= 1.20 times), (**d**= D/A= Dividend Portion or Payout= 30.00%), and (**Q**= Quoted Longterm Liabilities= 35,351 millions USD), then it's (**i**= Interest Portion planned), is:

$$i= [U+M-F-T-Ad-(Q+Vp/\{360[c-q]\})/I]/S$$
$$= [64,720+41,548-29,750-3,385$$
$$-7,899*0.3000-(35,351$$
$$+9,746*240/\{360[1.2000$$
$$-1.0100]\})/0.9900]/51,294$$
$$= 1.00\%$$

Corporate IFRS-GAAP (B/S-I/S), ISBN-13: **978-1720792789**, ISBN-10: **172079278X**

Law-7942:

If both (I= L/E= Leverage or Gearing Ratio= 99.00%), (i= I/S= Interest Portion= 1.00%), ($\$$= Sales or Revenues= 51,294 millions USD), (q= [C-P]/X= Quick or Acid Test Ratio= 1.01 times), (M= Margin of Contribution= 41,548 millions USD), (F= Fixed Cost= 29,740 millions USD), (U= Utilized or Starting Capital= 64,720 millions USD), (A= After Tax Income= 7,899 millions USD), (p= 360 P/V= Procured Inventory Days= 240 Days), (V= Variable Cost= 9,746 millions USD), (c= C/X= Current Ratio= 1.20 times), (d= D/A= Dividend Portion or Payout= 30.00%), and (Q= Quoted Longterm Liabilities= 35,351 millions USD), then it's (T= Tax Paid planned), is:

$$T= U+M-F-\$i-Ad-(Q+Vp/\{360[c-q]\})/I$$
$$= 64,720+41,548-29,750-51,294$$
$$*0.0100-7,899*0.3000$$
$$-(35,351+9,746*240/\{360$$
$$[1.2000-1.0100]\})/0.9900$$
$$= 3,385 \text{ millions USD}$$

Corporate IFRS-GAAP (B/S-I/S), ISBN-13: **978-1720792789**, ISBN-10: **172079278X**

Law-7943:

If both (I= L/E= Leverage or Gearing Ratio= 99.00%), (i= I/S= Interest Portion= 1.00%), ($\$$= Sales or Revenues= 51,294 millions USD), (q= [C-P]/X= Quick or Acid Test Ratio= 1.01 times), (M= Margin of Contribution= 41,548 millions USD), (F= Fixed Cost= 29,740 millions USD), (U= Utilized or Starting Capital= 64,720 millions USD), (T= Tax Paid= 3,385 millions USD), (p= 360 P/V= Procured Inventory Days= 240 Days), (V= Variable Cost= 9,746 millions USD), (c= C/X= Current Ratio= 1.20 times), (d= D/A= Dividend Portion or Payout= 30.00%), and (Q= Quoted Longterm Liabilities= 35,351 millions USD), then it's (A= After Tax Income planned), is:

$$A= [U+M-F-\$i-T-(Q+Vp/\{360[c-q]\})/I\,]/d$$

$$= [64,720+41,548-29,750-51,294$$
$$*0.0100-3,385-(35,351$$
$$+9,746*240/\{360[1.2000$$
$$-1.0100]\})/0.9900]/0.3000$$

$$= 7,899 \text{ millions USD}$$

Corporate IFRS-GAAP (B/S-I/S), ISBN-13: **978-1720792789**, ISBN-10: **172079278X**

<u>Law-7944</u>:

If both (**I**= L/E= Leverage or Gearing Ratio= <u>99.00%</u>), (**i**= I/S= Interest Portion= <u>1.00%</u>), (**S**= Sales or Revenues= <u>51,294</u> millions USD), (**q**= [C-P]/X= Quick or Acid Test Ratio= <u>1.01</u> times), (**M**= Margin of Contribution= <u>41,548</u> millions USD), (**F**= Fixed Cost= <u>29,740</u> millions USD), (**U**= Utilized or Starting Capital= <u>64,720</u> millions USD), (**T**= Tax Paid= <u>3,385</u> millions USD), (**p**= 360 P/V= Procured Inventory Days= <u>240</u> Days), (**V**= Variable Cost= <u>9,746</u> millions USD), (**c**= C/X= Current Ratio= <u>1.20</u> times), (**A**= After Tax Income= <u>7,899</u> millions USD), and (**Q**= Quoted Longterm Liabilities= <u>35,351</u> millions USD), then it's (**d**= Dividend Portion or Payout planned), is:

$$d= [U+M-F-Si-T-(Q+Vp/\{360[c-q]\})/I]/A$$
$$= [64,720+41,548-29,750-51,294$$
$$*0.0100-3,385-(35,351$$
$$+9,746*240/\{360[1.2000$$
$$-1.0100]\})/0.9900]/7,899$$
$$= \underline{30.00\%}$$

Corporate IFRS-GAAP (B/S-I/S), ISBN-13: **978-1720792789**, ISBN-10: **172079278X**

Law-7945:

If both ($\digamma$= L/E= Leverage or Gearing Ratio= 99.00%), (i= I/S= Interest Portion= 1.00%), (S= Sales or Revenues= 51,294 millions USD), (q= [C-P]/X= Quick or Acid Test Ratio= 1.01 times), (M= Margin of Contribution= 41,548 millions USD), (F= Fixed Cost= 29,740 millions USD), (U= Utilized or Starting Capital= 64,720 millions USD), (T= Tax Paid= 3,385 millions USD), (p= 360 P/V= Procured Inventory Days= 240 Days), (d= D/A= Dividend Portion or Payout= 30.00%), (c= C/X= Current Ratio= 1.20 times), (A= After Tax Income= 7,899 millions USD), and (Q= Quoted Longterm Liabilities= 35,351 millions USD), then it's (V= Variable Cost planned), is:

$$V= 360[c\text{-}q]\{\digamma[U+M\text{-}F\text{-}Si\text{-}T\text{-}Ad]\text{-}Q\}/p$$
$$= 360[1.2000\text{-}1.0100]\{0.9900[64,720$$
$$+41,548\text{-}29,750\text{-}51,294$$
$$*0.0100\text{-}3,385\text{-}7,899*0.3000]$$
$$-35,351\}/240$$
$$= 9,746 \text{ millions USD}$$

Corporate IFRS-GAAP (B/S-I/S), ISBN-13: **978-1720792789**, ISBN-10: **172079278X**

<u>Law-7946</u>:

If both (**I**= L/E= Leverage or Gearing Ratio= <u>99.00%</u>), (**i**= I/S= Interest Portion= <u>1.00%</u>), (**S**= Sales or Revenues= <u>51,294</u> millions USD), (**q**= [C-P]/X= Quick or Acid Test Ratio= <u>1.01</u> times), (**M**= Margin of Contribution= <u>41,548</u> millions USD), (**F**= Fixed Cost= <u>29,740</u> millions USD), (**U**= Utilized or Starting Capital= <u>64,720</u> millions USD), (**T**= Tax Paid= <u>3,385</u> millions USD), (**V**= Variable Cost= <u>9,746</u> millions USD), (**d**= D/A= Dividend Portion or Payout= <u>30.00%</u>), (**c**= C/X= Current Ratio= <u>1.20</u> times), (**A**= After Tax Income= <u>7,899</u> millions USD), and (**Q**= Quoted Longterm Liabilities= <u>35,351</u> millions USD), then it's (**p**= Procured Inventories Days planned), is:

$$p= 360[\textbf{c-q}]\{\textbf{I}\,[\textbf{U+M-F-Si-T-Ad}]\textbf{-Q}\}/\textbf{V}$$

$$= 360[1.2000\text{-}1.0100]\{0.9900[64,720$$
$$+41,548\text{-}29,750\text{-}51,294$$
$$*0.0100\text{-}3,385\text{-}7,899*0.3000]$$
$$\text{-}35,351\}/9,746$$

$$= \underline{240} \text{ days}$$

Corporate IFRS-GAAP (B/S-I/S), ISBN-13: **978-1720792789**, ISBN-10: **172079278X**

<u>Law-7947</u>:

If both (I= L/E= Leverage or Gearing Ratio= <u>99.00%</u>), (i= I/S= Interest Portion= <u>1.00%</u>), (S= Sales or Revenues= <u>51,294</u> millions USD), (q= [C-P]/X= Quick or Acid Test Ratio= <u>1.01</u> times), (M= Margin of Contribution= <u>41,548</u> millions USD), (F= Fixed Cost= <u>29,740</u> millions USD), (U= Utilized or Starting Capital= <u>64,720</u> millions USD), (T= Tax Paid= <u>3,385</u> millions USD), (V= Variable Cost= <u>9,746</u> millions USD), (d= D/A= Dividend Portion or Payout= <u>30.00%</u>), (p= 360P/V= Procured Inventory Days= <u>240</u> Days), (A= After Tax Income= <u>7,899</u> millions USD), and (Q= Quoted Longterm Liabilities= <u>35,351</u> millions USD), then it's (c= Current Ratio planned), is:

$$c= q+Vp/(360\{I\,[U+M-F-Si-T-Ad]-Q\})$$
$$= 1.0100+9,746*240/(360\{0.9900$$
$$[64,720+41,548-29,750$$
$$-51,294*0.0100-3,385$$
$$-7,899*0.3000]-35,351\})$$
$$= \underline{1.20} \text{ times}$$

Corporate IFRS-GAAP (B/S-I/S), ISBN-13: **978-1720792789**, ISBN-10: **172079278X**

Law-7948:

If both (**I**= L/E= Leverage or Gearing Ratio=
99.00%), (**i**= I/S= Interest Portion= 1.00%), (**S**= Sales
or Revenues= 51,294 millions USD), (**c**= C/X=
Current Ratio= 1.20 times), (**M**= Margin of
Contribution= 41,548 millions USD), (**F**= Fixed
Cost= 29,740 millions USD), (**U**= Utilized or Starting
Capital= 64,720 millions USD), (**T**= Tax Paid= 3,385
millions USD), (**V**= Variable Cost= 9,746 millions
USD), (**d**= D/A= Dividend Portion or Payout=
30.00%), (**p**= 360 P/V= Procured Inventory Days=
240 Days), (**A**= After Tax Income= 7,899 millions
USD), and (**Q**= Quoted Longterm Liabilities= 35,351
millions USD), then it's (**q**= Quick or Acid Test
Ratio planned), is:

$$q= c\text{-}Vp/(360\{I\,[U+M\text{-}F\text{-}Si\text{-}T\text{-}Ad]\text{-}Q\})$$
$$= 1.2000\text{-}9,746*240/(360\{0.9900$$
$$[64,720+41,548\text{-}29,750$$
$$\text{-}51,294*0.0100\text{-}3,385$$
$$\text{-}7,899*0.3000]\text{-}35,351\})$$
$$= \underline{1.01}\text{ times}$$

Corporate IFRS-GAAP (B/S-I/S), ISBN-13: **978-1720792789**, ISBN-10: **172079278X**

Law-7949:

If both (**I**= L/E= Leverage or Gearing Ratio= 99.00%), (**i**= I/S= Interest Portion= 1.00%), (**S**= Sales or Revenues= 51,294 millions USD), (**c**= C/X= Current Ratio= 1.20 times), (**M**= Margin of Contribution= 41,548 millions USD), (**F**= Fixed Cost= 29,740 millions USD), (**U**= Utilized or Starting Capital= 64,720 millions USD), (**T**= Tax Paid= 3,385 millions USD), (**v**= V/S= Variable Portion= 19.00%), (**d**= D/A= Dividend Portion or Payout= 30.00%), (**p**= 360 P/V= Procured Inventory Days= 240 Days), (**A**= After Tax Income= 7,899 millions USD), and (**q**= [C-P]/X= Quick or Acid Test Ratio= 1.01 times), then it's (**Q**= Quoted Longterm Liabilities planned), is:

$$Q = I[U+M-F-Si-T-Ad]-Svp/\{360[c-q]\}$$
$$= 0.9900[64,720+41,548-29,750$$
$$-51,294*0.0100-3,385$$
$$-7,899*0.3000]-51,294$$
$$*0.1900*240/\{360[1.2000$$
$$-1.0100]\}$$
$$= 35,351 \text{ millions USD}$$

Corporate IFRS-GAAP (B/S-I/S), ISBN-13: **978-1720792789**, ISBN-10: **172079278X**

Law-7950:

  If both (**Q**= Quoted Longterm Liabilities= 35,351
  millions USD), (**i**= I/S= Interest Portion= 1.00%), (**S**=
  Sales or Revenues= 51,294 millions USD), (**c**= C/X=
  Current Ratio= 1.20 times), (**M**= Margin of
  Contribution= 41,548 millions USD), (**F**= Fixed
  Cost= 29,740 millions USD), (**U**= Utilized or Starting
  Capital= 64,720 millions USD), (**T**= Tax Paid= 3,385
  millions USD), (**v**= V/S= Variable Portion= 19.00%),
  (**d**= D/A= Dividend Portion or Payout= 30.00%), (**p**=
  360 P/V= Procured Inventory Days= 240 Days), (**A**=
  After Tax Income= 7,899 millions USD), and (**q**=
  Quick or Acid Test Ratio= [C-P]/X= 1.01 times),
  then it's (**I** = Leverage or Gearing Ratio planned), is:

  I = (**Q**+**Svp**/{360[**c-q**]})/[**U**+**M**-**F**-**Si**-**T**-**Ad**]
  $\qquad$ = (35,351+51,294*0.1900*240/{360
  $\qquad\qquad$ [1.2000-1.0100]})/[64,720
  $\qquad\qquad$ +41,548-29,750-51,294
  $\qquad\qquad$ *0.0100-3,385-7,899*0.3000]
  $\qquad$ = 99.00%

<u>Law-7951</u>:

If both (**Q**= Quoted Longterm Liabilities= <u>35,351</u> millions USD), (**i**= I/S= Interest Portion= <u>1.00%</u>), (**S**= Sales or Revenues= <u>51,294</u> millions USD), (**c**= C/X= Current Ratio= <u>1.20</u> times), (**M**= Margin of Contribution= <u>41,548</u> millions USD), (**F**= Fixed Cost= <u>29,740</u> millions USD), (**l**= L/E= Leverage or Gearing Ratio= <u>99.00%</u>), (**T**= Tax Paid= <u>3,385</u> millions USD), (**v**= V/S= Variable Portion= <u>19.00%</u>), (**d**= D/A= Dividend Portion or Payout= <u>30.00%</u>), (**p**= 360 P/V= Procured Inventory Days= <u>240</u> Days), (**A**= After Tax Income= <u>7,899</u> millions USD), and (**q**= [C-P]/X= Quick or Acid Test Ratio= <u>1.01</u> times), then it's (**U**= Utilized or Starting Capital planned), is:

$$U= F+Si+T+Ad-M+(Q+Svp/\{360[c-q]\})/l$$
$$= 29,750+51,294*0.0100+3,385$$
$$+7,899*0.3000 -41,548$$
$$+(35,351+51,294*0.1900$$
$$*240/\{360[1.2000$$
$$-1.0100]\})/0.9900$$
$$= \underline{64,720} \text{ millions USD}$$

Corporate IFRS-GAAP (B/S-I/S), ISBN-13: **978-1720792789**, ISBN-10: **172079278X**

<u>Law-7952</u>:

If both (**Q**= Quoted Longterm Liabilities= 35,351 millions USD), (**i**= I/S= Interest Portion= 1.00%), (**$**= Sales or Revenues= 51,294 millions USD), (**c**= C/X= Current Ratio= 1.20 times), (**U**= Utilized or Starting Capital= 64,720 millions USD), (**F**= Fixed Cost= 29,740 millions USD), (**⌐**= L/E= Leverage or Gearing Ratio= 99.00%), (**T**= Tax Paid= 3,385 millions USD), (**v**= V/S= Variable Portion= 19.00%), (**d**= D/A= Dividend Portion or Payout= 30.00%), (**p**= 360 P/V= Procured Inventory Days= 240 Days), (**A**= After Tax Income= 7,899 millions USD), and (**q**= [C-P]/X= Quick or Acid Test Ratio= 1.01 times), then it's (**M**= Margin of Contribution planned), is:

$$M= F+Si+T+Ad-U+(Q+Svp/\{360[c-q]\})/I$$
$$= 29,750+51,294*0.0100+3,385$$
$$+7,899*0.3000-64,720$$
$$+(35,351+51,294*0.1900$$
$$*240/\{360[1.2000$$
$$-1.0100]\})/0.9900$$
$$= \underline{41,548} \text{ millions USD}$$

Corporate IFRS-GAAP (B/S-I/S), ISBN-13: **978-1720792789**, ISBN-10: **172079278X**

<u>Law-7953</u>:

If both (**Q**= Quoted Longterm Liabilities= <u>35,351</u> millions USD), (**i**= I/S= Interest Portion= <u>1.00%</u>), (**S**= Sales or Revenues= <u>51,294</u> millions USD), (**c**= C/X= Current Ratio= <u>1.20</u> times), (**U**= Utilized or Starting Capital= <u>64,720</u> millions USD), (**M**= Margin of Contribution= <u>41,548</u> millions USD), (**l**= L/E= Leverage or Gearing Ratio= <u>99.00%</u>), (**T**= Tax Paid= <u>3,385</u> millions USD), (**v**= V/S= Variable Portion= <u>19.00%</u>), (**d**= D/A= Dividend Portion or Payout= <u>30.00%</u>), (**p**= 360P/V= Procured Inventory Days= <u>240</u> Days), (**A**= After Tax Income= <u>7,899</u> millions USD), and (**q**= [C-P]/X= Quick or Acid Test Ratio= <u>1.01</u> times), then it's (**F**= Fixed Cost planned), is:

$$F= U+M-Si-T-Ad-(Q+Svp/\{360[c-q]\})/l$$
$$= 64,720+41,548-51,294*0.0100$$
$$-3,385-7,899*0.3000$$
$$-(35,351+51,294*0.1900$$
$$*240/\{360[1.2000$$
$$-1.0100]\})/0.9900$$
$$= \underline{29,750} \text{ millions USD}$$

Corporate IFRS-GAAP (B/S-I/S), ISBN-13: **978-1720792789**, ISBN-10: **172079278X**

Law-7954:

If both (**Q**= Quoted Longterm Liabilities= 35,351 millions USD), (**i**= I/S= Interest Portion= 1.00%), (**F**= Fixed Cost= 29,740 millions USD), (**c**= C/X= Current Ratio= 1.20 times), (**U**= Utilized or Starting Capital= 64,720 millions USD), (**M**= Margin of Contribution= 41,548 millions USD), (**l**= L/E= Leverage or Gearing Ratio= 99.00%), (**T**= Tax Paid= 3,385 millions USD), (**v**= V/S= Variable Portion= 19.00%), (**d**= D/A= Dividend Portion or Payout= 30.00%), (**p**= 360 P/V= Procured Inventory Days= 240 Days), (**A**= After Tax Income= 7,899 millions USD), and (**q**= [C-P]/X= Quick or Acid Test Ratio= 1.01 times), then it's (**$**= Sales or Revenues planned), is :

$$\$= [U+M-F-T-Ad-(Q+\$vp/\{360[c-q]\})/I]/i$$
$$= [64,720+41,548-29,750-3,385$$
$$-7,899*0.3000-(35,351$$
$$+51,294*0.1900*240$$
$$/\{360[1.2000-1.0100]\})$$
$$/0.9900]/0.0100$$
$$= 51,294 \text{ millions USD}$$

Corporate IFRS-GAAP (B/S-I/S), ISBN-13: **978-1720792789**, ISBN-10: **172079278X**

Law-7955:

If both (**Q**= Quoted Longterm Liabilities= <u>35,351</u> millions USD), (**$**= Sales or Revenues= <u>51,294</u> millions USD), (**F**= Fixed Cost= <u>29,740</u> millions USD), (**c**= C/X= Current Ratio= <u>1.20</u> times), (**U**= Utilized or Starting Capital= <u>64,720</u> millions USD), (**M**= Margin of Contribution= <u>41,548</u> millions USD), (**l**= L/E= Leverage or Gearing Ratio= <u>99.00%</u>), (**T**= Tax Paid= <u>3,385</u> millions USD), (**v**= V/S= Variable Portion= <u>19.00%</u>), (**d**= D/A= Dividend Portion or Payout= <u>30.00%</u>), (**p**= 360P/V= Procured Inventory Days= <u>240</u> Days), (**A**= After Tax Income= <u>7,899</u> millions USD), and (**q**= [C-P]/X= Quick or Acid Test Ratio= <u>1.01</u> times), then it's (**i** = Interest Portion planned), is:

$$i= [U+M-F-T-Ad-(Q+\$vp/\{360[c-q]\})/l\ ]/\$$$

$$
\begin{aligned}
&= [64,720+41,548-29,750-3,385 \\
&\quad -7,899*0.3000-(35,351 \\
&\quad +51,294*0.1900*240 \\
&\quad /\{360[1.2000-1.0100]\}) \\
&\quad /0.9900]/51,294
\end{aligned}
$$

$$= \underline{1.00\%}$$

Corporate IFRS-GAAP (B/S-I/S), ISBN-13: **978-1720792789**, ISBN-10: **172079278X**

Law-7956:

If both (**Q**= Quoted Longterm Liabilities= 35,351 millions USD), (**S**= Sales or Revenues= 51,294 millions USD), (**F**= Fixed Cost= 29,740 millions USD), (**c**= C/X= Current Ratio= 1.20 times), (**U**= Utilized or Starting Capital= 64,720 millions USD), (**M**= Margin of Contribution= 41,548 millions USD), (**l**= L/E= Leverage or Gearing Ratio= 99.00%), (**i**= I/S= Interest Portion= 1.00%), (**v**= V/S= Variable Portion= 19.00%), (**d**= D/A= Dividend Portion or Payout= 30.00%), (**p**= 360P/V= Procured Inventory Days= 240 Days), (**A**= After Tax Income= 7,899 millions USD), and (**q**= [C-P]/X= Quick or Acid Test Ratio= 1.01 times), then it's (**T** = Tax Paid planned), is:

$$T= U+M-F-Si-Ad-(Q+Svp/\{360[c-q]\})/l$$
$$= 64,720+41,548-29,750-51,294$$
$$*0.0100-7,899*0.3000$$
$$-(35,351+51,294*0.1900$$
$$*240/\{360[1.2000$$
$$-1.0100]\})/0.9900$$
$$= 3,385 \text{ millions USD}$$

Corporate IFRS-GAAP (B/S-I/S), ISBN-13: **978-1720792789**, ISBN-10: **172079278X**

Law-7957:

If both (**Q**= Quoted Longterm Liabilities= 35,351 millions USD), (**S**= Sales or Revenues= 51,294 millions USD), (**F**= Fixed Cost= 29,740 millions USD), (**c**= C/X= Current Ratio= 1.20 times), (**U**= Utilized or Starting Capital= 64,720 millions USD), (**M**= Margin of Contribution= 41,548 millions USD), (**l**= L/E= Leverage or Gearing Ratio= 99.00%), (**i**= I/S= Interest Portion= 1.00%), (**v**= V/S= Variable Portion= 19.00%), (**d**= D/A= Dividend Portion or Payout= 30.00%), (**p**= 360 P/V= Procured Inventory Days= 240 Days), (**T**= Tax Paid= 3,385 millions USD), and (**q**= [C-P]/X= Quick or Acid Test Ratio= 1.01 times), then it's (**A**= After Tax Income planned), is :

$$A= [U+M-F-Si-T-(Q+Svp/\{360[c-q]\})/l]/d$$
$$= [64,720+41,548-29,750-51,294$$
$$*0.0100-3,385-(35,351$$
$$+51,294*0.1900*240$$
$$/\{360[1.2000-1.0100]\})$$
$$/0.9900]/0.3000$$
$$= 7,899 \text{ millions USD}$$

Corporate IFRS-GAAP (B/S-I/S), ISBN-13: **978-1720792789**, ISBN-10: **172079278X**

<u>Law-7958</u>:

If both (**Q**= Quoted Longterm Liabilities= <u>35,351</u> millions USD), (**S**= Sales or Revenues= <u>51,294</u> millions USD), (**F**= Fixed Cost= <u>29,740</u> millions USD), (**c**= C/X= Current Ratio= <u>1.20</u> times), (**U**= Utilized or Starting Capital= <u>64,720</u> millions USD), (**M**= Margin of Contribution= <u>41,548</u> millions USD), (**l**= L/E= Leverage or Gearing Ratio= <u>99.00%</u>), (**i**= I/S= Interest Portion= <u>1.00%</u>), (**v**= V/S= Variable Portion= <u>19.00%</u>), (**A**= After Tax Income= <u>7,899</u> millions USD), (**p**= 360 P/V= Procured Inventory Days= <u>240</u> Days), (**T**= Tax Paid= <u>3,385</u> millions USD), and (**q**= [C-P]/X= Quick or Acid Test Ratio= <u>1.01</u> times), then it's (**d**= Dividend Portion or Payuot planned), is:

$$d= [U+M-F-Si-T-(Q+Svp/\{360[c-q]\})/l]/A$$

$$= [64{,}720+41{,}548-29{,}750-51{,}294$$
$$*0.0100-3{,}385-(35{,}351$$
$$+51{,}294*0.1900*240$$
$$/\{360[1.2000-1.0100]\})$$
$$/0.9900]/7{,}899$$

$$= \underline{30.00\%}$$

Corporate IFRS-GAAP (B/S-I/S), ISBN-13: **978-1720792789**, ISBN-10: **172079278X**

Law-7959:

If both (Q= Quoted Longterm Liabilities= 35,351 millions USD), (S= Sales or Revenues= 51,294 millions USD), (F= Fixed Cost= 29,740 millions USD), (c= C/X= Current Ratio= 1.20 times), (U= Utilized or Starting Capital= 64,720 millions USD), (M= Margin of Contribution= 41,548 millions USD), (I= L/E= Leverage or Gearing Ratio= 99.00%), (i= I/S= Interest Portion= 1.00%), (d= D/A= Dividend Portion or Payout= 30.00%), (A= After Tax Income= 7,899 millions USD), (p= 360 P/V= Procured Inventory Days= 240 Days), (T= Tax Paid= 3,385 millions USD), and (q= [C-P]/X= Quick or Acid Test Ratio= 1.01 times), then it's (v= Variable Portion planned), is:

$$v= 360[c\text{-}q]\{I\,[U+M\text{-}F\text{-}Si\text{-}T\text{-}Ad]\text{-}Q\}/[Sp]$$

$$= 360[1.2000\text{-}1.0100]\{0.9900$$
$$[64,720+41,548\text{-}29,750$$
$$-51,294*0.0100\text{-}3,385$$
$$-7,899*0.3000]\text{-}35,351\}$$
$$/[51,294*240]$$

$$= 19.00\%$$

Corporate IFRS-GAAP (B/S-I/S), ISBN-13: **978-1720792789**, ISBN-10: **172079278X**

<u>Law-7960</u>:

If both (**Q**= Quoted Longterm Liabilities= <u>35,351</u> millions USD), (**S**= Sales or Revenues= <u>51,294</u> millions USD), (**F**= Fixed Cost= <u>29,740</u> millions USD), (**c**= C/X= Current Ratio= <u>1.20</u> times), (**U**= Utilized or Starting Capital= <u>64,720</u> millions USD), (**M**= Margin of Contribution= <u>41,548</u> millions USD), (**I**= L/E= Leverage or Gearing Ratio= <u>99.00%</u>), (**i**= I/S= Interest Portion= <u>1.00%</u>), (**d**= D/A= Dividend Portion or Payout= <u>30.00%</u>), (**A**= After Tax Income= <u>7,899</u> millions USD), (**v**= V/S= Variable Portion= <u>19.00%</u>), (**T**= Tax Paid= <u>3,385</u> millions USD), and (**q**= [C-P]/X= Quick or Acid Test Ratio= <u>1.01</u> times), then it's (**p**= Procured Inventory Days planned), is:

$$p= 360[\mathbf{c\text{-}q}]\{\mathbf{I}\,[\mathbf{U+M\text{-}F\text{-}Si\text{-}T\text{-}Ad}]\text{-}\mathbf{Q}\}/[\mathbf{Sv}]$$
$$= 360[1.2000\text{-}1.0100]\{0.9900$$
$$[64,720+41,548\text{-}29,750$$
$$\text{-}51,294*0.0100\text{-}3,385$$
$$\text{-}7,899*0.3000]\text{-}35,351\}$$
$$/[51,294*0.1900]$$
$$= \underline{240}\text{ days}$$

Corporate IFRS-GAAP (B/S-I/S), ISBN-13: **978-1720792789**, ISBN-10: **172079278X**

Law-7961:

If both (Q= Quoted Longterm Liabilities= 35,351 millions USD), (S= Sales or Revenues= 51,294 millions USD), (F= Fixed Cost= 29,740 millions USD), (p= 360 P/V= Procured Inventory Days= 240 Days), (U= Utilized or Starting Capital= 64,720 millions USD), (M= Margin of Contribution= 41,548 millions USD), (I= L/E= Leverage or Gearing Ratio= 99.00%), (i= I/S= Interest Portion= 1.00%), (d= D/A= Dividend Portion or Payout= 30.00%), (A= After Tax Income= 7,899 millions USD), (v= V/S= Variable Portion= 19.00%), (T= Tax Paid= 3,385 millions USD), and (q= [C-P]/X= Quick or Acid Test Ratio= 1.01 times), then it's (c= Current Ratio planned), is:

$$c= q+Svp/(360\{I[U+M-F-Si-T-Ad]-Q\})$$

$$= 1.0100+51,294*0.1900*240(360$$
$$\{0.9900[64,720+41,548$$
$$-29,750-51,294*0.0100$$
$$-3,385-7,899*0.3000]$$
$$-35,351\})$$

$$= 1.20 \text{ times}$$

Corporate IFRS-GAAP (B/S-I/S), ISBN-13: **978-1720792789**, ISBN-10: **172079278X**

Law-7962:

If both (**Q**= Quoted Longterm Liabilities= 35,351 millions USD), (**S**= Sales or Revenues= 51,294 millions USD), (**F**= Fixed Cost= 29,740 millions USD), (**p**= 360 P/V= Procured Inventory Days= 240 Days), (**U**= Utilized or Starting Capital= 64,720 millions USD), (**M**= Margin of Contribution= 41,548 millions USD), (**l**= L/E= Leverage or Gearing Ratio= 99.00%), (**i**= I/S= Interest Portion= 1.00%), (**d**= D/A= Dividend Portion or Payout= 30.00%), (**A**= After Tax Income= 7,899 millions USD), (**v**= V/S= Variable Portion= 19.00%), (**T**= Tax Paid= 3,385 millions USD), and (**c**= C/X= Current Ratio= 1.20 times), then it's (**q** = Quick or Acid Test Ratio planned), is:

$$q= c\text{-}Svp/(360\{l\,[U+M\text{-}F\text{-}Si\text{-}T\text{-}Ad]\text{-}Q\})$$
$$= 1.2000\text{-}51,294*0.1900*240(360$$
$$\{0.9900[64,720+41,548$$
$$-29,750\text{-}51,294*0.0100$$
$$-3,385\text{-}7,899*0.3000]$$
$$-35,351\})$$
$$= \underline{1.01} \text{ times}$$

Law-7963:

If both (q= [C-P]/X= Quick or Acid Test Ratio= 1.01 times), (S'= Sales of Past Year= 48,851 millions USD), (s= [S/S']-1= Sales Growth= 5.00%), (S= Sales or Revenues= 51,294 millions USD), (F= Fixed Cost= 29,740 millions USD), (p= 360 P/V= Procured Inventory Days= 240 Days), (U= Utilized or Starting Capital= 64,720 millions USD), (M= Margin of Contribution= 41,548 millions USD), (I= L/E= Leverage or Gearing Ratio= 99.00%), (i= I/S= Interest Portion= 1.00%), (d= D/A= Dividend Portion or Payout= 30.00%), (A= After Tax Income= 7,899 millions USD), (v= V/S= Variable Portion= 19.00%), (T= Tax Paid= 3,385 millions USD), and (c= C/X= Current Ratio= 1.20 times), then it's (Q= Quoted Longterm Liabilities planned), is:

$$Q= I[U+M-F-Si-T-Ad]-S'vp[1+s]/\{360[c-q]\}$$
$$= 0.9900[64,720+41,548-29,750$$
$$-51,294*0.0100-3,385$$
$$-7,899*0.3000]- 48,851$$
$$*0.1900*240[1+0.0500]$$
$$\{360[1.2000-1.0100]\}$$
$$= 35,351 \text{ millions USD}$$

Corporate IFRS-GAAP (B/S-I/S), ISBN-13: **978-1720792789**, ISBN-10: **172079278X**

<u>Law-7964</u>:

If both (q= [C-P]/X= Quick or Acid Test Ratio= <u>1.01</u> times), (S'= Sales of Past Year= <u>48,851</u> millions USD), (s= [S/S']-1= Sales Growth= <u>5.00%</u>), (S= Sales or Revenues= <u>51,294</u> millions USD), (F= Fixed Cost= <u>29,740</u> millions USD), (p= 360 P/V= Procured Inventory Days= <u>240</u> Days), (U= Utilized or Starting Capital= <u>64,720</u> millions USD), (M= Margin of Contribution= <u>41,548</u> millions USD), (Q= Quoted Longterm Liabilities= <u>35,351</u> millions USD), (i= I/S= Interest Portion= <u>1.00%</u>), (d= D/A= Dividend Portion or Payout= <u>30.00%</u>), (A= After Tax Income= <u>7,899</u> millions USD), (v= V/S= Variable Portion= <u>19.00%</u>), (T= Tax Paid= <u>3,385</u> millions USD), and (c= C/X= Current Ratio= <u>1.20</u> times), then it's (I = Leverage or Gearing Ratio planned), is:

$$I = (Q+S'vp[1+s]/\{360[c-q]\})$$
$$/[U+M-F-Si-T-Ad]$$
$$= (35,351+48,851*0.1900*240$$
$$[1+0.0500]\{360[1.2000$$
$$-1.0100]\})/[64,720+41,548$$
$$-29,750-51,294*0.0100$$
$$-3,385-7,899*0.3000]$$
$$= \underline{99.00\%}$$

Corporate IFRS-GAAP (B/S-I/S), ISBN-13: **978-1720792789**, ISBN-10: **172079278X**

Law-7965:

If both (**q**= [C-P]/X= Quick or Acid Test Ratio= 1.01 times), (**S'**= Sales of Past Year= 48,851 millions USD), (**s**= [S/S']-1= Sales Growth= 5.00%), (**S**= Sales or Revenues= 51,294 millions USD), (**F**= Fixed Cost= 29,740 millions USD), (**p**= 360 P/V= Procured Inventory Days= 240 Days), (**I**= L/E= Leverage or Gearing Ratio= 99.00%), (**M**= Margin of Contribution= 41,548 millions USD), (**Q**= Quoted Longterm Liabilities= 35,351 millions USD), (**i**= I/S= Interest Portion= 1.00%), (**d**= D/A= Dividend Portion or Payout= 30.00%), (**A**= After Tax Income= 7,899 millions USD), (**v**= V/S= Variable Portion= 19.00%), (**T**= Tax Paid= 3,385 millions USD), and (**c**= C/X= Current Ratio= 1.20 times), then it's (**U**= Utilized or Starting Capital planned), is:

$$U= F+Si+T+Ad-M+(Q+S'vp[1+s]/\{360[c-q]\})/I$$

$$= 29,750+51,294*0.0100+3,385$$
$$+7,899*0.3000-41,548$$
$$+ (35,351+48,851*0.1900*$$
$$240[1+0.0500]\{360[1.2000$$
$$-1.0100]\})/0.9900$$

$$= 64,720 \text{ millions USD}$$

Corporate IFRS-GAAP (B/S-I/S), ISBN-13: **978-1720792789**, ISBN-10: **172079278X**

Law-7966:

If both (**q**= [C-P]/X= Quick or Acid Test Ratio= 1.01 times), (**S'**= Sales of Past Year= 48,851 millions USD), (**s**= [S/S']-1= Sales Growth= 5.00%), (**S**= Sales or Revenues= 51,294 millions USD), (**F**= Fixed Cost= 29,740 millions USD), (**p**= 360 P/V= Procured Inventory Days= 240 Days), (**l**= L/E= Leverage or Gearing Ratio= 99.00%), (**U**= Utilized or Starting Capital= 64,720 millions USD), (**Q**= Quoted Longterm Liabilities= 35,351 millions USD), (**i**= I/S= Interest Portion= 1.00%), (**d**= D/A= Dividend Portion or Payout= 30.00%), (**A**= After Tax Income= 7,899 millions USD), (**v**= V/S= Variable Portion= 19.00%), (**T**= Tax Paid= 3,385 millions USD), and (**c**= C/X= Current Ratio= 1.20 times), then it's (**M**= Margin of Contribution planned), is:

$$M= F+Si+T+Ad-U+(Q+S'vp[1+s]/\{360[c-q]\})/l$$

$$= 29,750+51,294*0.0100+3,385$$
$$+7,899*0.3000-64,720$$
$$+ (35,351+48,851*0.1900$$
$$*240[1+0.0500]\{360[1.2000$$
$$-1.0100]\})/0.9900$$

$$= 41,548 \text{ millions USD}$$

Corporate IFRS-GAAP (B/S-I/S), ISBN-13: **978-1720792789**, ISBN-10: **172079278X**

Law-7967:

If both (**q**= [C-P]/X= Quick or Acid Test Ratio= 1.01 times), (**S'**= Sales of Past Year= 48,851 millions USD), (**s**= [S/S']-1= Sales Growth= 5.00%), (**S**= Sales or Revenues= 51,294 millions USD), (**M**= Margin of Contribution= 41,548 millions USD), (**p**= 360 P/V= Procured Inventory Days= 240 Days), (**l**= L/E= Leverage or Gearing Ratio= 99.00%), (**U**= Utilized or Starting Capital= 64,720 millions USD), (**Q**= Quoted Longterm Liabilities= 35,351 millions USD), (**i**= I/S= Interest Portion= 1.00%), (**d**= D/A= Dividend Portion or Payout= 30.00%), (**A**= After Tax Income= 7,899 millions USD), (**v**= V/S= Variable Portion= 19.00%), (**T**= Tax Paid= 3,385 millions USD), and (**c**= C/X= Current Ratio= 1.20 times), then it's (**F**= Fixed Cost planned), is:

$$F= U+M-Si-T-Ad-(Q+S'vp[1+s]$$
$$/\{360[c-q]\})/l$$
$$= 64,720+41,548-51,294*0.0100$$
$$-3,385-7,899*0.3000-(35,351$$
$$+48,851*0.1900*240$$
$$[1+0.0500]\{360[1.2000$$
$$-1.0100]\})/0.9900$$
$$= 29,750 \text{ millions USD}$$

Corporate IFRS-GAAP (B/S-I/S), ISBN-13: **978-1720792789**, ISBN-10: **172079278X**

<u>Law-7968</u>:

If both (**q**= [C-P]/X= Quick or Acid Test Ratio= <u>1.01</u> times), (**S'**= Sales of Past Year= <u>48,851</u> millions USD), (**s**= [S/S']-1= Sales Growth= <u>5.00%</u>), (**F**= Fixed Cost= <u>29,740</u> millions USD), (**M**= Margin of Contribution= <u>41,548</u> millions USD), (**p**= 360 P/V= Procured Inventory Days= <u>240</u> Days), (**l**= L/E= Leverage or Gearing Ratio= <u>99.00%</u>), (**U**= Utilized or Starting Capital= <u>64,720</u> millions USD), (**Q**= Quoted Longterm Liabilities= <u>35,351</u> millions USD), (**i**= I/S= Interest Portion= <u>1.00%</u>), (**d**= D/A= Dividend Portion or Payout= <u>30.00%</u>), (**A**= After Tax Income= <u>7,899</u> millions USD), (**v**= V/S= Variable Portion= <u>19.00%</u>), (**T**= Tax Paid= <u>3,385</u> millions USD), and (**c**= C/X= Current Ratio= <u>1.20</u> times), then it's (**S**= Sales or Revenues planned), is:

$$S= [U+M-F-T-Ad-(Q+S'vp[1+s]$$
$$/\{360[c-q]\})/l\,]/i$$
$$= [64,720+41,548-29,750-3,385$$
$$-7,899*0.3000-(35,351$$
$$+48,851*0.1900*240$$
$$[1+0.0500]\{360[1.2000$$
$$-1.0100]\})/0.9900]/0.0100$$
$$= 29,750 \text{ millions USD}$$

Corporate IFRS-GAAP (B/S-I/S), ISBN-13: **978-1720792789**, ISBN-10: **172079278X**

Law-7969:

If both (q= [C-P]/X= Quick or Acid Test Ratio= 1.01 times), (S'= Sales of Past Year= 48,851 millions USD), (s= [S/S']-1= Sales Growth= 5.00%), (F= Fixed Cost= 29,740 millions USD), (M= Margin of Contribution= 41,548 millions USD), (p= 360 P/V= Procured Inventory Days= 240 Days), (I= L/E= Leverage or Gearing Ratio= 99.00%), (U= Utilized or Starting Capital= 64,720 millions USD), (Q= Quoted Longterm Liabilities= 35,351 millions USD), (S= Sales or Revenues= 51,294 millions USD), (d= D/A= Dividend Portion or Payout= 30.00%), (A= After Tax Income= 7,899 millions USD), (v= V/S= Variable Portion= 19.00%), (T= Tax Paid= 3,385 millions USD), and (c= C/X= Current Ratio= 1.20 times), then it's (i= Interest Portion planned), is:

i= [U+M-F-T-Ad-(Q+S'vp[1+s]
/{360[c-q]})/I]/S
= [64,720+41,548-29,750-3,385
-7,899*0.3000-(35,351
+48,851*0.1900*240
[1+0.0500]{360[1.2000
-1.0100]})/0.9900]/51.294

= 1.00%

Corporate IFRS-GAAP (B/S-I/S), ISBN-13: **978-1720792789**, ISBN-10: **172079278X**

<u>Law-7970</u>:

If both (**q**= [C-P]/X= Quick or Acid Test Ratio= <u>1.01</u> times), (**c**= C/X= Current Ratio= <u>1.20</u> times), (**S'**= Sales of Past Year= <u>48,851</u> millions USD), (**s**= [S/S']-1= Sales Growth= <u>5.00%</u>), (**F**= Fixed Cost= <u>29,740</u> millions USD), (**M**= Margin of Contribution= <u>41,548</u> millions USD), (**p**= 360 P/V= Procured Inventory Days= <u>240</u> Days), (**l**= L/E= Leverage or Gearing Ratio= <u>99.00%</u>), (**U**= Utilized or Starting Capital= <u>64,720</u> millions USD), (**Q**= Quoted Longterm Liabilities= <u>35,351</u> millions USD), (**S**= Sales or Revenues= <u>51,294</u> millions USD), (**d**= D/A= Dividend Portion or Payout= <u>30.00%</u>), (**A**= After Tax Income= <u>7,899</u> millions USD), (**v**= V/S= Variable Portion= <u>19.00%</u>), and (**i**= I/S= Interest Portion= <u>1.00%</u>), then it's (**T**= Tax Paid planned), is:

$$T= U+M-F-Si-Ad-(Q+S'vp[1+s]$$
$$/\{360[c-q]\})/l$$
$$= 64,720+41,548-29,750-51,294$$
$$*0.0100-7,899*0.3000$$
$$- (35,351+48,851*0.1900$$
$$*240[1+0.0500]\{360[1.2000$$
$$-1.0100]\})/0.9900$$
$$= \underline{3,385} \text{ millions USD}$$

Corporate IFRS-GAAP (B/S-I/S), ISBN-13: **978-1720792789**, ISBN-10: **172079278X**

Law-7971:

If both (**q**= [C-P]/X= Quick or Acid Test Ratio= 1.01 times), (**c**= C/X= Current Ratio= 1.20 times), (**S'**= Sales of Past Year= 48,851 millions USD), (**s**= [S/S']-1= Sales Growth= 5.00%), (**F**= Fixed Cost= 29,740 millions USD), (**M**= Margin of Contribution= 41,548 millions USD), (**p**= 360 P/V= Procured Inventory Days= 240 Days), (**l**= L/E= Leverage or Gearing Ratio= 99.00%), (**U**= Utilized or Starting Capital= 64,720 millions USD), (**Q**= Quoted Longterm Liabilities= 35,351 millions USD), (**S**= Sales or Revenues= 51,294 millions USD), (**d**= D/A= Dividend Portion or Payout= 30.00%), (**T**= Tax Paid= 3,385 millions USD), (**v**= V/S= Variable Portion= 19.00%), and (**i**= I/S= Interest Portion= 1.00%), then it's (**A**= After Tax Income planned), is:

$$A= [U+M-F-Si-T-(Q+S'vp[1+s] /\{360[c-q]\})/l]/d$$

$$= 64,720+41,548-29,750-51,294$$
$$*0.0100-3,385-(35,351$$
$$+48,851*0.1900*240$$
$$[1+0.0500]\{360[1.2000$$
$$-1.0100]\})/0.9900]/0.3000$$

$$= 7,899 \text{ millions USD}$$

Corporate IFRS-GAAP (B/S-I/S), ISBN-13: **978-1720792789**, ISBN-10: **172079278X**

Law-7972:

If both (**q**= [C-P]/X= Quick or Acid Test Ratio= 1.01 times), (**c**= C/X= Current Ratio= 1.20 times), (**S'**= Sales of Past Year= 48,851 millions USD), (**s**= [S/S']-1= Sales Growth= 5.00%), (**F**= Fixed Cost= 29,740 millions USD), (**M**= Margin of Contribution= 41,548 millions USD), (**p**= 360P/V= Procured Inventory Days= 240 Days), (**I**= L/E= Leverage or Gearing Ratio= 99.00%), (**U**= Utilized or Starting Capital= 64,720 millions USD), (**Q**= Quoted Longterm Liabilities= 35,351 millions USD), (**S**= Sales or Revenues= 51,294 millions USD), (**A**= After Tax Income= 7,899 millions USD), (**T**= Tax Paid= 3,385 millions USD), (**v**= V/S= Variable Portion= 19.00%), and (**i**= I/S= Interest Portion= 1.00%), then it's (**d**= Dividend Portion or payout planned), is :

$$d= [U+M-F-Si-T-(Q+S'vp[1+s]$$
$$/\{360[c-q]\})/I]/A$$
$$= [64,720+41,548-29,750-51,294$$
$$*0.0100-3,385-(35,351+48,851$$
$$*0.1900*240[1+0.0500]$$
$$\{360[1.2000-1.0100]\})$$
$$/0.9900]/7,899$$

$$= 30.00\%$$

Corporate IFRS-GAAP (B/S-I/S), ISBN-13: **978-1720792789**, ISBN-10: **172079278X**

Law-7973:

If both (**q**= [C-P]/X= Quick or Acid Test Ratio= 1.01 times), (**c**= C/X= Current Ratio= 1.20 times), (**d**= D/A= Dividend Portion or Payout= 30.00%), (**s**= [S/S']-1= Sales Growth= 5.00%), (**F**= Fixed Cost= 29,740 millions USD), (**M**= Margin of Contribution= 41,548 millions USD), (**p**= 360 P/V= Procured Inventory Days= 240 Days), (**l**= L/E= Leverage or Gearing Ratio= 99.00%), (**U**= Utilized or Starting Capital= 64,720 millions USD), (**Q**= Quoted Longterm Liabilities= 35,351 millions USD), (**S**= Sales or Revenues= 51,294 millions USD), (**A**= After Tax Income= 7,899 millions USD), (**T**= Tax Paid= 3,385 millions USD), (**v**= V/S= Variable Portion= 19.00%), and (**i**= I/S= Interest Portion= 1.00%), then it's (**S'**= Sales Past), must be:

$$S' = 360[c-q](l \{[U+M-F-Si-T-Ad]-Q\})$$
$$/\{vp[1+s]\}$$
$$= 360[1.2000-1.0100](0.9900$$
$$\{64,720+41,548-29,750$$
$$-51,294*0.0100-3,385$$
$$-7,899*0.3000]-35,351\})$$
$$/\{0.1900*240[1+0.0500]\}$$
$$= 48,851 \text{ millions USD}$$

Corporate IFRS-GAAP (B/S-I/S), ISBN-13: **978-1720792789**, ISBN-10: **172079278X**

Law-7974:

If both (**q**= [C-P]/X= Quick or Acid Test Ratio= 1.01 times), (**d**= D/A= Dividend Portion or Payout= 30.00%), (**c**= C/X= Current Ratio= 1.20 times), (**s**= [S/S']-1= Sales Growth= 5.00%), (**F**= Fixed Cost= 29,740 millions USD), (**M**= Margin of Contribution= 41,548 millions USD), (**S'**= Sales of Past Year= 48,851 millions USD), (**p**= 360 P/V= Procured Inventory Days= 240 Days), (**l**= L/E= Leverage or Gearing Ratio= 99.00%), (**U**= Utilized or Starting Capital= 64,720 millions USD), (**Q**= Quoted Longterm Liabilities= 35,351 millions USD), (**S**= Sales or Revenues= 51,294 millions USD), (**A**= After Tax Income= 7,899 millions USD), (**T**= Tax Paid= 3,385 millions USD), and (**i**= I/S= Interest Portion= 1.00%), then it's (**v**= Variable Portion planned), is:

$$v= 360[\mathbf{c\text{-}q}](\mathbf{l}\ \{[\mathbf{U\text{+}M\text{-}F\text{-}Si\text{-}T\text{-}Ad}]\text{-}\mathbf{Q}\})$$
$$/\{\mathbf{S'p}[1\text{+}\mathbf{s}]\}$$
$$= 360[1.2000\text{-}1.0100](0.9900$$
$$\{64,720\text{+}41,548\text{-}29,750$$
$$\text{-}51,294*0.0100\text{-}3,385$$
$$\text{-}7,899*0.3000]\text{-}35,351\})$$
$$/\{48,851*240[1\text{+}0.0500]\}$$
$$= 19.00\%$$

Corporate IFRS-GAAP (B/S-I/S), ISBN-13: **978-1720792789**, ISBN-10: **172079278X**

Law-7975:

If both (q= [C-P]/X= Quick or Acid Test Ratio= 1.01
times), (d= D/A= Dividend Portion or Payout=
30.00%), (c= C/X= Current Ratio= 1.20 times), (s=
[S/S']-1= Sales Growth= 5.00%), (F= Fixed Cost=
29,740 millions USD), (M= Margin of Contribution=
41,548 millions USD), (S'= Sales of Past Year=
48,851 millions USD), (v= V/S= Variable Portion=
19.00%), (l= L/E= Leverage or Gearing Ratio=
99.00%), (U= Utilized or Starting Capital= 64,720
millions USD), (Q= Quoted Longterm Liabilities=
35,351 millions USD), (S= Sales or Revenues=
51,294 millions USD), (A= After Tax Income= 7,899
millions USD), (T= Tax Paid= 3,385 millions USD),
and (i= I/S= Interest Portion= 1.00%), then it's (p=
Procured Inventory Days planned), is:

$$p= 360[c-q](l \{[U+M-F-Si-T-Ad]-Q\})$$
$$/\{S'v[1+s]\}$$
$$= 360[1.2000-1.0100](0.9900$$
$$\{64,720+41,548-29,750$$
$$-51,294*0.0100-3,385$$
$$-7,899*0.3000]-35,351\})$$
$$/\{48,851*0.1900$$
$$[1+0.0500]\}$$
$$= 240 \text{ days}$$

Corporate IFRS-GAAP (B/S-I/S), ISBN-13: **978-1720792789**, ISBN-10: **172079278X**

Law-7976:

If both (q= [C-P]/X= Quick or Acid Test Ratio= 1.01 times), (d= D/A= Dividend Portion or Payout= 30.00%), (c= C/X= Current Ratio= 1.20 times), (p= 360 P/V= Procured Inventory Days= 240 Days), (F= Fixed Cost= 29,740 millions USD), (M= Margin of Contribution= 41,548 millions USD), (S'= Sales of Past Year= 48,851 millions USD), (v= V/S= Variable Portion= 19.00%), (I= L/E= Leverage or Gearing Ratio= 99.00%), (U= Utilized or Starting Capital= 64,720 millions USD), (Q= Quoted Longterm Liabilities= 35,351 millions USD), (S= Sales or Revenues= 51,294 millions USD), (A= After Tax Income= 7,899 millions USD), (T= Tax Paid= 3,385 millions USD), and (i= I/S= Interest Portion= 1.00%), then it's (s= Sales Growth planned), is:

$$s = 360[c\text{-}q](I\{[U+M\text{-}F\text{-}Si\text{-}T\text{-}Ad]\text{-}Q\})$$
$$/[S'vp]\text{-}1$$
$$= 360[1.2000\text{-}1.0100](0.9900$$
$$\{64,720+41,548\text{-}29,750$$
$$\text{-}51,294*0.0100\text{-}3,385$$
$$\text{-}7,899*0.3000]\text{-}35,351\})$$
$$/[48,851*0.1900*240]\text{-}1$$
$$= 5.00\%$$

Corporate IFRS-GAAP (B/S-I/S), ISBN-13: **978-1720792789**, ISBN-10: **172079278X**

Law-7977:

If both (**q**= [C-P]/X= Quick or Acid Test Ratio= 1.01 times), (**d**= D/A=Dividend Portion or Payout= 30.00%), (**s**= [S/S']-1= Sales Growth= 5.00%), (**p**= 360 P/V= Procured Inventory Days= 240 Days), (**F**= Fixed Cost= 29,740 millions USD), (**M**= Margin of Contribution= 41,548 millions USD), (**S'**= Sales of Past Year= 48,851 millions USD), (**v**= V/S= Variable Portion= 19.00%), (**l**= L/E= Leverage or Gearing Ratio= 99.00%), (**U**= Utilized or Starting Capital= 64,720 millions USD), (**Q**= Quoted Longterm Liabilities= 35,351 millions USD), (**S**= Sales or Revenues= 51,294 millions USD), (**A**= After Tax Income= 7,899 millions USD), (**T**= Tax Paid= 3,385 millions USD), and (**i**= I/S= Interest Portion= 1.00%), then it's (**c**= Current Ratio planned), is:

$$c= q+S'vp[1+s]/(360\{l\,[U+M-F-Si-T-Ad]-Q\})$$
$$= 1.0100+48,851*0.1900*240$$
$$[1+0.0500]/(360\{0.9900$$
$$[64,720+41,548-29,750$$
$$-51,294*0.0100-3,385$$
$$-7,899*0.3000]-35,351\})$$

$$= 1.20 \text{ times}$$

Corporate IFRS-GAAP (B/S-I/S), ISBN-13: **978-1720792789**, ISBN-10: **172079278X**

Law-7978:

If both (**c**= C/X= Current Ratio= 1.20 times), (**d**= D/A= Dividend Portion or Payout= 30.00%), (**s**= [S/S']-1= Sales Growth= 5.00%), (**p**= 360 P/V= Procured Inventory Days= 240 Days), (**F**= Fixed Cost= 29,740 millions USD), (**M**= Margin of Contribution= 41,548 millions USD), (**$'**= Sales of Past Year= 48,851 millions USD), (**v**= V/S= Variable Portion= 19.00%), (**l**= L/E= Leverage or Gearing Ratio= 99.00%), (**U**= Utilized or Starting Capital= 64,720 millions USD), (**Q**= Quoted Longterm Liabilities= 35,351 millions USD), (**$**= Sales or Revenues= 51,294 millions USD), (**A**= After Tax Income= 7,899 millions USD), (**T**= Tax Paid= 3,385 millions USD), and (**i**= I/S= Interest Portion= 1.00%), then it's (**q**= Quick or Acid Test Ratio planned), is:

$$q= c\text{-}\$'vp[1+s]/(360\{\textbf{\textit{l}}\,[U+M\text{-}F\text{-}\$i\text{-}T\text{-}Ad]\text{-}Q\})$$

$$= 1.2000\text{-}48,851*0.1900*240$$
$$[1+0.0500]/(360\{0.9900$$
$$[64,720+41,548\text{-}29,750$$
$$\text{-}51,294*0.0100\text{-}3,385$$
$$\text{-}7,899*0.3000]\text{-}35,351\})$$

$$= 1.01 \text{ times}$$

Corporate IFRS-GAAP (B/S-I/S), ISBN-13: **978-1720792789**, ISBN-10: **172079278X**

<u>Law-7979</u>:

If both (**F**= Fixed Cost= <u>29,740</u> millions USD), (**M**= Margin of Contribution= <u>41,548</u> millions USD), (**I**= L/E= Leverage or Gearing Ratio= <u>99.00%</u>), (**U**= Utilized or Starting Capital= <u>64,720</u> millions USD), (**S**= Sales or Revenues= <u>51,294</u> millions USD), (**X**= Xpress or Current Debt= <u>34,196</u> millions USD), (**t**= T/B= Tax Rate= <u>30.00%</u>), (**D**= Dividend Paid= <u>2,370</u> millions USD), and (**i**= I/S= Interest Portion= <u>1.00%</u>), then it's (**Q**= Quoted Longterm Liabilities planned), is:

$$Q= I\{U+[M-F-Si][1-t]-D\}-X$$
$$= 0.9900\{64,720+[41,548-29,750$$
$$-51,294*0.0100][1-0.3000]$$
$$-2,370\}-34,196$$
$$= \underline{35,351} \text{ millions USD}$$

Corporate IFRS-GAAP (B/S-I/S), ISBN-13: **978-1720792789**, ISBN-10: **172079278X**

<u>Law-7980</u>:

If both (**F**= Fixed Cost= <u>29,740</u> millions USD), (**M**= Margin of Contribution= <u>41,548</u> millions USD), (**Q**= Quoted Longterm Liabilities= <u>35,351</u> millions USD), (**U**= Utilized or Starting Capital= <u>64,720</u> millions USD), (**$**= Sales or Revenues= <u>51,294</u> millions USD), (**X**= Xpress or Current Debt= <u>34,196</u> millions USD), (**t**= T/B= Tax Rate= <u>30.00%</u>), (**D**= Dividend Paid= <u>2,370</u> millions USD), and (**i**= I/S= Interest Portion= <u>1.00%</u>), then it's (**I** = Leverage or Gearing Ratio planned), is:

$$I = [Q+X]/\{U+[M-F-Si][1-t]-D\}$$
$$= [35,351+34,196]/\{64,720+[41,548$$
$$-29,750-51,294*0.0100]$$
$$[1-0.3000]-2,370\}$$
$$= \underline{99.00\%}$$

Corporate IFRS-GAAP (B/S-I/S), ISBN-13: **978-1720792789**, ISBN-10: **172079278X**

Law-7981:

If both (**F**= Fixed Cost= 29,740 millions USD), (**M**= Margin of Contribution= 41,548 millions USD), (**I**= L/E= Leverage or Gearing Ratio= 99.00%), (**Q**= Quoted Longterm Liabilities= 35,351 millions USD), (**$**= Sales or Revenues= 51,294 millions USD), (**X**= Xpress or Current Debt= 34,196 millions USD), (**t**= T/B= Tax Rate= 30.00%), (**D**= Dividend Paid= 2,370 millions USD), and (**i**= I/S= Interest Portion= 1.00%), then it's (**U**= Utilized or Starting Capital planned), is:

$$\textbf{U}= \textbf{D}+[\textbf{Q}+\textbf{X}]/\textbf{I} -[\textbf{M}-\textbf{F}-\textbf{\$i}][1-\textbf{t}]$$
$$= 2,370+[35,351+34,196]/0.9900$$
$$-[41,548-29,750-51,294$$
$$*0.0100][1-0.3000]$$
$$= 64,720 \text{ millions USD}$$

Corporate IFRS-GAAP (B/S-I/S), ISBN-13: **978-1720792789**, ISBN-10: **172079278X**

Law-7982:

If both (**F**= Fixed Cost= 29,740 millions USD), (**U**=
Utilized or Starting Capital= 64,720 millions USD),
(**I**= L/E= Leverage or Gearing Ratio= 99.00%), (**Q**=
Quoted Longterm Liabilities= 35,351 millions USD),
(**S**= Sales or Revenues= 51,294 millions USD), (**X**=
Xpress or Current Debt= 34,196 millions USD), (**t**=
T/B= Tax Rate= 30.00%), (**D**= Dividend Paid= 2,370
millions USD), and (**i**= I/S= Interest Portion= 1.00%),
then it's (**M**= Margin of Contribution planned), is:

$$\mathbf{M= F+Si+\{D-U+[Q+X]/I\}/[1-t]}$$
$$= 29{,}750+51{,}294*0.0100+\{2{,}370$$
$$-64{,}720+[35{,}351+34{,}196]$$
$$/0.9900\}/[1-0.3000]$$
$$= \underline{41{,}548} \text{ millions USD}$$

Corporate IFRS-GAAP (B/S-I/S), ISBN-13: **978-1720792789**, ISBN-10: **172079278X**

<u>Law-7983</u>:

If both (**M**= Margin of Contribution= <u>41,548</u> millions USD), (**U**= Utilized or Starting Capital= <u>64,720</u> millions USD), (**I**= L/E= Leverage or Gearing Ratio= <u>99.00%</u>), (**Q**= Quoted Longterm Liabilities= <u>35,351</u> millions USD), (**S**= Sales or Revenues= <u>51,294</u> millions USD), (**X**= Xpress or Current Debt= <u>34,196</u> millions USD), (**t**= T/B= Tax Rate= <u>30.00%</u>), (**D**= Dividend Paid= <u>2,370</u> millions USD), and (**i**= I/S= Interest Portion= <u>1.00%</u>), then it's (**F**= Fixed Cost planned), is:

$$F= M\text{-}Si\text{-}\{D\text{-}U+[Q+X]/I\}/[1\text{-}t]$$
$$= 41,548\text{-}51,294*0.0100\text{-}\{2,370$$
$$-64,720+[35,351+34,196]$$
$$/0.9900\}/[1\text{-}0.3000]$$
$$= \underline{29,750} \text{ millions USD}$$

Corporate IFRS-GAAP (B/S-I/S), ISBN-13: **978-1720792789**, ISBN-10: **172079278X**

Law-7984:

If both (**M**= Margin of Contribution= 41,548 millions USD), (**U**= Utilized or Starting Capital= 64,720 millions USD), (**I**= L/E= Leverage or Gearing Ratio= 99.00%), (**Q**= Quoted Longterm Liabilities= 35,351 millions USD), (**F**= Fixed Cost= 29,740 millions USD), (**X**= Xpress or Current Debt= 34,196 millions USD), (**t**= T/B= Tax Rate= 30.00%), (**D**= Dividend Paid= 2,370 millions USD), and (**i**= I/S= Interest Portion= 1.00%), then it's (**S**= Sales or Revenues planned), is:

$$\textsf{S}= (\textsf{M-F-}\{\textsf{D-U+}[\textsf{Q+X}]/\textsf{I}\}/[1-\textsf{t}])/\textsf{i}$$
$$= 41,548-29,750-\{2,370-64,720$$
$$+[35,351+34,196]/0.9900\}$$
$$/[1-0.3000])/\ 0.0100$$
$$= \underline{51,294} \text{ millions USD}$$

Corporate IFRS-GAAP (B/S-I/S), ISBN-13: **978-1720792789**, ISBN-10: **172079278X**

Law-7985:

If both (**M**= Margin of Contribution= 41,548 millions USD), (**U**= Utilized or Starting Capital= 64,720 millions USD), (**I**= L/E= Leverage or Gearing Ratio= 99.00%), (**Q**= Quoted Longterm Liabilities= 35,351 millions USD), (**F**= Fixed Cost= 29,740 millions USD), (**X**= Xpress or Current Debt= 34,196 millions USD), (**t**= T/B= Tax Rate= 30.00%), (**D**= Dividend Paid= 2,370 millions USD), and (**S**= Sales or Revenues= 51,294 millions USD), then it's (**i**= Interest Portion planned), is:

$$i= (M-F-\{D-U+[Q+X]/I\}/[1-t])/S$$
$$= 41,548-29,750-\{2,370-64,720$$
$$+[35,351+34,196]/0.9900\}$$
$$/[1-0.3000])/51,294$$

$$= 1.00\%$$

Corporate IFRS-GAAP (B/S-I/S), ISBN-13: **978-1720792789**, ISBN-10: **172079278X**

Law-7986:

If both (**M**= Margin of Contribution= 41,548 millions USD), (**U**= Utilized or Starting Capital= 64,720 millions USD), (**I**= L/E= Leverage or Gearing Ratio= 99.00%), (**Q**= Quoted Longterm Liabilities= 35,351 millions USD), (**F**= Fixed Cost= 29,740 millions USD), (**X**= Xpress or Current Debt= 34,196 millions USD), (**i**= I/S= Interest Portion= 1.00%), (**D**= Dividend Paid= 2,370 millions USD), and (**$**= Sales or Revenues= 51,294 millions USD), then it's (**t**= Tax Rate planned), is:

$$t = 1 - \{D-U+[Q+X]/I\}/[M-F-\$i]$$
$$= 1 - \{2,370-64,720+[35,351+34,196]$$
$$/0.9900\}/[41,548-29,750$$
$$-51,294*0.0100]$$
$$= 30.00\%$$

Corporate IFRS-GAAP (B/S-I/S), ISBN-13: **978-1720792789**, ISBN-10: **172079278X**

Law-7987:

If both (**M**= Margin of Contribution= 41,548 millions USD), (**U**= Utilized or Starting Capital= 64,720 millions USD), (**I**= L/E= Leverage or Gearing Ratio= 99.00%), (**Q**= Quoted Longterm Liabilities= 35,351 millions USD), (**F**= Fixed Cost= 29,740 millions USD), (**X**= Xpress or Current Debt= 34,196 millions USD), (**i**= I/S= Interest Portion= 1.00%), (**t**= T/B= Tax Rate= 30.00%), and (**$**= Sales or Revenues= 51,294 millions USD), then it's (**D**= Dividend Paid planned), is:

$$D= U+[M-F-\$i][1-t]-[Q+X]/I$$
$$= 64,720+[41,548-29,750-51,294$$
$$*0.0100][1-0.3000]-[35,351$$
$$+34,196]/0.9900$$
$$= 2,370 \text{ millions USD}$$

Corporate IFRS-GAAP (B/S-I/S), ISBN-13: **978-1720792789**, ISBN-10: **172079278X**

<u>Law-7988</u>:

If both (**M**= Margin of Contribution= <u>41,548</u> millions USD), (**U**= Utilized or Starting Capital= <u>64,720</u> millions USD), (**I**= L/E= Leverage or Gearing Ratio= <u>99.00%</u>), (**Q**= Quoted Longterm Liabilities= <u>35,351</u> millions USD), (**F**= Fixed Cost= <u>29,740</u> millions USD), (**D**= Dividend Paid= <u>2,370</u> millions USD), (**i**= I/S= Interest Portion= <u>1.00%</u>), (**t**= T/B= Tax Rate= <u>30.00%</u>), and (**S**= Sales or Revenues= <u>51,294</u> millions USD), then it's (**X**= Xpress or Current Debt planned), is:

$$X = I\{U+[M-F-Si][1-t]-D\}-Q$$
$$= 0.9900\{64,720+[41,548-29,750$$
$$-51,294*0.0100][1-0.3000]$$
$$-2,370\}-35,351$$
$$= \underline{34,196} \text{ millions USD}$$

Corporate IFRS-GAAP (B/S-I/S), ISBN-13: **978-1720792789**, ISBN-10: **172079278X**

Law-7989:

If both (**M**= Margin of Contribution= 41,548 millions USD), (**U**= Utilized or Starting Capital= 64,720 millions USD), (**I**= L/E= Leverage or Gearing Ratio= 99.00%), (**c**= C/X= Current Ratio= 1.20 times), (**q**= [C-P]/X= Quick or Acid Test Ratio= 1.01 times), (**P**= Procured Inventories= 6,497 millions USD), (**F**= Fixed Cost= 29,740 millions USD), (**D**= Dividend Paid= 2,370 millions USD), (**i**= I/S= Interest Portion= 1.00%), (**t**= T/B= Tax Rate= 30.00%), and (**S**= Sales or Revenues= 51,294 millions USD), then it's (**Q**= Quoted Longterm Liabilities planned), is:

$$Q= I \{U+[M-F-Si][1-t]-D\}-P/[c-q]$$
$$= 0.9900\{64,720+[41,548-29,750$$
$$-51,294*0.0100][1-0.3000]$$
$$-2,370\}-6,497$$
$$/[1.2000-1.0100]$$
$$= 35,351 \text{ millions USD}$$

Corporate IFRS-GAAP (B/S-I/S), ISBN-13: **978-1720792789**, ISBN-10: **172079278X**

Law-7990:

If both (**M**= Margin of Contribution= 41,548 millions
USD), (**U**= Utilized or Starting Capital= 64,720
millions USD), (**Q**= Quoted Longterm Liabilities=
35,351 millions USD), (**c**= C/X= Current Ratio= 1.20
times), (**q**= [C-P]/X= Quick or Acid Test Ratio= 1.01
times), (**P**= Procured Inventories= 6,497 millions
USD), (**F**= Fixed Cost= 29,740 millions USD), (**D**=
Dividend Paid= 2,370 millions USD), (**i**= I/S=
Interest Portion= 1.00%), (**t**= T/B= Tax Rate=
30.00%), and (**$**= Sales or Revenues= 51,294 millions
USD), then it's (**I** = Leverage or Gearing Ratio
planned), is:

$$I = \{Q+P/[c-q]\}/\{U+[M-F-\$i][1-t]-D\}$$
$$= \{35,351+6,497/[1.2000-1.0100]\}$$
$$/\{64,720+[41,548-29,750$$
$$-51,294*0.0100][1-0.3000]$$
$$-2,370\}$$
$$= 99.00\%$$

Corporate IFRS-GAAP (B/S-I/S), ISBN-13: **978-1720792789**, ISBN-10: **172079278X**

Law-7991:

If both (**M**= Margin of Contribution= 41,548 millions USD), (**I**= L/E= Leverage or Gearing Ratio= 99.00%), (**Q**= Quoted Longterm Liabilities= 35,351 millions USD), (**c**= C/X= Current Ratio= 1.20 times), (**q**= Quick or Acid Test Ratio= [C-P]/X= 1.01 times), (**P**= Procured Inventories= 6,497 millions USD), (**F**= Fixed Cost= 29,740 millions USD), (**D**= Dividend Paid= 2,370 millions USD), (**i**= I/S= Interest Portion= 1.00%), (**t**= T/B= Tax Rate= 30.00%), and (**S**= Sales or Revenues= 51,294 millions USD), then it's (**U**= Utilized or Starting Capital planned), is:

$$U= D+\{Q+P/[c-q]\}/I-[M-F-Si][1-t]$$
$$= 2,370+\{35,351+6,497/[1.2000$$
$$-1.0100]\}/0.9900-[41,548$$
$$-29,750-51,294*0.0100]$$
$$[1-0.3000]$$
$$= 64,720 \text{ millions USD}$$

Corporate IFRS-GAAP (B/S-I/S), ISBN-13: **978-1720792789**, ISBN-10: **172079278X**

Law-7992:

If both (**U**= Utilized or Starting Capital= 64,720 millions USD), (**⌐**= L/E= Leverage or Gearing Ratio= 99.00%), (**Q**= Quoted Longterm Liabilities= 35,351 millions USD), (**c**= C/X= Current Ratio= 1.20 times), (**q**= [C-P]/X= Quick or Acid Test Ratio= 1.01 times), (**P**= Procured Inventories= 6,497 millions USD), (**F**= Fixed Cost= 29,740 millions USD), (**D**= Dividend Paid= 2,370 millions USD), (**i**= I/S= Interest Portion= 1.00%), (**t**= T/B= Tax Rate= 30.00%), and (**$**= Sales or Revenues= 51,294 millions USD), then it's (**M**= Margin of Contribution planned), is:

$$M= F+\$i+(D-U+\{Q+P/[c-q]\}/\mathbf{I})/[1-t]$$

$$= 29,750+51,294*0.0100+(2,370$$
$$-64,720+\{35,351+6,497$$
$$/[1.2000-1.0100]\}/0.9900)$$
$$/[1-0.3000]$$

$$= \underline{41,548} \text{ millions USD}$$

Corporate IFRS-GAAP (B/S-I/S), ISBN-13: **978-1720792789**, ISBN-10: **172079278X**

Law-7993:

If both (**U**= Utilized or Starting Capital= 64,720
millions USD), (**l**= L/E= Leverage or Gearing Ratio=
99.00%), (**Q**= Quoted Longterm Liabilities= 35,351
millions USD), (**c**= C/X= Current Ratio= 1.20 times),
(**q**= [C-P]/X= Quick or Acid Test Ratio= 1.01 times),
(**P**= Procured Inventories= 6,497 millions USD), (**M**=
Margin of Contribution= 41,548 millions USD), (**D**=
Dividend Paid= 2,370 millions USD), (**i**= I/S=
Interest Portion= 1.00%), (**t**= T/B= Tax Rate=
30.00%), and (**$**= Sales or Revenues= 51,294 millions
USD), then it's (**F**= Fixed Cost planned), is:

$$F= M\text{-}\$i\text{-}(D\text{-}U+\{Q+P/[c\text{-}q]\}/l)/[1\text{-}t]$$

$$= 41{,}548\text{-}51{,}294*0.0100\text{-}(2{,}370$$
$$-64{,}720+\{35{,}351+6{,}497$$
$$/[1.2000\text{-}1.0100]\}/0.9900)$$
$$/[1\text{-}0.3000]$$

$$= \underline{29{,}750} \text{ millions USD}$$

Corporate IFRS-GAAP (B/S-I/S), ISBN-13: **978-1720792789**, ISBN-10: **172079278X**

Law-7994:

If both (**U**= Utilized or Starting Capital= 64,720
millions USD), (**⊩**= L/E= Leverage or Gearing Ratio=
99.00%), (**Q**= Quoted Longterm Liabilities= 35,351
millions USD), (**c**= C/X= Current Ratio= 1.20 times),
(**q**= [C-P]/X= Quick or Acid Test Ratio= 1.01 times),
(**P**= Procured Inventories= 6,497 millions USD), (**M**=
Margin of Contribution= 41,548 millions USD), (**D**=
Dividend Paid= 2,370 millions USD), (**i**= I/S=
Interest Portion= 1.00%), (**t**= T/B= Tax Rate=
30.00%), and (**F**= Fixed Cost= 29,740 millions USD),
then it's (**$**= Sales or Revenues planned), is:

$$\$= [M-F-(D-U+\{Q+P/[c-q]\}/\mathit{l})/[1-t]]/i$$
$$= [41{,}548\text{-}29{,}750\text{-}(2{,}370\text{-}64{,}720$$
$$+\{35{,}351\text{+}6{,}497/[1.2000$$
$$-1.0100]\}/0.9900)$$
$$/[1\text{-}0.3000]]/0.0100$$
$$= 51{,}294 \text{ millions USD}$$

Corporate IFRS-GAAP (B/S-I/S), ISBN-13: **978-1720792789**, ISBN-10: **172079278X**

Law-7995:

If both (**U**= Utilized or Starting Capital= 64,720
millions USD), (**I**= L/E= Leverage or Gearing Ratio=
99.00%), (**Q**= Quoted Longterm Liabilities= 35,351
millions USD), (**c**= C/X= Current Ratio= 1.20 times),
(**q**= [C-P]/X= Quick or Acid Test Ratio= 1.01 times),
(**P**= Procured Inventories= 6,497 millions USD), (**M**=
Margin of Contribution= 41,548 millions USD), (**D**=
Dividend Paid= 2,370 millions USD), (**$**= Sales or
Revenues= 51,294 millions USD), (**t**= T/B= Tax
Rate= 30.00%), and (**F**= Fixed Cost= 29,740 millions
USD), then it's (**i**= Interest Portion planned), is:

$$i= [M-F-(D-U+\{Q+P/[c-q]\}/I)/[1-t]]/\$$$
$$= [41,548-29,750-(2,370-64,720$$
$$+\{35,351+6,497/[1.2000$$
$$-1.0100]\}/0.9900)$$
$$/[1-0.3000]]/51,294$$
$$= 1.00\%$$

Corporate IFRS-GAAP (B/S-I/S), ISBN-13: **978-1720792789**, ISBN-10: **172079278X**

Law-7996:

If both (**U**= Utilized or Starting Capital= 64,720 millions USD), (**I**= L/E= Leverage or Gearing Ratio= 99.00%), (**Q**= Quoted Longterm Liabilities= 35,351 millions USD), (**c**= C/X= Current Ratio= 1.20 times), (**q**= [C-P]/X= Quick or Acid Test Ratio= 1.01 times), (**P**= Procured Inventories= 6,497 millions USD), (**M**= Margin of Contribution= 41,548 millions USD), (**D**= Dividend Paid= 2,370 millions USD), (**$**= Sales or Revenues= 51,294 millions USD), (**i**= I/S= Interest Portion= 1.00%), and (**F**= Fixed Cost= 29,740 millions USD), then it's (**t**= Tax Rate planned), is:

$$t= 1-(\mathbf{D}-\mathbf{U}+\{\mathbf{Q}+\mathbf{P}/[\mathbf{c}-\mathbf{q}]\}/\mathbf{I})/[\mathbf{M}-\mathbf{F}-\mathbf{\$i}]$$

$$= 1-(2,370-64,720+\{35,351+6,497$$
$$/[1.2000-1.0100]\}/0.9900)$$
$$/[41,548-29,750-51,294$$
$$*0.0100]$$

$$= 30.00\%$$

Corporate IFRS-GAAP (B/S-I/S), ISBN-13: **978-1720792789**, ISBN-10: **172079278X**

Law-7997:

If both (**U**= Utilized or Starting Capital= 64,720 millions USD), (**F**= L/E= Leverage or Gearing Ratio= 99.00%), (**Q**= Quoted Longterm Liabilities= 35,351 millions USD), (**c**= C/X= Current Ratio= 1.20 times), (**q**= [C-P]/X= Quick or Acid Test Ratio= 1.01 times), (**P**= Procured Inventories= 6,497 millions USD), (**M**= Margin of Contribution= 41,548 millions USD), (**t**= T/B= Tax Rate= 30.00%), (**$**= Sales or Revenues= 51,294 millions USD), (**i**= I/S= Interest Portion= 1.00%), and (**F**= Fixed Cost= 29,740 millions USD), then it's (**D**= Dividend Paid planned), is:

$$D= U+[M\text{-}F\text{-}\$i][1\text{-}t]\text{-}\{Q+P/[c\text{-}q]\}/F$$
$$= 64,720+[41,548\text{-}29,750\text{-}51,294$$
$$*0.0100][1\text{-}0.3000]\text{-}\{35,351$$
$$+6,497/[1.2000\text{-}1.0100]\}$$
$$/0.9900$$
$$= \underline{2,370} \text{ millions USD}$$

Corporate IFRS-GAAP (B/S-I/S), ISBN-13: **978-1720792789**, ISBN-10: **172079278X**

Law-7998:

If both (**U**= Utilized or Starting Capital= 64,720 millions USD), (**F**= L/E= Leverage or Gearing Ratio= 99.00%), (**Q**= Quoted Longterm Liabilities= 35,351 millions USD), (**c**= C/X= Current Ratio= 1.20 times), (**q**= [C-P]/X= Quick or Acid Test Ratio= 1.01 times), (**D**= Dividend Paid= 2,370 millions USD), (**M**= Margin of Contribution= 41,548 millions USD), (**t**= T/B= Tax Rate= 30.00%), (**$**= Sales or Revenues= 51,294 millions USD), (**i**= I/S= Interest Portion= 1.00%), and (**F**= Fixed Cost= 29,740 millions USD), then it's (**P**= Procured Inventories planned), is:

$$P= [c-q](F \{U+[M-F-Si][1-t]-D\}-Q)$$
$$= [1.2000-1.0100](0.9900\{64,720$$
$$+[41,548-29,750-51,294$$
$$*0.0100][1-0.3000]-2,370\}$$
$$-35,351)$$
$$= 6,497 \text{ millions USD}$$

Corporate IFRS-GAAP (B/S-I/S), ISBN-13: **978-1720792789**, ISBN-10: **172079278X**

Law-7999:

If both (**U**= Utilized or Starting Capital= 64,720 millions USD), (**Ɩ**= L/E= Leverage or Gearing Ratio= 99.00%), (**Q**= Quoted Longterm Liabilities= 35,351 millions USD), (**P**= Procured Inventories= 6,497 millions USD), (**q**= [C-P]/X= Quick or Acid Test Ratio= 1.01 times), (**D**= Dividend Paid= 2,370 millions USD), (**M**= Margin of Contribution= 41,548 millions USD), (**t**= T/B= Tax Rate= 30.00%), (**S**= Sales or Revenues= 51,294 millions USD), (**i**= I/S= Interest Portion= 1.00%), and (**F**= Fixed Cost= 29,740 millions USD), then it's (**c**= Current Ratio planned), is:

$$c= q+P/(Ɩ\{U+[M-F-Si][1-t]-D\}-Q)$$
$$= 1.0100+6,497/(0.9900\{64,720$$
$$+[41,548-29,750-51,294$$
$$*0.0100][1-0.3000]-2,370\}$$
$$-35,351)$$
$$= 1.20 \text{ times}$$

Corporate IFRS-GAAP (B/S-I/S), ISBN-13: **978-1720792789**, ISBN-10: **172079278X**

<u>Law-8000</u>:

If both (**U**= Utilized or Starting Capital= <u>64,720</u> millions USD), (**I**= L/E= Leverage or Gearing Ratio= <u>99.00%</u>), (**Q**= Quoted Longterm Liabilities= <u>35,351</u> millions USD), (**P**= Procured Inventories= <u>6,497</u> millions USD), (**c**= C/X= Current Ratio= <u>1.20</u> times), (**D**= Dividend Paid= <u>2,370</u> millions USD), (**M**= Margin of Contribution= <u>41,548</u> millions USD), (**t**= T/B= Tax Rate= <u>30.00%</u>), (**$**= Sales or Revenues= <u>51,294</u> millions USD), (**i**= I/S= Interest Portion= <u>1.00%</u>), and (**F**= Fixed Cost= <u>29,740</u> millions USD), then it's (**q**= Quick or Acid Test Ratio planned), is:

$$q= c-P/(I \{U+[M-F-\$i][1-t]-D\}-Q)$$

$$= 1.2000-6,497/(0.9900\{64,720$$
$$+[41,548-29,750-51,294$$
$$*0.0100][1-0.3000]-2,370\}$$
$$-35,351)$$

$$= \underline{1.01} \text{ times}$$

Corporate IFRS-GAAP (B/S-I/S), ISBN-13: **978-1720792789**, ISBN-10: **172079278X**

Finance Construction-11, *Tim Asikin, Steve Asikin, Indra Senihardja*

The Rest will be Continued in
The Series of Book

Finance Construction 12

Corporate IFRS-GAAP (B/S-I/S) Engineering
Technologies No. 8,001-8,500 of 111,111 Laws

At
http: //www.Amazon.Com

Corporate IFRS-GAAP (B/S-I/S), ISBN-13: **978-1720792789**, ISBN-10: **172079278X**

REFERENCES:

A. Books:

1) **Amin, A.R. & Asikin, S.**(2011a- 110 pages), *CEO Time Matrix: Productivity Management Skill that Increase CEO Effectiveness by 12400%*, Seattle (US): Amazon Createspace. http://www.amazon.com/CEO-Time-Matrix-Productivity-Effectiveness/dp/1461066069/ref=sr_1_1?s=books&ie=UTF8&qid=1387091880&sr=1-1&keywords=ceo+time+matrix. ISBN-13: 978-1461066064, ISBN-10: 1461066069

2) **Amin, A.R. & Asikin, S.** (2014- 76 pages), *No Urgency CEO: A High-Tech Book in Modern Management*, Seattle (US): Amazon Createspace. http://www.amazon.com/No-Urgency-CEO-Hi-Tech-Management/dp/1489509356/ref=sr_1_1?s=books&ie=UTF8&qid=1396440688&sr=1-1&keywords=no+urgency. ISBN-13: 978-1489509352, ISBN-10: 1489509356

3} **Amin, A.R. & Asikin, S.** (2011b- 292 pages), *The Celestial Management: Spiritual Wisdom for All Human Beings*, Seattle (US): Amazon Createspace. http://www.amazon.com/Celestial-Management-Islamic-Wisdom-Beings/dp/1456450700/ref=sr_1_1?s=books&ie=UTF8&qid=1387092036&sr=1-1&keywords=celestial+management. ISBN-13: 978-1456450700, ISBN-10: 1456450700

4) **Asikin, S.** (2009- 548 pages), *Arigato, Obrigado... (Thanks): Portuguese-Japan 1550-1647 Historic Novel*, Seattle (US): Amazon Createspace. http://www.amazon.com/Arigato-Obrigado-Thanks-Portuguese-Japan-1550-1647/dp/1453786368/ref=sr_1_1?s=books&ie=UTF8&qid=1396442162&sr=1-1&keywords=arigato. ISBN-13: 978-1453786369, ISBN-10: 1453786368

5} **Asikin, S.** (2013a- 824 pages), *Econometric Rex: 888 Marketing Promo 6666 Six Sigma Parametric Theories (of 10 Data)*, Seattle (US): Amazon Createspace. http://www.amazon.co.uk/Econometric-Rex-Marketing-Parametric-Theories/dp/1482745259, ISBN-13: 978-1482745252, ISBN-10: 1482745259

6) **Asikin, S.** (2010b- 224 pages), *FINANCE Architecture: VISUAL art, for Corporate Finance's Beauty, Safety and Simplicity*, Seattle (US): Amazon Createspace. http://www.amazon.co.uk/FINANCE-Architecture-Corporate-Finances-Simplicity/dp/1453829547/ref=sr_1_14?s=books&ie=UTF8&qid=1385371842&sr=1-14&keywords=finance+architecture. ISBN-13: 978-1453829547, ISBN-10: 1453829547

7) **Asikin, S.** (2010c- 150 pages), *Geometric FINANCE: Useful concepts that work better than Economic Noble Laureates'* Seattle (US): Amazon Createspace. http://www.amazon.co.uk/Magic-Geometric-FINANCE-concepts-Laureates/dp/1453790284/ref=tmm_pap_title_0, ISBN-13: 978-1453790281, ISBN-10: 1453790284

Corporate IFRS-GAAP (B/S-I/S), ISBN-13: **978-1720792789**, ISBN-10: **172079278X**

8) **Asikin, S.** (2010a- 290 pages), *Investment ALGEBRA: Strategic Profitable Comprehensive Business Finance Engineering Technology*, Seattle (US): Amazon Createspace. http://www.amazon.co.uk/Investment-ALGEBRA-Profitable-Comprehensive-Engineering/dp/1453856811/ref=tmm_pap_title_0?ie=UTF8&qid=1385371309&sr=1-1, ISBN-13: 978-1453856819, ISBN-10: 1453856811

9) **Asikin, S.** (2011a- 352 pages), *State Economics: Comprehensive Macro-Micro Economics' Simple Fiscal-Monetary Export-Import Accouting, Integrated Supply-Demand Managerial ... Mathematical Engineering Visual* . Seattle (US): Amazon Createspace. http://www.amazon.co.uk/STATE-ECONOMICS-Steve-Asikin-ebook/dp/B00G89Q5SS, http://www.amazon.co.uk State-Economics-Steve-Asikin-ebook/dp/B00BBPLQDI, ISBN-13: 978-1461066095, ISBN-10: 1461066093

10) **Asikin, S.** (2013b- 260 pages), *Technoeconomistat: Economic &Statistical Technology for Financial Planning*, Seattle (US): Amazon Createspace. http://www.amazon.co.uk/Technoeconomistat-Economic-Statistical-Technology-Financial/dp/1482304244, ISBN-13: 978-1482304244, ISBN-10: 1482304244

11) **Asikin, S.** (2011b- 540 pages), *Terima Kasih: Novel Internasional Sejarah Jepang-Portugis 1550-1647*, Seattle (US): Amazon Createspace. http://www.amazon.com/Terima-Kasih-Internasional-Jepang-Portugis-1550-1647/dp/146351106X/ref=sr_1_1?s=books&ie=UTF8&qid=1396442361&sr=1-1&keywords=terima, ISBN-13: 978-1463511067, ISBN-10: 146351106X

12) **Asikin, S.** (2010d- 294 pages), *TREASURY Planology: Strategic Profitable Comprehensive Business Finance Engineering Technology*, Seattle (US): Amazon Createspace. http://www.amazon.co.uk/TREASURY-PLANOLOGY-Steve-Asikin-ebook/dp/B00G88MTFW/ref=sr_1_1?s=books&ie=UTF8&qid=1385403690&sr=1-1&keywords=Treasury+Planology. ISBN-13: 978-1453891476, ISBN-10: 1453891471

13) **Asikin, S.** (2014a- 190 pages), *Calculus Accountancy: Leibnitz Newton Pacioli's Polynomial Quantitative Finance Risk Modelling Optimization*, Seattle (US): Amazon Createspace. http://www.amazon.com/Calculus-Accountancy-Polynomial-Quantitative-Optimization/dp/1495463028/ref=sr_1_1?s=books&ie=UTF8&qid=1396441338&sr=1-1&keywords=calculus+accountancy, ISBN-13: 978-1495463020, ISBN-10: 1495463028

14) **Asikin, S.**(2014b- 318 pages), *No Parametric: No Parametric: Hyperbolic Octahedron Central 3-Dimensional Orthogonal Risk Distribution Topology*, Seattle (US): Amazon Createspace. http://www.amazon.com/Parametric-Hyperbolic-Octahedron-3-Dimensional-Distribution/dp/1496089731/ref=sr_1_1?s=books&ie=UTF8&qid=1396441543&sr=1-1&keywords=No+Parametric, ISBN-13: 978-1496089731, ISBN-10: 1496089731

Corporate IFRS-GAAP (B/S-I/S), ISBN-13: **978-1720792789**, ISBN-10: **172079278X**

15) **Asikin, S.** (2014c- 138 pages), *Quick Help:* Religion vs Real-Legion *Philosophy,* Seattle (US): Amazon Createspace.
http://www.amazon.com/Quick-Help-Religion-Real-Legion-Philosophy/dp/1497466520/ref=sr_1_2?s=books&ie=UTF8&qid=1396441740&sr=1-2&keywords=quick+help, ISBN-13: 978-1497466524, ISBN-10: 1497466520

16) **Asikin, S.** (2014d- 824 pages), *Anne of Denmark:* King James I the Bible, Historic Plays Drama, Seattle (US): Amazon Createspace.
http://www.amazon.com/Anne-Denmark-James-Bible-Drama/dp/149299734X/ref=sr_1_1?s=books&ie=UTF8&qid=1396441857&sr=1-1&keywords=anne+asikin, ISBN-13: 978-1492997344, ISBN-10: 149299734X

17) **Asikin, S.** (2014e- 184 pages), *Constructive Spells:* Positive Spiritual Mentality Personal Development: Seattle (US): Amazon Createspace.
http://www.amazon.com/Constructive-Spells-Spiritual-Mentality-Development/dp/1497498899/ref=sr_1_1?s=books&ie=UTF8&qid=1397664512&sr=1-1&keywords=constructive+spells; ISBN-13: 978-1497498891, ISBN-10: 1497498899

18) **Asikin, S.** (2014f- 466 pages), *Elizabeth of Russia:* Another Virgin Queen Gloariana: A Romanov Plays Drama, Seattle (US): Amazon Createspace.
http://www.amazon.com/Elizabeth-RUSSIA-Another-Gloriana-Romanov/dp/1505542480/ref=sr_1_4?s=books&ie=UTF8&qid=1423888482&sr=1-4&keywords=elizabeth+russia; ISBN-13: 978-1505542486, ISBN-10: 1505542480

19) **Asikin, S.** (2014f- 708 pages), *Shinmen Munisai:*Miyamoto Musashi's Father, A Sword Samurai Plays Drama, Seattle (US): Amazon Createspace. http://www.amazon.com/Shinmen-Munisai-Miyamoto-Musashis-Samurai/dp/1507857799/ref=sr_1_1?s=books&ie=UTF8&qid=1428094280&sr=1-1&keywords=munisai; ISBN-13: 978-1507857793, ISBN-10: 1507857799

20) **Asikin, S.** (2015a-767 pages), *Math Fin Law-1: Mathematical Financial Law for Public Listed Firms Rule 1-4441,* Seattle (US): Amazon Createspace.
http://www.amazon.com/Math-Fin-Law-Mathematical-Financial-ebook/dp/B00WLNX3Y4/ref=asap_bc?ie=UTF8; ISBN-13: 978-1511792219, ISBN-10: 1511792213

21) **Asikin, S.** (2015b-813 pages), *Math Fin Law-2: Mathematical Financial Law for Public Listed Firms Rule 4442-8712,* Seattle (US): Amazon Createspace. http://www.amazon.com/Math-Fin-Law-Mathematical-No-4442-8712/dp/1512285080/ref=sr_1_1?s=books&ie=UTF8&qid=1432732199&sr=1-1&keywords=math+fin+law+2; ISBN-13: 978-1512285086, ISBN-10: 1512285080

22) **Asikin, S.** (2015c-813 pages), *Math Fin Law-3: Mathematical Financial Law for Public Listed Firms Rule 8713-12575,* Seattle (US): Amazon Createspace. http://www.amazon.com/Math-Fin-Law-Mathematical-8712-12575/dp/1514276763/ref=sr_1_1?s=books&ie=UTF8&qid=1433982149&sr=1-1&keywords=math+fin+law+3, ISBN-13: 978-1514276761. ISBN-10: 1514276763

Corporate IFRS-GAAP (B/S-I/S), ISBN-13: **978-1720792789**, ISBN-10: **172079278X**

23) **Asikin, S.** (2015d-813 pages), *Math Fin Law-4: Mathematical Financial Law for Public Listed Firms Rule 12576-16333,* Seattle (US): Amazon Createspace. http://www.amazon.com/Math-Fin-Law-Mathematical-12576-16333/dp/1514685132/ref=sr_1_1?s=books&ie=UTF8&qid=1435393628&sr=1-1&keywords=math+fin+law+4, ISBN-13: 978-1514685136, ISBN-10: 1514685132

24) **Asikin, S.** (2015d-816 pages), *Math Fin Law-5: Mathematical Financial Law for Public Listed Firms Rule 16334-19904,* Seattle (US): Amazon Createspace. http://www.amazon.com/Math-Fin-Law-Mathematical-No-16334-19904/dp/1515073009/ref=sr_1_1?s=books&ie=UTF8&qid=1372246936&sr=1-1&keywords=math+fin+law+5. ISBN-13: 978-1515073000, ISBN-10: 1515073009

25) **Asikin, S.** (2015d-728 pages), *Math Fin Law-6: Mathematical Financial Law for Public Listed Firms Rule 19905-23237,* Seattle (US): Amazon Createspace. https://www.amazon.com/Math-Fin-Law-Mathematical-No-19905-23238-ebook/dp/B0129ILVNK/ref=sr_1_2?ie=UTF8&qid=1478017624&sr=8-2&keywords=Math+Fin+Law+Asikin. ISBN-13: 978-1515158691, ISBN-10: 1515158691

26) **Asikin, S.** (2016a-728 pages), *Math Fin Law-7: Mathematical Financial Law for Public Listed Firms Rule 23238-27115,* Seattle (US): Amazon Createspace. https://www.amazon.com/Math-Fin-Law-Mathematical-No-23238-27115/dp/1508518521/ref=la_B00428V6JU_1_27?s=books. ISBN-13: 978-1508518525, ISBN-10: 1508518521

27) **Asikin, S.** (2016b-728 pages), *Math Fin Law-8: Mathematical Financial Law for Public Listed Firms Rule 27116-30330,* Seattle (US): Amazon Createspace. https://www.amazon.com/dp/1541091884, https://www.amazon.com/Math-Fin-Law-Mathematical-No-27116-30330/dp/1541091884/ref=la_B00428V6JU_1_27?s=books&ie=UTF8&qid=1479398538&sr=1-27&refinements=p_82%3AB00428V6JU, ISBN-13: 978-1508518525, ISBN-10: 1508518521

28) **Asikin, S.** (2017a-828 pages), *Math Fin Law-9: Mathematical Financial Law for Public Listed Firms Rule 30330-33373,* Seattle (US): Amazon Createspace. https://www.amazon.com/Math-Fin-Law-Mathematical-30331-33373-ebook/dp/B01MRA8HZQ/ref=la_B00428V6JU_1_4?s=books&ie=UTF8&qid=1485921040&sr=1-4. ISBN-13: 978-1541092433, ISBN-10: 1541092430

29) **Asikin, S.** (2017b-823 pages), *Math Fin Law-10: Mathematical Financial Law for Public Listed Firms Rule 33374-36242,* Seattle (US): Amazon Createspace. https://www.amazon.com/s/ref=nb_sb_noss?url=search-alias%3Dstripbooks&field-keywords=math+fin+law+10. https://www.amazon.com/s/ref=nb_sb_noss?url=search-alias%3Dstripbooks&field-keywords=math+fin+law+10, ISBN-13: 978-1541092761, ISBN-10: 1541092767

Corporate IFRS-GAAP (B/S-I/S), ISBN-13: **978-1720792789**, ISBN-10: **172079278X**

30) **Asikin, S**. (2017c-823 pages), *Math Finance Law-11: Mathematical Financial Law for Public Listed Firms Rule 36243-39158*, Seattle (US): Amazon Createspace. https://www.amazon.com/s/ref=nb_sb_noss?url=search-alias%3Dstripbooks&field-keywords=math+finance+law+11, https://www.amazon.com/dp/1541215265/ref=sr_1_1?s=books&ie=UTF8&qid=14880299786 sr=1-1&keywords=math+finance+law, ISBN-13: 978-1541215269. ISBN-10: 1541215265

31) **Asikin, S**. (2017d-823 pages), *Math Finance Law-12: Mathematical Financial Law for Public Listed Firms Rule 30150-42152*, Seattle (US): Amazon Createspace. https://www.amazon.com/s/ref=nb_sb_noss?url=search-alias%3Dstripbooks&field-keywords=math+finance+law+12 https://www.amazon.com/Math-Finance-Law-Mathematical-39159-42152/dp/1541215516/ref=sr_1_2?s=books&ie=UTF8&qid=14909903428sr=1-2&keywords=math+finance+law+12. ISBN-13: 978-1541215511, ISBN-10: 1541215516

32) **Asikin, S**. (2017e-824 pages), *Math Finance Law-13: Mathematical Financial Law for Public Listed Firms Rule 42153-44938*, Seattle (US): Amazon Createspace. https://www.amazon.com/dp/1542675219/ref=sr_1_1?s=books&ie=UTF8&qid=1491146227&s r=1-1&keywords=math+finance+law+13 . ISBN-13: 978-1542675215, ISBN-10: 1542675219

33) **Asikin, S**. (2017f-824 pages), *Math Finance Law-14: Mathematical Financial Law for Public Listed Firms Rule 44939-47745*, Seattle (US): Amazon Createspace. https://www.amazon.com/dp/1542675219/ref=sr_1_1?s=books&ie=UTF8&qid=1491146227&s r=1-1&keywords=math+finance+law+14 . ISBN-13: 978-1544628967, ISBN-10: 154462896X

34) **Asikin, S**. (2017g-810 pages), *Math Finance Law-15: Mathematical Financial Law for Public Listed Firms Rule 47746-50465*, Seattle (US): Amazon Createspace. https://www.amazon.com/dp/1545118647/ref=sr_1_1?s=books&ie=UTF8&qid=1493626425&s r=1-1&keywords=math+finance+law+15, . ISBN-13: 978-1545118641, ISBN-10: 1545118647

35) **Asikin, S**. (2017h-816 pages), *Math Finance Law 16: Mathematical Financial Law Public Listed Firm Rule No.50,466-53,320*, Seattle (US), Amazon CreateSpace, https://www.amazon.com/dp/1545320179/ref=sr_1_1?s=books&ie=UTF8&qid=14956926956 sr=1-1&keywords=math+finance+law+16. ISBN-13: 978-1545320174, ISBN-10: 1545320179

Corporate IFRS-GAAP (B/S-I/S), ISBN-13: **978-1720792789**, ISBN-10: **172079278X**

36) **Asikin, S**. (2017i-809 pages), *Math Finance Law 17*: *Mathematical Financial Law Public Listed Firm Rule No.53,231-56,124*, Seattle (US), Amazon CreateSpace, https://www.amazon.com/dp/1545320179/ref=sr_1_1?s=books&ie=UTF8&qid=14956926956 sr=1-1&keywords=math+finance+law+17, ISBN-13: 978-1545486474, ISBN-10: 1545486476

37) **Asikin, S**. (2017j-815 pages), *Math Finance Law 18*: *Mathematical Financial Law Public Listed Firm Rule No.56,125-58,942*, Seattle (US), Amazon CreateSpace, https://www.amazon.com/dp/1545320179/ref=sr_1_1?s=books&ie=UTF8&qid=14956926956 sr=1-1&keywords=math+finance+law+18, ISBN-13: 978-1546912361, ISBN-10: 1546912363

38) **Asikin, S. (**2014-47 pages), *Doing Business in Comprehensive Interactive Interdisciplinary Micro-Macro Eco nomic International Accounting Architecture of Serial Well Known General Great Theories*, Proceedings in Finance and Risk Perspective 13, ACRN Oxford Publishing House, UK, p.197-243, https://books.google.com/books?id=4e4KAwAAQBAJ&pg=PA219&lpg=PA219&dq=steve+asiki n+oxford&source=bl&ots=qVq9KyAmem&sig=ABdBliwuzUwRiglsuLimuJ5ZdW4&hl=en&sa=X&s qi=2&ved=0ahUKEwjZkMrF7KjTAhUDSiYKHREDB- qQ6AEILjAC#v=onepage&q=steve%20asikin%20oxford&f=false, ISBN 978-3-9503518-1-1.

39) **Asikin, T.& Asikin, S**. (2013- 828 pages), *448 Poems*: *KARAOKE Song Book: 24 Hours Story of 30 Languages*, Seattle (US): Amazon Createspace. http://www.amazon.com/448-Poems-KARAOKE-Song- Book/dp/1480034363/ref=sr_1_1?s=books&ie=UTF8&qid=13964420346sr=1- 1&keywords=448+poems, ISBN-13: 978-1480034365, ISBN-10: 1480034363

40) **Asikin, T, Asikin, S, Senihardja, I.** (2017a- 508 pages), *Finance Construction 1*: *Corporate IFRS-GAAP (B/S-I/S) Engineering Technologies No. 1-1,000 of 111,111 Laws*, Seattle (US): Amazon Createspace. https://www.amazon.com/s/ref=nb_sb_noss?url=search- alias%3Dstripbooks&field-keywords=finance+construction+1+asikin. ISBN- 13: 978-1544059525, ISBN-10: 1544059523

41) **Asikin, T, Asikin, S, Senihardja, I.** (2017b- 526 pages), *Finance Construction 2*: *Corporate IFRS-GAAP (B/S-I/S) Engineering Technologies No. 1,001-2,000 of 111,111 Laws*, Seattle (US): Amazon Createspace. https://www.amazon.com/s/ref=nb_sb_noss?url=search- alias%3Dstripbooks&field-keywords=finance+construction+2+asikin. ISBN- 13: 978-1542402415, ISBN-10: 1542402417

Corporate IFRS-GAAP (B/S-I/S), ISBN-13: **978-1720792789**, ISBN-10: **172079278X**

42) **Asikin, T, Asikin, S, Senihardja, I.** (2017c- 738 pages), *Finance Construction 3: Corporate IFRS-GAAP (B/S-I/S) Engineering Technologies No. 2,001-3,000 of 111,111 Laws*, Seattle (US): Amazon Createspace. https://www.amazon.com/s/ref=nb_sb_noss?url=search-alias%3Dstripbooks&field-keywords=finance+construction+3+asikin. ISBN-13: 978-1975999292, ISBN-10: 1975999290

43) **Asikin, T, Asikin, S, Senihardja, I.** (2018a- 656 pages), *Finance Construction 4: Corporate IFRS-GAAP (B/S-I/S) Engineering Technologies No. 3,001-4,000 of 111,111 Laws*, Seattle (US): Amazon Createspace. https://www.amazon.com/s/ref=nb_sb_noss?url=search-alias%3Dstripbooks&field-keywords=finance+construction+4+asikin. ISBN-13: 978-1977659187, ISBN-10: 1977659187

44) **Asikin, T, Asikin, S, Senihardja, I.** (2018b- 656 pages), *Finance Construction 5: Corporate IFRS-GAAP (B/S-I/S) Engineering Technologies No. 4,001-5,000 of 111,111 Laws*, Seattle (US): Amazon Createspace. https://www.amazon.com/s/ref=nb_sb_noss?url=search-alias%3Dstripbooks&field-keywords=finance+construction+5+asikin. ISBN-978-1983566141, ISBN-10: 1983566144

45) **Asikin, T, Asikin, S, Senihardja, I.** (2018c- 524 pages), *Finance Construction 6: Corporate IFRS-GAAP (B/S-I/S) Engineering Technologies No. 5,001-5,500 of 111,111 Laws*, Seattle (US): Amazon Createspace. https://www.amazon.com/s/ref=nb_sb_noss?url=search-alias%3Dstripbooks&field-keywords=finance+construction+6+asikin. ISBN-978-1983566332, ISBN-10: 1983566330

46) **Asikin, T, Asikin, S, Senihardja, I.** (2018d- 532 pages), *Finance Construction 7: Corporate IFRS-GAAP (B/S-I/S) Engineering Technologies No. 5,501-6,000 of 111,111 Laws*, Seattle (US): Amazon Createspace. https://www.amazon.com/s/ref=nb_sb_noss?url=search-alias%3Dstripbooks&field-keywords=finance+construction+7+asikin. ISBN-978-1987571509, ISBN-10: 1987571509

47) **Asikin, T, Asikin, S, Senihardja, I.** (2018e- 534 pages), *Finance Construction 8: Corporate IFRS-GAAP (B/S-I/S) Engineering Technologies No. 6,001-6,500 of 111,111 Laws*, Seattle (US): Amazon Createspace. https://www.amazon.com/s/ref=nb_sb_noss?url=search-alias%3Dstripbooks&field-keywords=finance+construction+8+asikin. ISBN-13: 978-1718712300, ISBN-10: 1718712308

48) **Asikin, T, Asikin, S, Senihardja, I.** (2018f- 536 pages), *Finance Construction 9: Corporate IFRS-GAAP (B/S-I/S) Engineering Technologies No. 6,501-7,000 of 111,111 Laws*, Seattle (US): Amazon Createspace. https://www.amazon.com/s/ref=nb_sb_noss?url=search-alias%3Dstripbooks&field-keywords=finance+construction+9+asikin. ISBN-13: 978-1718784055, ISBN-10: 1718784058

Corporate IFRS-GAAP (B/S-I/S), ISBN-13: **978-1720792789**, ISBN-10: **172079278X**

49) **Asikin, T, Asikin, S, Senihardja, I.** (2018g- 536 pages), *Finance Construction 10: Corporate IFRS-GAAP (B/S-I/S) Engineering Technologies No. 7,001-7,500 of 111,111 Laws*, Seattle (US): Amazon Createspace. https://www.amazon.com/s/ref=nb_sb_noss?url=search-alias%3Dstripbooks&field-keywords=finance+construction+10+asikin. ISBN-13: 978-1719046626, ISBN-10: 171904662X

50) **Brigham, EF & Gapenski, LC,** . (1988- Textbook), *Financial Management: Theory and Practice (Study Guide)*, NY (US): Dryden Press. http:/www.amazon.com/Financial-Management-Theory-Practice-Study/dp/0030125391/ref=sr_1_1?s=books&ie=UTF8&qid=14293808356sr=1-1 ;

51) **Chandonet, RE,** . (2011-32 pages), *Balancesheet Management, Trends & Strategies*, NY (US): Sandler O'Neil. https://www.ncbankers.org/uploads/File/Meetings/Presentations/2011CONV-Chandonnet.pdf;

52) **Cornuejols G & Tutuncu, R,** . (2006-358 pages), *Optimization Methods in Finance*, Cambridge (UK): Cambridge University Press. http://www.math.ku.dk/~rolf/CT_FinOpt.pdf;

53) **Crook, M.** (2015-6 pages), *Optimization Methods in Finance*, UBS FS (Swiss): CIO WM Research. https://www.ubs.com/content/dam/WealthManagementAmericas/documents/investment-strategy-insights-2015-01-09-balance-sheet-optimization.pdf;

54) **Fee, Greg.** (2012-127 pages), *Catalan's Constant: (Ramanujan's Formula}: Catalan Constant to 300,000 digits*, Seattle (US): Amazon Kindle. http://www.Amazon.com/gp/aw/d/B008ZZ5IEW/ref=mp_sa_1_4?ie=UTF8%Qqid=1 4843678186sr=8-4&pi=AC_SX236_SY340_QL65&keywords=ramanujan&dpPl=1&dpID=51FqlwDZpRL&ref=plSrch

55) **Gardenfors, Peter.** (2014) *The Geometry of Meaning: Semantics Based on Conceptual Spaces*. Cambridge, Mass.: MIT Press. ISBN 0-262-02678-3.

56) **Gardenfors, Peter.** (2000), *Conceptual Spaces: The Geometry of Thought*. Cambridge, Mass.: MIT Press. ISBN 0-262-07199-1.

57) **Kanigel, Robert.** (2013-465 pages), *The Man Who Knew Infinity:A Life of the Genius Ramanujan*, Washington (US): Washington Square Press&Amazon Kindle. http://www.Amazon.com/gp/aw/d/B008BW4VEGM/ref=pd_aw_sim351_1?ie=UTF86 psc=1&refRID=VYN345GVDBJ6ATDQ35W2&dpID=9IZgpc4ETdL

58) **Limlingan, Victor S.** (1993-1 page), *The Limlingan Financial Model*, Manila: Asian Manager. http://limlingan.com/images/article.gif

Corporate IFRS-GAAP (B/S-I/S), ISBN-13: **978-1720792789**, ISBN-10: **172079278X**

59) **Newton, Isaac.** (2015-631 pages), *Principia:The Mathematical Principles of Natural Philosophy (Active Content)*, London (UK): RSM&Amazon Kindle.
http://www.Amazon.com/gp/aw/sitb/B01ISFJSJ0?ref=sib_dp_aw_kb_udp

60) **Nikisa, Evets.** (2014g- 828 pages), *Wonderful Live Spells: 144-Steps Recreative Nice Children Story, Succesful Development:* Seattle (US): Amazon Createspace. http://www.amazon.com/WONDERFUL-Live-Spells-Recreative-Development/dp/1499248423/ref=sr_1_1?s=books&ie=UTF8&qid=1418698379&sr=1-1&keywords=wonderful+asikin, ISBN-13: 978-1499248425, ISBN-10: 1499248423

61) **Nielson, SS.** (2007- 430 pages), *Practical Financial Optimization: Decision Making for Financial Engineers:* NY (US): Wiley.
https://books.google.com/books/about/Practical_Financial_Optimization.html?id=XkwLAAAACAAJ

62) **Niswonger CR, Fess PE & Warren CE.** (1993-Textbook), *Accounting Principles:* Seattle (US): South Western Publishing Co.
http://www.amazon.com/Accounting-Principles-Chapters-C-Rollin-Niswonger/dp/0538818603

63) **Omerod, P**. (1993-Textbook), (1994), *The Death of Ecoomics*, St Martin's Press, London, . http://www.amazon.com/Accounting-Principles-Chapters-C-Rollin-Niswonger/dp/0538818603, ISBN-13: 978-0312134648, ISBN-10: 0312134649

64) **Paccioli, L. (1989-Textbook)**, *Summa de Arithmetica Geometria Proportioni et Proportionalita*, Venice 1494.
http://www.amazon.com/Arithmetica-Geometria-Proportioni-Proportionalita-Venice/dp/B00T4HXTIW/ref=sr_1_fkmr0_2?s=books&ie=UTF8&qid=1429597966&sr=1-2-fkmr0&keywords=summa+mhematica+arithmetica

65) **Persson, Torsten & Tabellini, Guido** (2000), *Political Economics – Explaining Economic Policy*, MIT Press.
https://mitpress.mit.edu/authors/torsten-persson

66) **Persson, Torsten & Tabellini, Guido** (2003), *The Economic Effects of Constitutions*, MIT Press. https://mitpress.mit.edu/authors/torsten-persson

67) **Persson, Torsten & Tabellini, Guido** (2000), *Monetary and Fiscal Policy –1 Credibility*, MIT Press. https://mitpress.mit.edu/authors/torsten-persson

68) **Persson, Torsten & Tabellini, Guido** (2003), *Monetary and Fiscal Policy –2 Politics*, MIT Press. https://mitpress.mit.edu/authors/torsten-persson

69) **Puts, J** (2012-102 pages), *Balancesheet Optimization Under Basel III*, Amsterdam (Netherlads): Vrije Universiteit.
http://www.math.vu.nl/~sbhulai/papers/thesis-puts.pdf

70) **Timmermans, K** (2012-30 pages), *Balancesheet Optimization Under Basel III*, Amsterdam (Netherlads): ING Investor Day.
https://www.ing.com/web/file?uuid=09c803d9-79a1-47ec-bfe5...owner...

Corporate IFRS-GAAP (B/S-I/S), ISBN-13: **978-1720792789**, ISBN-10: **172079278X**

71) Weston JF, Copeland TE & Shastri, K. (2004-Textbook), *Financial Theory and Corporate Policy:* NY (US): Addison Wesley.
http:/www.amazon.com/Financial-Corporate-Copeland-Kuldeep-Paperback/dp/B008YSUE4M/ref=sr_1_1?s=books&ie=UTF8&qid=1429381700&sr=1-1&keywords=weston+copeland

72) Zenios SA. (2002-350 pages), *Financial Optimization:* Cambridge (UK): Cambridge University Press.
https://books.google.com/books/about/Financial_Optimization.html?id=-FJdqjZ0qSAC. .

Corporate IFRS-GAAP (B/S-I/S), ISBN-13: **978-1720792789**, ISBN-10: **172079278X**

B. Intellectual Properties:

1)<u>**Indonesian Banker Certification 1993**</u>, *Sertifikasi Bankir Indonesia 1993*;
2)<u>**MATHFINPLAN**</u> (Mathematical Financial Planning);
    Perencanaan Keuangan Matematis
3)<u>**BANK MATHPLAN**</u> (Bank's Mathematical Planning);
    Perencanaan Matematis Bank
4)<u>**TECHNO-ECONOMISTAT**</u> (Technology for Economical Statistics);
    Teknologi Statistik untuk Ekonomi
5)<u>**FINANCIAL GEOMETRY**</u> (IT Program); *Ilmu Geometri Keuangan*
6)<u>**BAK DOBEL**</u> (Car Gasoline Tank: Left and Right Side Fuel Fill-in).
    Tangki Mobil yang bisa diisi dari Kiri atau Kanan
7)<u>**Mister GO-CHENG**</u> (Cheap Food Franchising). *Waralaba Pangan Murah*

Corporate IFRS-GAAP (B/S-I/S), ISBN-13: **978-1720792789**, ISBN-10: **172079278X**

C. SCIENTIFIC PAPERS:

1)**Eichhornia Crassipes**, as Anti Polutant and Fertile Sediment, *Pemanfaatan Eceng Gondok untuk Menetralkan Polusi dan Pengurukan Subur pada Genangan Air,* LKIR, Lomba Karya Ilmiah Remaja, LIPI (1979)

2)**Archimedes Catesian Buoyancy Law**, *Penyimpangan Hukum Archimedes pada Model Pelampung Cartesian,* Lomba Karya Ilmiah Remaja, Depdikbud-RI (1980)

3)**SURAMADU**, Surabaya-Madura Inter Island Bridge, Design Supervisor, *Sumarsono Sumali Engineering Contractor,* Tambak-V, Tandes, Surabaya (1982),

4)**GAZOLINE Taxi-Pool Depot**, Jemur Handayani, Surabaya, *Sumarsono Sumali Engineering Contractor,* Tambak-V, Tandes, Surabaya (1982),

5)**SOEKARNO-HATTA**, Jakarta International Airport, Garuda Maintenance Facilities, *Junior Architect,* As Bulit Drawings, Coinstruction Management, Aerospatiale-Encona Engineering (1983-1985),

6)**ADISUCIPTO**, Yogya International Airport, *Junior Architect,* Preliminary Technical Drawings, Budiono Surasno-Encona Engineering (1986),

7)**BLOK-M PLAZA**, "New Garden Hall", Kebayoran Baru Supermall, *Preliminary Design Architect,* Danisworo-James Ferry- Jasa Ferry-Encona (1986),

8)**NORTH JAKARTA PUBLIC LIBRARY**, Semper, Tanjung Priok, *Chief Architect,* Agung Darma Sarana-Asep Hermawan (1987),

9)**The Banker's Trainer**: A Manual for The Training Head in Banking. *Profesi Bankir Pelatih, Tuntunan bagi Kepala Diklat Perbankan,* (1200 Pages, 1993),

10)**Indonesian Bankers Certification**, a Strategic Move for the National Banking Improvement. *Sertifikasi Bankir Indonesia, Langkah Strategis Pembenahan Perbankan Nasional,* (KOMPAS, 20 Juli 1993).

11)**Indonesian Rupiah's Exchange Rate**, on a Quiet but Steady Movement. *Nilai Tukar Rupiah, Diam-diam Meningkat Pesat,* (MANAJEMEN, Jan-Feb 1994).

12)**The Bank's Credit**, for the Best Practices. *Kredit Praktis* (200 pages, 1994),

13)**Down to Earth Credit Meeting Credo**, *Kiat Rapat Pemberian Kredit yang Realistik.* (MANAJEMEN, Sep-Oct 1994).

14)**SCHIPHOL**, Amsterdam International Airport, 1st Anthony Fokker Weg, Sloten, *Airport Engineer,* Samsung- Fokker-Beurlin, BV (1996),

15)**Ultrapeneurship: The Secret of Starting Business Without Capital**. Ultrapreneur: *Kiat Usaha Tanpa Modal,* (MANAJEMEN, Nov-Dec 1996),

16)**Recipe from the Corporate Raider's Kitchen**, *Menguak Dapur Corporate Raiders,* (MANAJEMEN, Jan-Feb 1997).

17)**The 10-Tips and Tricks of Engineered Conglomeration**. *Sepuluh Rekayasa Keuangan Konglomerat,* (MANAJEMEN, Mar-Apr 1997),

18)**The Management of Succession in Five Parts of the World**. *Manajemen Suksesi di Lima bagian Dunia,* (MANAJEMEN, Mei-Jun 1997),

Corporate IFRS-GAAP (B/S-I/S), ISBN-13: **978-1720792789**, ISBN-10: **172079278X**

19)**Auditing**: The Global Language for the 2003's Human Resources. *Audit : Bahasa Global untuk SDM 2003,* (MANAJEMEN, Jul-Aug 1997),

20)**The Elected Parliament 1997-2002**, and the Hope of Professionals. *DPR 1997–2002 dan Harapan Profesional,* (MANAJEMEN, Sep-Okt 1997),

21)**Disappearance of World's Soft Currencies**, and Proposition for Global Economic Thinking. *Lenyapnya Mata Uang Lunak Dunia dan Usulan Pemikiran Ekonomi Global,* (MANAJEMEN, Nov-Dec 1997),

22)**The Human Resources and Its Accountability**, in Critical Situation. *Akuntabilitas Sumberdaya Manusia di Masa Krisis,* (MANAJEMEN, Jan-Feb 1998), '

23)**The Investment Economic Model**, and the Failure of Growth Economic Theories. *Model Ekonomi Investasi dan Kegagalan Teori Ekonomi Pertumbuhan,* (MANAJEMEN, Nov 1998),

24)**The Exact Equilibrium of Capital Budgeting**, *Persamaan-persamaan akurat dalam Penganggaran Modal Perusahaan.* (MANAGEMENT EXPOSE, November 1998),

25)**The Corruptive Economic Model**, and Its Solution for Long Term Debt's Problem. *Model Ekonomi Korupsi dan Penyelesaian Utang Jangka Panjang,* (MANAJEMEN, Dec 1998),

26)**The Self Equity Budgeting**, and Operations Research in Investments. *Penganggaran Modal Sendiri dan Riset Operasi dalam Investasi,* (MANAJEMEN, Jan 1999),

27)**Actual Impact of Inflation**, and the Need of Geometric Approach in Statistic. *Pengaruh Riil Inflasi dan Perlunya Statistika Geometri,* (MANAJEMEN, Feb 1999),

28)**Multiplication of Money**, to Overcome the Crisis. *Melipat Gandakan Uang untuk Mengusir Krisis,* (MANAJEMEN, Apr 1999),

29)**The Interactive Mathematical Model**, for Multi Constraints Financial Budgeting, *Model Matematika Ekonomi Interaktif bagi Penganggaran Keuangan Multi Kendala.* (SCRIPTA ECONOMICA, Vol. 2, April 1999)

30)**Application of Ethics**, for the Management Accounting Practices. *Aplikasi Etika dalam Praktek Akuntansi Manajemen,* (MANAJEMEN, May 1999),

31)**The Curriculum of Magistrate of Management in Finance**: Its Needs and Practices. *Kurikulum Magister Manajemen Keuangan 1988-1998: Antara Kebutuhan dan Praktek,* (MANAJEMEN, Jun 1999),

32)**The Attractiveness**, Personal Power and Management. *Pesona Pribadi: "Personal Power" dan Manajemen,* (MANAJEMEN, Jul 1999),

33)**The Market Research**, for Magistrate Degree in Management. *Soal Jawab Riset Pasar untuk Pascasarjana Manajemen* (126 Pages, 1999),

34)**The Thesis Manual**, Question and Answer for Magistrate Degree in Management. *Soal Jawab Tugas Akhir untuk Pascasarjana Manajemen* (126 Pages, 1999),

35)**The Risk Management**, Question and Answers for Magistrate Degree in Management. *Soal Jawab Manajemen Resiko untuk Pascasarjana Manajemen* (126 Pages, 1999),

Corporate IFRS-GAAP (B/S-I/S), ISBN-13: **978-1720792789**, ISBN-10: **172079278X**

36)**The Credit Manual**, Question and Answers for Magistrate Degree in Management. *Soal Jawab Manajemen Kredit untuk Pascasarjana Manajemen* (126 Pages, 1999),

37)**Enhancing the Morality Structure**. *Solidkan Struktur Moralitas*, (MANAJEMEN, Oct 1999)

38)**Performance of PT. Astra Otoparts**, Plc, in its Press Releases and Public Financial Reports. *Kinerja PT. Astra Otoparts dalam Siaran Pers dan Laporan Keuangan Publik*, (MANAJEMEN, Oct 1999)

39)**The Management of Conflicts**, for the New Indonesian's "Rainbow Colored" Cabinet. *Manajemen Konflik dan Kabinet Pelangi*, (MANAJEMEN, Dec 1999),

40)**Technoeconometric**: An Engineering Elaboration to Matrix Multi Variate Econometrics, *Tekno Ekonometrik sebagai sumbangan teknologi untuk Ekonometrika Matrik Multi Variat.* (MANAGEMENT EXPOSE, November 1999),

41)**Technoeconometric** : A Mathematical Shortcut for Non Linear Bivariate Econometrics, *Tekno Ekonometrik sebagai Usulan Jalan Pintas Matematik untuk Ekonometrika non Linear Bivariat.* (SCRIPTA ECONOMICA, Vol. 2, No. 3, Desember 1999),

42)**Developing "New Image"**, for the New Indonesia. *Membangun Citra Indonesia Baru*, (MANAJEMEN, Jan 2000),

43)**Segregations of Duties**, the Needs of Both Indonesian Governments and Private Sectors. *Pemerintah dan Swasta Perlu Berbagi Tugas*, MANAJEMEN, Mar 2000),

44)**The 812.500 Financial Models**, for the Autonomous Country Governments. *Tentang 812.500 Model Keuangan Otonomi Daerah*, (MANAJEMEN, Mar 2000),

45)**The Forecast of 10 Most Attractive Share Values**, in the Jakarta Stock Exchange, April 2000. *Prakiraan Harga Sepuluh Saham Unggulan di Bursa Efek Jakarta, April 2000*, (MANAJEMEN, Mar 2000),

46)**Let Us Merchandise the Intelectual Properties**!. *Jual "Intelectual Property" Saja!* (MANAJEMEN, Apr 2000),

47)**The Forecast of ten Most Attractive Share Values**, in the Jakarta Stock Exchange May 2000. *Prakiraan Harga Sepuluh Saham Unggulan di Bursa Efek Jakarta Mei 2000*, (MANAJEMEN, Apr 2000),

48)**HEATHROW**, London International Airport, Second Runway, *Q-Basic Algorithmic Engineer*, Beasley-Krishnamoorthy, Imperial College, Cambridge, Post Doctoral Project (2000),

49)**Why the Banking Sector Should be Healthy**, Before Others?. *Mengapa Perbankan Harus Sehat?* (MANAJEMEN, Mei 2000),

50)**Recycling of Wastes as a Negative Costing**. *Memanfaatkan Sampah atau Negative Costing*, (MANAJEMEN, Agt 2000),

51)**The Guidance for Indonesia Corporate Restructuring**. *Kiat Restrukturisasi Perusahaan Indonesia*, (MANAJEMEN, Agt 2000),

Corporate IFRS-GAAP (B/S-I/S), ISBN-13: **978-1720792789**, ISBN-10: **172079278X**

52)<u>Redefinition of the Ethical Comprehension</u>. *Pemaknaan Kembali atas Pemahaman Etika*, (MANAJEMEN, Sep 2000),

53)<u>The Critical View the 1999's Indonesians Central Bank Financial Statement</u>. *Analisa Laporan Keuangan Bank Indonesia 1999*, (MANAJEMEN, Oct 2000),

54)<u>Becoming an "Achiever"</u>, not Just Common "Worker". *Bukan Sekadar "Kerja", Tapi Menghasilkan Karya!* (MANAJEMEN, Feb 2001),

55)<u>Ten Steps to Real Wealth!</u> *Sepuluh Langkah Menjadi Kaya*, (MANAJEMEN, Jun 2001),

56)<u>Creative Financial Breakthrough</u>, *Seminar Sehari tentang Terobosan Keuangan yang Kreatif*, (Gramedia-PPM dan MANAJEMEN, Jun 2001),

57)<u>The Marketing Practices of Financial Softwares</u>, for Publics Listed Corporation. *Praktek Pemasaran Program Perangkat Lunak untuk Keuangan Perusahaan Publik* (Univ Satyagama, Jakarta Juni 24, 2001).

58)<u>The Mathematical Financial Planning</u>, *Perancangan Keuangan Secara Eksak dengan Matematik.* (MANAJEMEN, Jul 2001),

59)<u>The Real Research and Development, not Rework and Destruction</u>. *Litbang yang Bukan "Sulit Berkembang",* (MANAJEMEN, Jul 2001),

60)<u>The Treatise on the Advantages and Disadvantages</u>, of Revolutionary Learning Methods. *Baik Buruknya: Revolusi Cara Belajar*, (MANAJEMEN, Jul 2001),

61)<u>Better Make an Indonesian "Corruption Prevention" Body</u>, not Just a "Corruption Watch". *Buat Indonesian Corruption Prevention (ICP), Bukan Indonesian Corruption Watch, ICW!* (MANAJEMEN, Okt 2001),

62)<u>Preparation of the Worlds Business Leaders</u>, for our Century and the Century After. *Mempersiapkan Manusia-manusia yang akan menjadi Khafilah Bisnis Dunia Abad ini dan Abad Mendatang.* (Univ Satyagama – Gregorio Araneta Univ, Jakarta, Oktober 27, 2001).

63)<u>The Good Corporate Governance</u>, is not just a Morale Condition. *Kepemerintahan Perusahaan yang Baik, Bukan Sekedar Moralitas,* (MANAJEMEN, Nov 2001),

64)<u>Corporate Healthiness and Cleanliness as a Challenge</u>. *Bersih, Sehat Perusahaan, sebuah Tantangan* (MANAJEMEN, Jan 2002),

65)<u>Developing the Cultural Base for Future Competitions</u>. *Budaya untuk Bersaing di Masa Depan* (MANAJEMEN, Jan 2002),

66)<u>"Catapult", "Bullet" or "Ballistic Missiles" Approach</u>, for Corporate Healthiness and Cleanliness. *Pola "Ketapel", "Peluru Biasa" atau "Rudal" untuk Keuangan Perusahaan yang Bersih Sehat.* (MANAJEMEN, Feb 2002),

67)<u>The Mathematical Invesment Solution Mode</u>l, for Indonesian Foreign Debt and Its Corruption, *Solusi Matematika Investasi bagi Penyelesaian Masalah Utang Luar Negeri dan Korupsi di Indonesia.* (MANAGEMENT EXPOSE, Mar 2002),

Corporate IFRS-GAAP (B/S-I/S), ISBN-13: **978-1720792789**, ISBN-10: **172079278X**

68)**The Attitude, Skills and Knowledge (ASK)**, in Educational System Approach. *Sikap, Keterampilan dan Pengetahuan Menurut Pendekatan Sistemik Pendidikan,* (MANAJEMEN, Apr 2002),
69)**Strategic Treatise on Regional Autonomy**, in Futurization, Digitalization, Securitization and Globalization Era in Accordance to the Awakening of the Mercantile Nations. *Kajian Strategik Otonomi Daerah : Era Futurisasi, Digitalisasi, Strukturisasi, dan Globalisasi dalam Kebangkitan Negara-Negara Dagang.* (Training for Provincial Government Regional Management, Indonesian Association of Provincial Governments, *Pelatihan Pengembangan Manajemen Pemerintah Daerah bagi Perangkat Pemerintah Propinsi, Angkatan II,* (APPSI), June 27, 2002.
70)**The Suggestion for the Doctoral Course**, for Islamic Financial Management System. *Usulan-usulan Perbaikan dalam Kurikulum Doktor Manajemen Keuangan Islam.* (December, 17-19, 2002, Canadian International Development Agency - CIDA, ADSGM, Universiti Utara Malaysia, Kedah, Da'arul Sintok).
71)**The Defensive Strategy for the Market Leader**, a Nokia Cellular Telephone's Case. *Strategi Bertahan Sang Pemimpin Pasar dengan Menggunakan Multi Merk, Kasus Telepon Seluler Nokia.* (March, 12, 2004, Univ Tarumanegara, Jakarta).

Corporate IFRS-GAAP (B/S-I/S), ISBN-13: **978-1720792789**, ISBN-10: **172079278X**

Corporate IFRS-GAAP (B/S-I/S), ISBN-13: **978-1720792789**, ISBN-10: **172079278X**

Finance Construction-11, *Tim Asikin, Steve Asikin, Indra Senihardja*

Corporate IFRS-GAAP (B/S-I/S), ISBN-13: **978-1720792789**, ISBN-10: **172079278X**

Corporate IFRS-GAAP (B/S-I/S), ISBN-13: **978-1720792789**, ISBN-10: **172079278X**

Corporate IFRS-GAAP (B/S-I/S), ISBN-13: **978-1720792789**, ISBN-10: **172079278X**

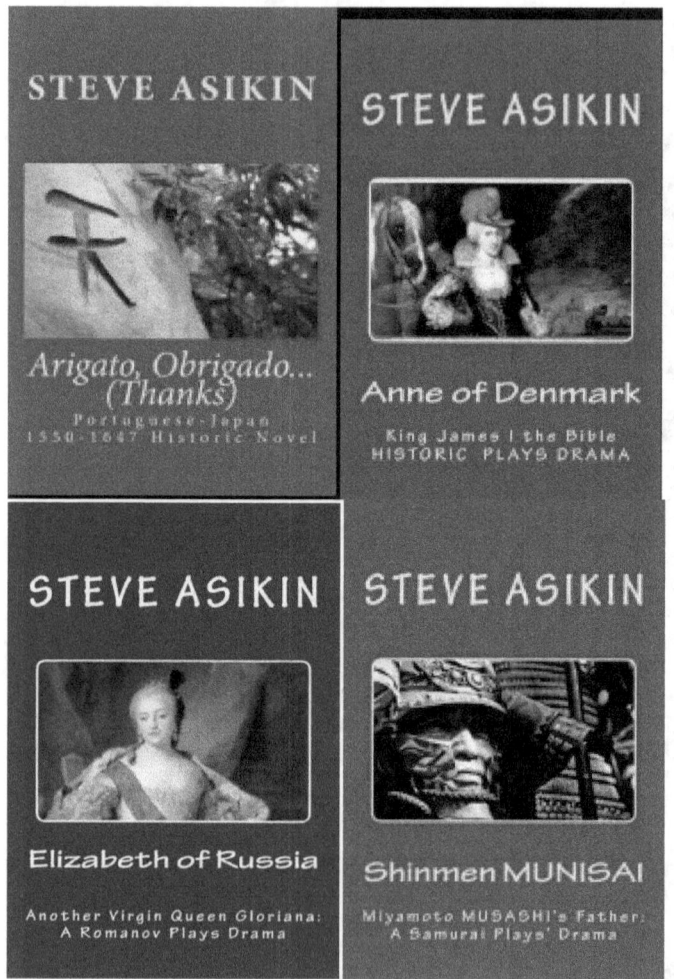

Corporate IFRS-GAAP (B/S-I/S), ISBN-13: **978-1720792789**, ISBN-10: **172079278X**

Corporate IFRS-GAAP (B/S-I/S), ISBN-13: **978-1720792789**, ISBN-10: **172079278X**

Corporate IFRS-GAAP (B/S-I/S), ISBN-13: **978-1720792789**, ISBN-10: **172079278X**

Corporate IFRS-GAAP (B/S-I/S), ISBN-13: **978-1720792789**, ISBN-10: **172079278X**

Finance Construction-11, *Tim Asikin, Steve Asikin, Indra Senihardja*

Corporate IFRS-GAAP (B/S-I/S), ISBN-13: **978-1720792789**, ISBN-10: **172079278X**

Finance Construction-11, *Tim Asikin, Steve Asikin, Indra Senihardja*

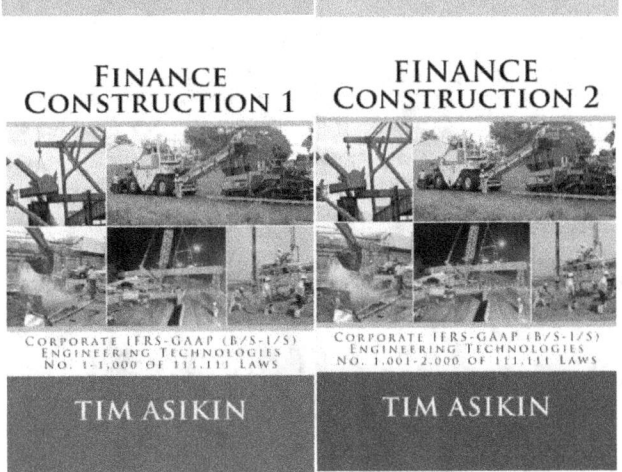

Corporate IFRS-GAAP (B/S-I/S), ISBN-13: **978-1720792789**, ISBN-10: **172079278X**

Finance Construction-11, *Tim Asikin, Steve Asikin, Indra Senihardja*

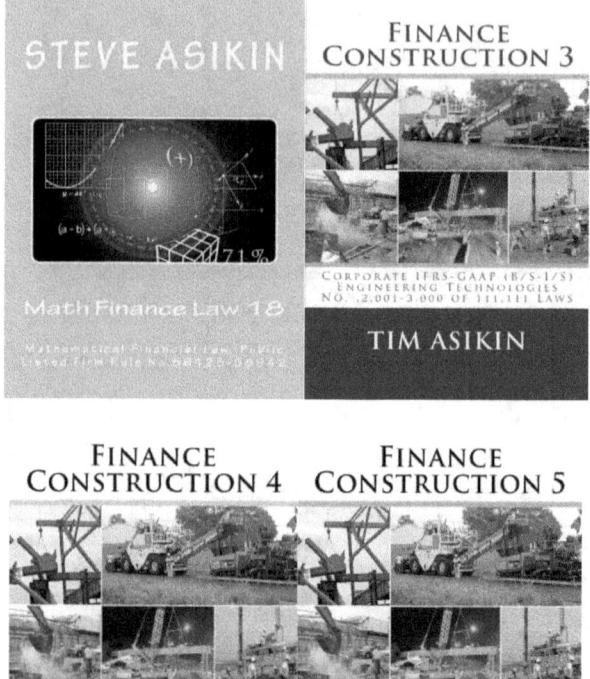

Corporate IFRS-GAAP (B/S-I/S), ISBN-13: **978-1720792789**, ISBN-10: **172079278X**

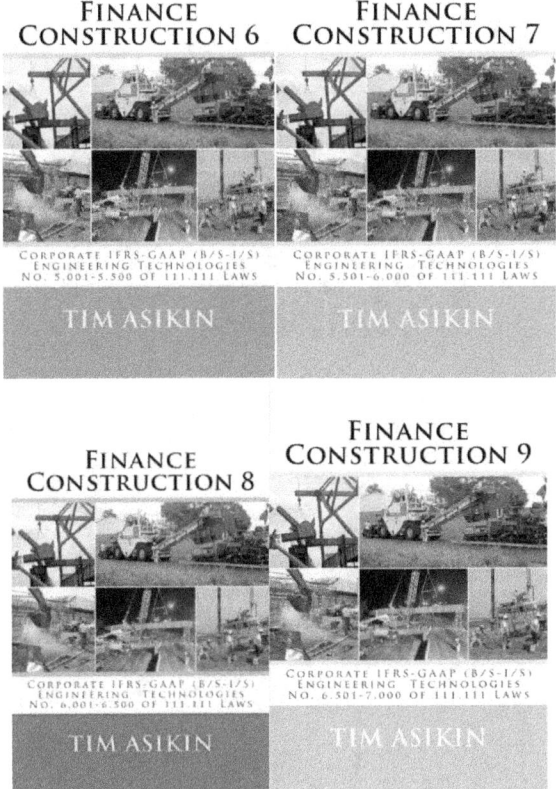

Corporate IFRS-GAAP (B/S-I/S), ISBN-13: **978-1720792789**, ISBN-10: **172079278X**

FINANCE
CONSTRUCTION 10

CORPORATE IFRS-GAAP (B/S-I/S)
ENGINEERING - TECHNOLOGIES
NO. 7.001-7.500 OF 111.111 LAWS

TIM ASIKIN

Corporate IFRS-GAAP (B/S-I/S), ISBN-13: **978-1720792789**, ISBN-10: **172079278X**